VIRTUE AND MEDICINE

PHILOSOPHY AND MEDICINE

Editors:

H. TRISTRAM ENGELHARDT, JR.

The Center for Ethics, Medicine, and Public Issues,
Baylor College of Medicine, Houston, Texas, U.S.A.

STUART F. SPICKER

School of Medicine,
University of Connecticut Health Center, Farmington, Connecticut, U.S.A.

VOLUME 17

VIRTUE AND MEDICINE

Explorations in the Character of Medicine

Edited by

EARL E. SHELP

Center for Ethics, Medicine, and Public Issues
Baylor College of Medicine, and
Institute of Religion, Houston, Texas, U.S.A.

D. REIDEL PUBLISHING COMPANY

A MEMBER OF THE KLUWER ACADEMIC PUBLISHERS GROUP

DORDRECHT / BOSTON / LANCASTER

Library of Congress Cataloging in Publication Data

Main entry under title:

Virtue and medicine.

(Philosophy and medicine; v. 17)
Bibliography: p.
Includes index.
1. Medical ethics. 2. Nursing ethics. 3. Medicine–Philosophy.
4. Virtue. I. Shelp, Earl E., 1947– . II. Series.
R724.V48 1984 174′.2 85-1836
ISBN 90-277-1808-3

Publishing by D. Reidel Publishing Company,
P.O. Box 17, 3300 AA Dordrecht, Holland.

Sold and distributed in the U.S.A. and Canada
by Kluwer Academic Publishers,
190 Old Derby Street, Hingham, MA 02043, U.S.A.

In all other countries, sold and distributed
by Kluwer Academic Publishers Group,
P.O. Box 322, 3300 AH Dordrecht, Holland.

Printed in The Netherlands.

TABLE OF CONTENTS

SECTION IV / CRITIQUE

EARL E. SHELP

INTRODUCTION

Interest in theories of virtue and the place of virtues in the moral life continues to grow. Nicolai Hartmann [7], George F. Thomas [20], G. E. M. Anscombe [1], and G. H. von Wright [21], for example, called to our attention decades ago that virtue had become a neglected topic in modern ethics. The challenge implicit in these sorts of reminders to rediscover the contribution that the notion of virtue can make to moral reasoning, moral character, and moral judgment has not gone unattended. Arthur Dyck [3], P. T. Geach [5], Josef Pieper [16], David Harned [6], and, most notably, Stanley Hauerwas [8–11], in the theological community, have analyzed or utilized in their work virtue-based theories of morality. Philosophical probings have come from Lawrance Becker [2], Philippa Foot [4], Edmund Pincoffs [17], James Wallace [22], and most notably, Alasdair MacIntyre [12–14]. Drawing upon and revising mainly ancient and medieval sources, these and other commentators have ignited what appears to be the beginning of a sustained examination of virtue.

In addition to a theoretical consideration of the value of virtue to moral decision-making in general, applications of the virtues and virtue ethics to specific areas of activity are beginning to appear. The practice of medicine has been a prime candidate for these exercises, given medicine's reliance on virtuous character and conduct as distinguishing traits of medical practitioners. Several types of studies have appeared. Alasdair MacIntyre has looked at how virtues function in different contexts within a moral pluralism [13]. G. E. Pence has called for a greater emphasis on character traits, rather than an emphasis on analyses of rights, duties, or principles, as pathways to solutions to medical-moral dilemmas [15]. Earl Shelp has examined the meaning and relevance of a specific virtue, courage, for patients and physicians in the moral community constituted as the patient-physician relationship [18, 19]. These analytical and applied studies of virtue ethics have tended to provoke fresh debate about the contribution of virtue-based methods of ethics to the moral life, in general, and moral medical practice, in particular.

The focus in bioethics on right and wrong conduct has grown out of a methodological commitment to the primacy of moral rules and principles as resources for moral guidance. Deontological and consequentialist methods of

Earl E. Shelp (ed.), Virtue and Medicine, vii–xx.

moral decision making have tended to de-emphasize a concern for moral 'being' in favor of a concentration on moral 'doing'. It could be argued that the renewed interest in virtue reflects, on the part of some people, an uneasiness with the sufficiency of moral analysis in which the relevance and contribution of notions of virtue and character are slighted. The call to return to virtue, on the one hand, may be a summons to displace the focus on conduct in favor of a focus on character. On the other hand, the call to virtue may be a plea for balance and/or comprehensiveness in moral reflection.

The contribution virtue theory and the intellectual, moral, nonmoral, and theological virutes ultimately will make to the philosophy of medicine, bioethics, and the delivery of health care is unknown. There can be little doubt, however, that a virtue-based approach to these areas of inquiry will be different from consequentialist or deontological ones. Evidence of these divisions can be found in the essays in this volume. Diverse understandings of the definition of virtue, the classification of virtues, and the application of virtue to medicine are presented. These investigations, together with historical studies and evaluative essays provide an orientation to the current status of the examination of virtue and identifies topics for further investigation.

The essays are presented in four sections. The first section contains four historical analyses of understandings of the virtues, their application to medicine, and their expression in clinical encounters. The second section provides a theoretical discussion of virtue theory, including a proposed taxonomy for the virtues, an analysis of virtue from a Christian theological perspective, and a critique of Alasdair MacIntyre's account of virtue. Following these conceptual explorations, seven essays speak directly to the relation of virtue and the virtues to a broad understanding of medicine. This penultimate section begins with three examinations of the meaning, embodiment, and normativeness of notions of virtue for medical practice. The virtuous physician, nurse, and patient then are considered prior to an application of virtue theory to health policy. The volume concludes with a final, fourth section which contains three evaluations of the promise and problems of efforts to settle moral disputes in medicine by appeal to virtue-based methods of analysis and decision-making.

The historical section begins with two essays jointly authored by Gary Ferngren and Darrel Amundsen. One chapter is focused on pre-Christian antiquity. The second chapter reviews the first through the sixteenth centuries. The study of the pre-Christian era traces the development and interconnection of virtue, health, and medicine in classical civilization from Homer to the time of Christ. Homer's notion of *arete* or excellence dominated the first half of this development, and the excellences most sought were the masculine ones of courage, loyalty, magnanimity, and patriotism. Moral virtues received, by comparison, little attention. Health came to be extolled as the paramount virtue because it was considered the *sine qua non* of happiness. The Greeks opposed extending life if it would mean prolonged, hopeless suffering, and suicide was considered honorable in cases of grim prognosis. Philosophers, such as Plato and Aristotle, developed theories of moral virtue by analogy with theories of the healthy body. Even the personal virtue or etiquette of the physician was defined in terms of health, for the physician was to be 'of healthy appearance and suitable weight'. Little is said during this period of the moral basis of medicine. The essential requirement for practice was excellence in 'the act'. The virtuous physician was not expected to be philanthropic, a requirement largely a result of later Christian influence. Nevertheless, by the third century B.C.E., a foreshadowing of Christian influence began to affect attitudes toward virtue and health. Mainly due to the influence of Stoicism, Epicureanism, and Cynicism, emphasis on physical health as *arete* began to decline. At the same time the 'gentler' virtues – kindness, charity, forgiveness – came to be emphasized. This spirit reached the practice of medicine, which subsequently came to be viewed as a vehicle for compassion. Thus, according to the authors, late classical shifts in conceptions of the virtues and of medicine anticipated and prepared the way for the subsequent influences of Christianity.

The historical review by Darrel Amundsen and Gary Ferngren is extended through the sixteenth century in their second contribution. They first examine the general topic of virtue as it developed from the era of the New Testament to the Protestant Reformation. Texts from the New Testament, especially Pauline works; the Church Fathers, Ambrose, Boethius, and Augustine; Aquinas, representing the classic Roman Catholic position; personalities of the Renaissance; and Luther and Calvin are examined. Particular attention is given to the Reformation dispute over the relative values of faith and works, as interpreted in the terminology of virtue. Following this conceptual study, the relation between medicine and virtue in a Christian context is explored. They review answers to the question, 'Are the practice and use of medicine consistent with the demands of Christian virtue?' Behind this query is the worry about whether illness is a result of God's will, in which case submission to physical distress is virtuous for the Christian and medical intervention is vicious. The chapter concludes with a survey of conceptions of the virtuous Christian in the roles of patient and of physician.

Dietrich von Engelhardt extends the historical inquiry into the Enlightenment in Germany. The author examines German Enlightenment thought on ethics and medicine, with particular emphasis on the place of the virtues. He identifies three centers or foci of interest for medical ethical writers of the period, *viz*., the physician, the patient, and the state. In three substantive sections the contribution made to each by physicians, philosophers, theologians and jurists is surveyed. Throughout the chapter, emphasis is placed on the impact which eighteenth century changes in social structure, such as the transition from an aristocratic to a bourgeois society, had on the contemporary conceptions of ethics and medicine.

A review of medical morality during the eighteenth and nineteenth centuries in England and America concludes the historical section. Laurence McCullough finds precedents during this period for modern attempts to locate virtue in medical ethics. John Gregory (18th century) is seen as a philosophical forebearer of virtue-based Anglo-American medical ethics. Gregory based his account of the physician's moral responsibility in sympathy, understood as a disposition to experience the feelings of another. Sympathy led to benevolence for the physician because it directed the physician's attention to the best interests of the patient. His system possessed a high degree of completeness, since it comprised an account of the physician's moral duties and the virtues and etiquette appropriate to the role. Anglo-American medical ethics in the century following Gregory's work saw a dissolution of his comprehensive, sympathy-based approach. Pivotal to this dissolution were the work of Thomas Percival and the original (1848) Code of Medical Ethics of the AMA. For they had the effect, first of destroying the systematic connection between duties, virtues, and etiquette established by Gregory; and, second, of grounding the moral duties of physicians in the power, importance, and prestige of the profession, rather than the best interests of the patient.

The second grouping of essays analyze virtue philosophically and theologically. Bernard Gert initiates the inquiry by providing an account of the virtues (and vices) which takes its lead from Hobbes. According to Gert, there is a fundamental cleavage between the personal virtues (vices) and the moral virtues (vices). Personal virtues (vices) are character traits which all (no) rational persons want for themselves. Moral virtues (vices) are character traits which all impartial rational persons want everyone (no one) to have. Gert's account of personal virtues (vices) is situational, in the sense that virtue is defined in terms of what it would be reasonable to expect a person to do in a given situation. For example, in a situation where, because of fear or danger, it would be reasonable to expect a person to act irrationally, and that person nevertheless acts rationally, the virtue of courage has been displayed. Similarly for the vices. The concept of acting (failing to act) rationally plays a role in the definition of all the personal virtues (vices). What distinguishes them, according to Gert, is the nature of the situation in which the rational or irrational action occurs.

Gert takes issue with the view that acting rationally is just acting to maximize the satisfaction of one's desires. He urges that rationality is not coextensive with self-interest; that it is also rational to be interested in the good of others. His definition of the moral virtues casts them as character traits which, when had by everyone (or nearly everyone), would tend to result in the least suffering overall. These virtues, like the personal ones, are distinguished from one another situationally. Unlike the personal ones, the moral virtues are coordinated with rules. Nevertheless they do not involve blind or exceptionless following of rules. Rather, in Gert's view, having the appropriate moral virtue means that one will be disposed to act in accord with the coordinated moral rule in all cases *except* those in which one could publicly advocate violation of the rule.

One of the central problems for Gert is the resolution of apparent conflicts of virtues. As an example, Gert considers the physician who is deciding whether to inform a patient of a grim diagnosis. It seems the virtue of truthfulness inclines the physician to tell, while the virtue of kindness inclines him not to tell. Gert resolves such problems by suggesting that given situations often provide opportunities for exemplifying more than one virtue, though it is not possible to exemplify them all simultaneously.

Edmund Pincoffs has concerns similar to Gert but pursues a different path to resolve them. Pincoffs's major purpose is to provide a definition of virtue and to take first-steps toward a taxonomy of the virtues that avoids a reductionism that, in his judgment, characterizes many other attempts in this direction. His definition of virtue is functional, i.e., "determinable dispositions to behavior which can serve as grounds for preference/avoidance in human choice, not of acts, but of persons." The liberality of this definition is developed by comparing it with other, reductive ones. The virtues then are classified as instrumental and non-instrumental, aesthetic, ameliorating, mediating, temperamental, and moral. He concludes with specific criticisms against the theories of G. H. von Wright, Lester Hunt, Maurice Mandelbaum, Alasdair MacIntyre and James Wallace.

The work of Alasdair MacIntyre is subjected to further analysis by Kai Nielsen. Nielsen begins by distinguishing *Kantian* and *utilitarian* moral theories, which take the fundamental task of moral theory to be formulation and justification of basic principles of conduct to guide human choice; and *virtue*-ethics, which seeks to define the ultimate ends or goods of human life and to characterize, in terms of those goods, what it is to be a good person. Nielsen views MacIntyre's *After Virtue* to be the most significant step in the direction of providing a contemporary virtue-based ethic. On MacIntyre's account, virtue-ethics requires a reasonably determinate conception of the good of *a* human life conceived of as a unity. The unity of a life is achieved when the human agent to some extent understands one's actions as part of an ordered narrative sequence. What is best for the individual is whatever one could do in order to live out that narrative and bring it to completion. The 'good for man', considered abstractly, would then be what all of the individual goods must have in common. Moral virtue, accordingly, is defined in terms of this abstract good of humanity. Nielsen's criticism turns largely on his claim that seeing one's life as a narrative unity in this way is neither a necessary nor a sufficient condition for one's acting in ways considered virtuous. Moreover, he argues, a given life is susceptible to narration in many different, perhaps incompatible ways. Probably no one is in a position to decide which of possible narratives is correct. But this indeterminancy, according to Nielsen seems to leave MacIntyre's account of the virtues open to the same charges of liberalism and individualism which he presses against prevailing deontological and utilitarian theories.

Gilbert Meilaender completes the second section of the volume with a theological analysis of the virtues. Meilaender observes a *prima facie* tension between Old Testament and New Testament understandings of virtue. Old Testament passages, notably from the Psalms, are cited which imply that humanity is naturally capable of righteous (i.e., virtuous) actions, and that these are meritorious in God's eyes. On the other hand, New Testament passages, notably from Paul's epistles, are cited which imply that humanity is naturally incapable of righteousness, and that whatever merit people have in God's eyes results from God's willingness to 'count Christ's virtue as our own'. Meilaender's goal is to relax this apparent tension through a theological analysis of human virtue from the perspective of Luther's dispute with Latomus. The key to an understanding of Luther's position is the recognition that, for him, virtue is not exemplified piecemeal in individual actions, or at least not exclusively so, but rather and primarily in the unified character of the actor. Thus two concepts of virtue are operative in Luther's view. On the one hand, virtues are characterized as individual and, in a sense, unintegrated improvements to one's life achieved by rote through small, disciplined steps. The main trouble with this concept of virtue is that, from such isolated improvements, it is impossible to construct a unified virtuous self. Though individual actions may be virtuous, they do not guarantee the existence of a virtuous person. But, according to Luther, we face God's judgment not as bundles of actions, but rather as whole persons or selves. Thus Luther, on the other hand, advances the concept of virtue as the quality of a redeemed self, where the redemption is, of course, understood as the result of God's free gift. The unified, virtuous person is therefore possible after all, but only as a creation of God through grace. Nor are the two concepts of virtue unrelated: a virtuous self, in the second sense, is likely to be the author of many virtuous acts in the first sense. In this way Meilaender, via Luther, presents a New Testament understanding of virtue as the completion and perfection of a characteristically Old Testament notion, rather than as the antithesis of the latter.

Turning from more restricted conceptual analyses, the third grouping of essays are more directly concerned with the relation of virtue to medicine. Marx Wartofsky is concerned with the general question of the normative basis for a critique of medical virtue. His point of departure is the recognition that medicine is an historical practice subject to pervasive changes which may occur either gradually or, as has been true in this century, with great rapidity. This historic variability affects all aspects of the practice: legal, political, economic, moral, and technical. Wartofsky inquires, first, whether there are any characteristics which have been considered medically virtuous by practitioners and public alike, regardless of the particular state of historic development enjoyed by the practice at a given time; and, secondly, if there are any such historically durable conceptions of medical virtue, whether they are the 'right' or morally correct ones. In pursuing his analysis, the author seeks to avoid what he takes to be two common errors of normative studies of this kind: the error of ahistorical or *a priori* 'essentialism', and the error of historical relativism. His own dialectical method represents an attempt to find a middle ground between these two unacceptable extremes. Wartofsky admits that problems can confront his program, and he admits that it may not entirely sidestep the undesirable aspects of alternative approaches. Yet he argues that it is, nonetheless, the most promising course available.

Edmund Erde's essay speaks to the virtues of medicine. Operating largely within MacIntyre's Aristotelian notions of 'virtue' and 'practice', Erde distinguishes three senses of 'the virtues of medicine': (a) virtues as orientations of character which the institutions of organized medicine demand of its practitioners; (b) virtues as advantages which organized medicine offers to society; (c) virtue as the orientation of character which constitutes being a physician.

The major portion of his essay is devoted to an examination of these senses, particularly (c). According to (c), medicine is a trait or character-orientation of the physician according to which he/she chooses, in particular cases, to act in accord with the best technical knowledge in order to achieve specific ends such as the health and well-being of the patient. This characterization of the virtue of medicine imputes to it elements of both intellectual and moral virtue, in the Aristotelian senses of those terms, and Erde argues for the legitimacy of this hybrid. He construes the virtues – including the virtues of medicine as given in (a)–(c) – analogously to N. R. Hanson's construal of empirical concepts. As our visual concepts (and other sensory concepts) are ways of our being-in-the-physical-world, so moral virtues are instructions or guidelines for our being-in-the-social-world. A lengthy penultimate section discusses the possibility of virtue-conflicts, as well as some ways in which the virtues of medicine may be exaggerated in certain directions to the point of vice.

Next a physician considers the topics of virtue and medicine. Allen Dyer argues that two distinct approaches to medical ethics, viz., bioethics and professional ethics, are presently handicapped by their failure to take explicit account of the concept of virtue. Bioethics (or medical ethics proper) is seen as inappropriately abstract and aloof from the realities of the clinical setting. Ethics comes to be viewed as a rival discipline to medicine, the province of professional 'ethicists'. This trend toward specialization, Dyer claims, has had the effect of depersonalizing not only ethics, but medicine as well. It implies that the physician is a scientific technician whose clinical judgments may be made in a value-neutral fashion. Professional ethics, on the other hand, is characterized by Dyer as "reflection upon the moral standards of the professional group by which its members define their identity as professionals and by which they determine standards of inclusion or exclusion." He argues that some concept of virtue already operates here – at least at a tacit level of awareness – especially in medical 'gatekeeping' decisions. He feels that, by becoming more explicit about the place of virtues in the profession, their legitimacy and importance would be further enhanced.

A possible objection to an emphasis on virtue can be derived from a Weberian sociological analysis of the medical profession. On such an analysis, groups such as the AMA are seen simply as trade organizations which attempt to monopolize markets and promote self-interest. Dyer believes that virtue, as a trust-inducing trait of the physician, would be viewed by a Weberian as yet another way of placing the consumer in the hands of the professional, and of further monopolistic interests such as price-fixing. Dyer attempts to describe a more optimistic role than this for virtues in the medical context. He provides a sketch of the psychological reasons for the centrality of two such virtues, trustworthiness and confidentiality, and ends with a brief note on some virtues appropriate to the patient.

Shifting the discussion from the profession of medicine to the practitioner of medicine, Edmund Pellegrino characterizes the virtuous physician in terms of a classical-medieval understanding of 'virtue' and a theory of the ends of medicine. The author presents a brief historical sketch of the development of the concept of virtue from Socrates to Aquinas, and adopts the Thomistic version: virtues are habits and dispositions that enable a person to reason well (intellectual virtues) and to act in accordance with right reason (moral virtues). Thus, moral virtues are conceived as character traits according to which individuals tend habitually to seek moral perfection, i.e., to do the good. Pellegrino's theory of the good or end of medicine identifies that good with the good of the patient as expressed in a particular healing action. Thus the virtuous physician is one who is habitually disposed to act to promote the patient's good through healing actions, and who is to be relied on consistently to place the patient's good above the physician's own. Pellegrino argues that, for this reason, the virtuous physician practices at a higher level of moral sensitivity than does the physician practicing merely within the strictures of a rights- or duty-based ethic. For such ethics, according to Pellegrino, are best seen as setting only the minimum standards for medical practice – a necessary function in a morally pluralistic society, but one which is exceeded regularly by the virtuous practitioner.

The notion of the virtuous nurse is examined by Martin Benjamin and Joy Curtis. They, like other contributors to the volume, draw upon MacIntyre's account of virtues, practices, and integrated lives. They apply MacIntyre's analysis to the practice of nursing: first, in an historical attempt to understand Florence Nightingale's conception of nursing and its connection with ideals of Christian service, Enlightenment rationalism, and the virtues of Victorian womanhood; and, second, in an attempt to determine to what extent Nightingale's model of the 'good and intelligent nurse' is viable for the modern practice of nursing. Benjamin and Curtis point out that, even in her own time, a tension was present among Nightingale's ideals. For, on the one hand, her conception of the good *nurse* drew heavily on such Enlightenment themes as clear-headedness, independent thinking, and self-reliance; while her conception of the good *woman* – which every good nurse was expected to be – was informed with traditional Victorian qualities such as intellectual subordination to and reliance upon (male) physicians. This tension has become more pronounced in the twentieth century as widespread changes in the political, economic and sexual roles of women has forced re-evaluation of many predominantly female endeavors, including nursing. The authors conclude by surveying some of the results of this modern re-thinking of the virtues of nursing.

Turning from the profession of medicine and the providers of medical care, the final two essays explore the relevance of virtue to patients and to public health. Karen Lebacqz, in an essay on "The Virtuous Patient," understands virtue quite broadly as 'any kind of perfection'. Lebacqz asks whether there are perfections of character which the estate or circumstances of being a patient calls forth. The circumstances of being a patient include the stresses of pain, discomfort, physical limitation, loss of autonomy, violation of privacy, vulnerability, and erosion of self-concept. Traditionally the 'good' patient has been conceived of as long-suffering, compliant, obedient, conscientious – in short, one who 'makes the best of everything'. Lebacqz challenges this conception, and proposes instead fortitude, prudence and hope as the chief virtues of the patient.

Marc Lappé's study of "Virtue and Public Health" concludes the third section of the collection. Lappé attempts to fill a void in contemporary medical-ethical studies. He observes that there has been a tendency to apply concepts of virtue to individuals only, overlooking the possibility of their application to collectives. A natural setting for a discussion of virtuous collective endeavor is the area of public health. A persistent theme of Lappé is that, to count as virtuous, an action must entail some abrogation, or at least down-playing, of the interests of the actor – some element, that is, of selflessness. Thus, virtue in public health requires that the collective's response to individual needs and vulnerabilities transcend, to some extent, considerations of cost-effectiveness and be characterized by extraordinary effort, generosity, and altruism. As with individual virtue, institutional or collective virtue must be characterized in terms of the goods or goals sought by institutional action. Lappé discusses three interdependent public health goals whose achievement requires both individual and collective action: health promotion, health maintenance, and health protection. He attempts to define the boundaries of collective/institutional responsibility in the virtuous – as opposed to the minimal, cost-constrained – pursuit of each of these goals.

The final section of essays contains three responses to the discussion of virtue ethics in general and to the linkage of virtue to medicine in this volume. Tom Beauchamp is concerned to forestall what he considers the over-emphasis which philosophers and others tend to attach to their preferred forms of moral theory. Many of the essays in this volume, he claims, over value virtue-theory, vis-à-vis rights-or-duty-based theories, as the correct approach to medical ethics; while in other places the converse overvaluation is the order of the day. To counteract this tendency, which he feels ill serves the purposes of doing medical ethics in the first place, Beauchamp argues for two principal theses. The first is that virtues, rights and duties, and the theories that accompany each, need not and ought not be viewed as competitors for the title of best moral entity or viewpoint. Rather, virtues, rights, and duties are to be seen as correspondents and correlatives of one another in the sense that, for any instance of one of the three categories, an appropriately corresponding instance of the others either exists or can be constructed. Beauchamp is frank in his evaluation of the difficulties with this view, but argues that most of the difficulties are surmountable. His second thesis is that virutes, rights, and duties "are all instruments, or means" to the ends of life which it is the business of moral theory to recognize and prosecute. Beauchamp's consequentialism allows equal importance to each of these types of instruments as viewed from the overall standpoint of the ultimate goods to be attained; while, viewed from the standpoint of appropriateness within a given context or social situation, it allows that one might be preferred to the others. Moreover, it is entirely likely that, even within a given context (such as the practice of medicine), some features of the ultimate goods of life will best be achieved by reliance on persons of good character (virtue), while other features will best be achieved by emphasis on the more juridical instruments of rights/duties. For these reasons Beauchamp counsels toleration rather than parochialism in the construction of ethical theories.

Robert Veatch takes a less tolerant view in his critique. He clearly demarcates the field of discussion through a definition of virtue as praiseworthy character traits for persons in given social roles. He thereby distinguishes virtues both from principles of right action and from right actions themselves, correcting what he sees as an occasional confusion and conflation of these concepts by writers on virtue. He then proceeds to argue for the difficulty and question the desirability of developing a theory of virtue for medicine. These remarks are intended to serve as a corrective to the exuberant optimism, evident in many of the volume's essays, regarding virtue theory as the best candidate for an ethics of medicine.

Veatch's criticism seeks to establish four main points. The less controversial of these are, first, that no one set of virtues is appropriate to every social role; and, secondly, that even within practices, such as medicine, the variety of conceivable professional and lay roles testifies against the existence of a unique set of character traits which we could call *the* virtue of medicine. These points serve as a reminder of the great difficulty to be encountered in an attempt to formulate a virtue theory for medicine; the *desirability* of such a theory is called in question by the more controversial of Veatch's claims. These are, thirdly, that even where a definitive set of virtues can be agreed on for a given social role, the possession of these virtues by persons in that role will not guarantee right action and may, if it engenders in the possessors a sense of moral over-confidence, actually result in wrong actions; and, fourthly, that in the context of what the author calls 'stranger madicine' a theory of virtue is superfluous, since the desired moral standards of practice are more reliably secured via a theory of rights and/or duties. Stranger medicine – the practice of medicine among people who are 'essentially strangers to one another' – is perhaps most clearly exemplified in the emergency room of a big-city hospital, but is, according to Veatch, to a greater or lesser degree the model on which nearly all medical care is provided in today's impersonal urban setting. Hence stranger medicine represents the form of medical practice with which most persons, both as patients and as health care providers, are likely to encounter in an urbanized society. So to say that virtue theory is unnecessary in this setting is greatly to diminish the attractiveness of such theory as the preferred model for an ethics of medicine. Veatch's concluding section describes a medical setting within which virtue theory might actually be the preferred type of ethical theory, but, as he points out in his introduction, practically no health care is today delivered in such a setting.

The volume ends with a contribution by Stanley Hauerwas, one of the more active evangelists in the revival of virtue. Pointing to the abstract aloofness, in the clinical setting, of consequentialist and deontological normative ethical theories, Hauerwas finds reason to be generally sanguine about the recent heightening of interest in the virtues on the part of those engaged in medical ethics. For though he does not view an account of virtue as a candidate to *replace* normative medical ethics, Hauerwas does see this 'turn to virtue' as a sign that many moral philosophers and theologians have begun to recognize a major flaw in the ethical tradition of the past few centuries. Hauerwas's general description of that flaw cites the failure of deontologists and consequentialists alike to recognize common goods pertaining to all members of society precisely as a result of their being members of one society. The only goods recognized have been those pertaining to individuals *simpliciter*, with the result that the important ethical questions have tended to take the form 'What ought the individual in this situation to do?' The turn to virtue, with its emphasis on the common goods of practices and its descriptions, in terms of those goods, of the kinds of people we ought to be, is considered by Hauerwas to represent an important corrective of the presuppositions on which modern medical ethics – and modern ethics in general – has been carried out. Because the subject of virtue has received so little attention in ethical literature over the past several hundred years, the author believes that 'almost everything remains to be done'. In the latter half of the essay he presents an agenda of important questions which an account of virtue must answer if it is to provide the desired degree of supplementation to more traditional approaches to medical ethics.

The disagreements and disputes that are joined through these essays suggest that the scholarly debate about virtue is far from over. Given this theoretical turmoil, there can be little surprise that contemporary applications of virtue theory to the philosophy of medicine and bioethics have been met with suspicion. The 'family quarrels' among advocates of virtue ethics indicates that consensus regarding the definition, classification, source, and expression of virtue will not be reached anytime soon. This admission makes no judgment regarding the worthiness of the effort. It only takes account of the internal debate. The challenge by critics constitutes a second barrier to a robust acceptance of virtue ethics over against deontological and consequentialist approaches. Tom Beauchamp may be correct that a flight to virtue is soon followed by a flight from virtue. He may be incorrect with regard to the current revival of interest in virtue theory. Time will tell.

There is a frustration among some medical clinicians and ethicists with ruled-based and consequentialist ethics for their failure to chart an unambiguous and clear moral path through complex medical cases. The disagreements and disputes within and between these schools of thought are no less intense than those that surround virtue ethics. They too are seemingly unable to forge a consensus regarding right and wrong conduct in specific instances. An appeal to virtue theory may provide some relief for the frustration. It may not. We shall not know until all sides are heard and proposals tested. It is hoped that this collection will make a contribution to the dialogue.

It has been a pleasure to compile this volume. The contributors generously have given themselves to their assignments. Each author should know of my gratitude to them. The editors of the series, H. Tristram Engelhardt, Jr., and Stuart F. Spicker, whose counsel and assistance were at my beck and call are due a special word of thanks. Susan M. Engelhardt's careful proofreading kept countless errors from the text. Mrs. Audrey Laymance worked skillfully in typing the manuscript. My research assistant, Jay Jones, pursued references and typos with a vengence. His assistance with the introduction also should be acknowledged. These people, collectively and individually, deserve and have my thanks for their labor on behalf of this project. They are the ones who made it possible. I am happy to have been the integrating part of it.

BIBLIOGRAPHY

[1] Anscombe, G. E. M.: 1958, 'Modern Moral Philosophy', *Philosophy* **33**, 1–19.
[2] Becker, L. C.: 1975, 'The Neglect of Virtue', *Ethics* **85**, 110–122.
[3] Dyck, A. J.: 1973, 'A Unified Theory of Virtue and Obligation', *Journal of Religious Ethics* **1** (Fall), 37–53.
[4] Foot, P.: 1978, *Virtues and Vices*, University of California Press, Berkeley.
[5] Geach, P. T.: 1977, *The Virtues*, Cambridge University Press, Cambridge.
[6] Harned, D. B.: 1973, *Faith and Virtue*, Pilgrim Press, Philadelphia.
[7] Hartmann, N.: 1932, *Ethics*, Vol. II, S. Cort (trans.), Macmillan, New York.
[8] Hauerwas, S.: 1974, *Vision and Virtue*, Fides Publishing, Notre Dame, Indiana.
[9] Hauerwas, S.: 1975, *Character and the Christian Life*, Trinity University Press, San Antonio, Texas.
[10] Hauerwas, S.: 1977, *Truthfulness and Tragedy*, University of Notre Dame Press, Notre Dame, Indiana.
[11] Hauerwas, S.: 1981, *A Community of Character*, University of Notre Dame Press, Notre Dame, Indiana.
[12] MacIntyre, A.: 1975, 'How Virtues Become Vices: Values, Medicine and Social Context', in H. T. Engelhardt, Jr. and S. F. Spicker (eds.), *Evaluation and Explanation in the Biomedical Sciences*, D. Reidel Publishing Co., Dordrecht, pp. 97–111.
[13] MacIntyre, A.: 1981, *After Virtue*, University of Notre Dame Press, Notre Dame, Indiana.
[14] MacIntyre, A.: 1981, 'The Nature of Virtue', *Hastings Center Report* **11** (April), 27–34.
[15] Pence, G. E.: 1980, *Ethical Options in Medicine*, Medical Economics Press, Oradell, New Jersey.
[16] Pieper, J.: 1966, *The Four Cardinal Virtues*, University of Notre Dame Press, Notre Dame, Indiana.
[17] Pincoffs, E.: 1971, 'Quandary Ethics', *Mind* **80**, 552–571.
[18] Shelp, E. E.: 1983, 'Courage and Tragedy in Clinical Medicine', *The Journal of Medicine and Philosophy* **8**, 417–429.
[19] Shelp, E. E.: 1984, 'Courage: A Neglected Virtue in the Patient–Physician Relationship', *Social Science and Medicine* **18**, 351–360.
[20] Thomas, G. F.: 1955, *Christian Ethics and Moral Philosophy*, Charles Scribner's Sons, New York.
[21] von Wright, G. H.: 1963, *The Varieties of Goodness*, Humanities Press, New York.
[22] Wallace, J. D.: 1978, *Virtues and Vices*, Cornell University Press, Ithaca, New York.

SECTION I

HISTORICAL ANALYSES

GARY B. FERNGREN AND DARREL W. AMUNDSEN

VIRTUE AND HEALTH/MEDICINE IN PRE-CHRISTIAN ANTIQUITY

THE ORIGINS OF CLASSICAL VIRTUE

In current usage the word 'virtue' retains little of its etymological meaning. Derived from the Latin *virtus* ('valor or manliness'), it now usually refers to moral integrity or uprightness. In popular usage the kind of moral behavior which it and its adjective, 'virtuous', convey is sexual as, for instance, in the phrase, 'a virtuous woman'. A definite influence in the development of this specialized and etymologically-remote nuance is the emphasis of Christianity on sexual purity. Even in its broader sense of moral integrity or uprightness generally, 'virtue' today typically has connotations that reflect Christian rather than classical values. These Christian values are qualities that are associated with the triad of the Christian graces of faith, hope, and love. The Apostle Paul describes the "fruit of the Spirit" as "love, joy, peace, longsuffering, gentleness, goodness, faith, meekness, temperance" (Galatians 5:22–23, AV). It is these qualities that have produced the Christian ideal of the virtuous individual. Yet they were for the most part undervalued by the classical world in providing the basis for moral conduct. Lecky observed that whereas Christianity has stressed those virtues traditionally regarded as feminine, the classical world admired the virtues that were thought to be distinctively masculine ([19], Vol. 2, pp. 361–363). The qualities that were valued in paganism included courage, loyalty, magnanimity, and patriotism. They were part of the heroic ideal that dominated classical culture for a thousand years. The marked difference between the classical and Christian concepts of virtue is fundamental to understanding the relationship between health/medicine and virtue in the ancient world.

The Greek word for virtue is *arete*, but the word was used to refer to a much broader range of meanings than those that denote moral qualities.[1] The basic meaning of the word is 'excellence' and it is employed to refer to the quality or proficiency of men, gods, animals, and things. In Homer *arete* is not used with reference to moral behavior but to describe efficiency or fulfillment of a natural function. Insofar as they fulfill their function weapons and horses possess *arete*. There can be an *arete* of feet, of fighting, of shoemaking, or of the mind. The word is used in the plural as well as the

Earl E. Shelp (ed.), Virtue and Medicine, 3–22.

singular to indicate that there is a variety of 'excellences'. There is a set of human qualities that *arete* may describe: these are the virtues possessed by the Homeric hero and they are the qualities of an aristocratic world. They are not available to everyone; indeed, they are limited to a small number of those who enjoy certain advantages with which they are endowed by birth and inheritance. Hence the ordinary person has no *arete*. The characteristics of the Homeric hero are those suited to the warrior: skill, cunning, courage, self-reliance, loyalty, love of one's friends and hatred of one's enemies, courtesy, generosity, and hospitality. These values are intensely individualistic and frankly competitive. Distinctions between individuals are emphasized. Great importance is placed on personal attainment. One is expected to realize one's nature and to bring it to perfection. The goal is to win renown by one's own achievements and to distinguish oneself above one's fellows. Opportunity for renown is found not only in warfare but in competitive festivals that provide opportunity to compete in athletic contests, music, and poetry. Victory in a contest of valor, whether in battle or in games, is the test of manly virtue. Thus the hero's activity consists in repeated striving for supremacy. The object is to win honor and glory: nothing mattered more in the heroic outlook than fame and it was eagerly sought by those who had a claim to *arete*. "To strive always for excellence (*arete*) and to surpass all others" (*Iliad* 6.208; cp. 11.784): this was the desire of the Homeric nobility.

The heroic code enshrined in the concept of *arete* was based on honor rather than morality. It stressed nobility, achievement, and reputation. The motives were happiness, material advantage, and the desire for glory. The 'good man' who possesses *arete* was not defined as one who exhibits proper moral behavior. The sanction in Homeric and early Greek society was not a moral code but 'what people will say', i.e., public opinion and public disapproval. The qualities that were admired were those that were necessary for the preservation of life, namely, skill and courage; while the basis of *arete* was physical strength and prowess, which were the essential marks of the warrior. The moral virtues are seldom commended or immoral actions condemned. Praise and blame are given for honor and dishonor, not for violations of moral standards.

Homer was regarded in Plato's time (ca 429–347 B.C.) as the educator of all Greece (Plato, *Republic* 606E) and classical Greek ethics was derived from the aristocratic ethical code found in Homer. The old concepts had to undergo much transformation in order to meet the needs and changing circumstances of later generations. But they showed a remarkable persistence. The aristocratic ideals of Greek culture remained long after the disappearance of the world that gave birth to them, much as certain values of chivalry have continued for centuries after the disappearance of knighthood in Europe. In particular, the physical basis of Greek culture, which found its origins in the need of the warrior to possess strength and prowess, remained an essential component of the idea of a balanced and controlled personality. And while *arete* developed a much broader meaning in classical times as its characteristics were adapted to the developments of Greek ethical thinking, it retained the essential shape that it had been given by Homer in the eighth century B.C. And that shape was a distinctively and rigorously masculine one in which there was little appreciation of the quieter virtues.

HEALTH AS A VIRTUE

The Greeks considered health to be both essential to *arete* and an aspect of *arete*. It was thus a virtue and an indicator of virtue and the *sine qua non* of the good life. This theme is iterated often in Greek literature. According to an old Attic drinking-song, health leads the list of the four best goods:

For a man health is the first and best possession,
Second best to be born with shapely beauty,
And the third is wealth honestly won,
Fourth are the days of youth spent in delight with friends
([7], pp. 103–104).[2]

The same theme is echoed by Sophocles:

The fairest thing of all is to be just;
The best to live without disease; most sweet
Power to win each day the heart's desire
(Fr. 236 Pearson; [7], p. 104).

Much later Sextus Empiricus, a physician and philosopher of the Sceptical school (*fl. ca* 200 A.D.), says that to ordinary people health is the *summum bonum*, the highest good (*Adv. Mathem.* 11.49; [10], p. 357). The Greeks believed that without good health nothing else in life can be enjoyed. "Neither wealth nor anything else is of any value without health," writes the author of the pseudo-Hippocratic *Regimen* (3, ch. 69; [10], p. 380). "When health is absent," according to Herophilus, "wisdom cannot reveal itself, art cannot become manifest, strength cannot fight, wealth becomes useless, and intelligence cannot be made use of" (*ap.* Sextus Empiricus, *Adv. Mathem.* 11.50;

[10], p. 358). In a society that valued physical culture as the very basis of *arete*, health was the indispensable requisite of happiness. A Sicyonian poet named Ariphron, about 400 B.C., composed a beautiful paean to Health (Hygieia) that expresses the high value that the Greeks assigned to it:

Health, best of the Blessed Ones to men,
May I dwell with you for the rest of my days,
And may you be kind and stay with me.
For if there is any joy in wealth or in children,
Or in royal rule which makes men like the gods,
Or in the desires which we hunt
With Aphrodite's secret snares,
Or if men have any other delight
From the gods or respite from their labours,
With you, blessed Health,
All things are strong and shine with the converse of the Graces,
And without you no man is happy (*ap.* Athenaeus, *Deipnosophistae* 15.701F; [7], pp. 104–105).

Although the desire to remove disease and preserve health is the motivation behind any kind of medical tradition, it was especially prominent in Greek society, given the significance placed on the body, and perhaps this accounts for the importance given to medicine as a part of general education. Yet health cannot be maintained forever and medicine is restricted in what it can do to alleviate physical suffering. The Greeks realized this and they mourned the passing of youth and dreaded the coming of old age, which brought with it the decline of physical powers and made a person a burden both to himself and to others. The Greeks did not regard it as desirable to prolong life when health is so permanently impaired that the patient cannot spend his life free of suffering. Plato argues in the *Republic* that no medical treatment ought to be given to one who cannot continue in his occupation and who is of no use to himself or to society (*Republic* 406A–407D). And Plutarch quotes with approval the words of Euripides:

I hate the men who would prolong their lives
By foods and drinks and charms of magic art,
Perverting nature's course to keep off death;
They ought, when they no longer serve the land,
To quit this life, and clear the way for youth
(*Consolatio ad Apollonium*, trans. Babbitt; [10], p. 382).

The desire to end a life that had become unendurable rather than prolong it led many in the classical world to resort to suicide. Suicide was considered by many philosophical schools of antiquity to be an honorable act for those faced with painful or chronic diseases.[3] Physicians, moreover, are advised in the medical literature not to undertake the treatment of diseases that are regarded as hopeless.[4] The Christian belief that pain and suffering had a redemptive purpose was unknown to the classical world and so basic was health to the enjoyment of life that a life without it was not thought to be worth living.[5]

If health was, for most Greeks, the greatest of the virtues, it is not surprising that they devoted a great deal of attention to preserving it. As an essential component of *arete* physical culture was an important part of the life of what the Greeks called the *kalos kagathos*, the cultivated gentleman, who represented in classical times the ideal of the human personality. The phrase means 'a fine (i.e., handsome) and good man'. To the Greeks good looks were essential to a cultivated man. Plato reflects popular opinion when he ranks health as the highest of the physical graces, physical beauty as second, and physical strength as third.[6] Like health, good looks are a gift given by the gods to those whom they love. They are not allotted to everyone. Yet, according to Aristotle, lack of physical beauty makes it difficult for a person to be counted happy: "for a man is scarcely happy if he is very ugly to look at, or of low birth, or solitary and childless" (*Nichomachean Ethics* 1099a 35–37, trans. Thomson). Herodotus too reflects common Greek opinion when he makes Solon say to Croesus that the following blessings characterize the happy man: "he is whole of limb, a stranger to disease, free from misfortune, happy in his children, and comely to look upon" (1.32.6, trans, Rawlinson). The Greeks greatly admired physical beauty and frequently commented on it. They believed that in beauty men and women resemble the gods. They particularly admired the body of young men: in the games athletes performed naked and statues depicted young men in the nude (by contrast young women in the fifth century B.C. were depicted clothed). The human form was idealized and glorified in sculpture and marked by proportion, dignity, harmony, and restraint.

The 'cult of the body', which is so characteristic a feature of Greek life, explains the importance of athletic contests to the Greeks. The panhellenic games were religious festivals held in honor of a Greek god. But they provided an opportunity to display and to view the *arete* of the contestants. Finely-trained young athletes at the peak of their physical grace and powers competed for victory and fame. The best athletes that Greece could produce were observed as they displayed flawless physique, skill, poise, and endurance, and the victors were accorded almost superhuman honors and were treated as heroes for the rest of their lives. If the Greeks tended to place too much emphasis on the cultivation of the body, at least by our standards, it was because they believed that it was a necessary part of the developed and balanced personality. *Mens sana in corpore sano*, 'a sound mind in a sound body', was a Greek ideal that was in turn passed on to the Romans.[7] The Greeks thought it as perverse to develop the mind and neglect the body as it was to cultivate the body and deprive the mind. While they did not believe that a sound body would produce a sound mind ([10], p. 356), they did believe that there is a relationship between the two and that only in a healthy body can there dwell a fully healthy mind. This characteristically Greek view is attributed by Diogenes Laertius to Diogenes the Cynic (ca 400–ca 325 B.C.):

> He used to affirm that training was of two kinds, mental and bodily: the latter being that whereby, with constant exercise, perceptions are formed such as secure freedom of movement for virtuous deeds; and the one half of this training is incomplete without the other, good health and strength being just as much included among the essential things, whether for body or soul. And he would adduce indisputable evidence to show how easily from gymnastic training we arrive at virtue. For in the manual crafts and other arts it can be seen that the craftsmen develop extraordinary manual skill through practice. Again, take the case of flute-players and of athletes: what surpassing skill they acquire by their own incessant toil; and, if they had transferred their efforts to the training of the mind, how certainly their labours would not have been unprofitable or ineffective (6.70, trans. Hicks).

It is this principle that produced the celebrated Greek contempt of the manual trades. It was held that leading sedentary lives and working inside all day weakened the body and as a consequence the soul as well.[8] According to Aristotle, "a task and also an art or a science must be deemed vulgar if it renders the body or soul or mind of free men useless for the employments and actions of virtue (*arete*). Hence we entitle vulgar all such arts as deteriorate the condition of the body, and also the industries that earn wages; for they make the mind preoccupied and degraded" (*Politics* 1337b 1). A Greek 'gentleman' expected to spend much of his time out of doors working on his farm or conducting his business in the agora, or exercising in the gymnasium. In Xenophon's *Oeconomicus* Ischomachos, who represents a typical *kalos kagathos*, is concerned to maintain his health and strength and he describes to Socrates how he deliberately arranges his daily schedule to obtain a moderate diet and sufficient exercise (*Oeconomicus* 11.8, 12, and 19). A combination of physical and mental activity characterized the daily routine that the Greeks of the classical era thought suitable to produce a healthy and well-rounded life. It was a particularly Greek kind of life, shaped by Greek ideals. In creating a culture of the body and the soul, the Greeks were attempting to design a pattern of living that was suited to the *whole* personality.

HEALTH AS AN ANALOGUE OF VIRTUE

Arete became the central ideal of Greek culture in the classical period and it came to provide the moral imperative for individual conduct. The word began to take on a broader meaning in the late fifth century before Christ with the activity of the sophists, a group of itinerant teachers who claimed to be able to teach *arete*, by which they meant the ability to conduct one's personal affairs efficiently. The term was used by Socrates with reference to excellence of conduct in a moral sense. Afterwards it was used by philosophers to refer to right conduct and hence began to approximate our idea of virtue as 'moral behavior'. We find it used in this sense by Plato and Aristotle. This is apparent in the development of the four cardinal virtues. The virtues – wisdom, temperance, courage, and justice – appear for the first time in Plato's *Republic*.[9] They were apparently already traditional in Plato's day, but in some form they probably go back at least to the sixth century B.C. Similar virtues are mentioned by Aeschylus and Pindar in the early fifth century. There appears to have been no single canonical list, but Plato selected qualities that were inherited from traditional wisdom and made them the primary or cardinal (from the Latin *cardo*, 'hinge') virtues of conduct.[10] It has been suggested that they developed originally to describe the four sides of the human personality (physical, aesthetic, moral, and intellectual). Plato owes much to his teacher Socrates who had attempted to go beyond the mere enumeration of individual virtues by framing a general definition of *arete*. For Plato they describe a life of virtuous activity under the rule of reason. He makes the four virtues primary ones because they correspond to the parts of the soul. In the Platonic scheme the soul is composed of three parts.[11] The first is the intellect, which gives to man the power to think and deliberate. The second is the will or the spirited part of the personality. The third is the appetite, which has a natural desire for physical satisfaction. Corresponding to these elements of the tri-partite soul are the appropriate virtues: wisdom, which is appropriate to the intellect; courage, which is appropriate to the will; and temperance, which is appropriate to the appetite. The fourth of the virtues, justice, is not related to one of the parts of the soul. Rather it is conceived as the virtue that regulates the others. Justice in the soul results from each part of the soul fulfilling its proper function. Virtue arises from a harmony of the constituent parts of the soul, which are organized for the best performance of living. Since for Plato virtue is one, all four of the virtues overlap and it is impossible to have one virtue without having them all. An individual whose soul is well ordered will be wise, since his reason will be in command; brave, as his will enables him to carry out what reason prescribes; and temperate, as his appetites are kept under control by his reason. Plato has Socrates compare the harmony that is produced by virtue in the soul with the harmony of the parts of a healthy body:

> And the creation of health is the institution of a natural order and government of one by another in the parts of the body; and the creation of disease is the production of a state of things at variance with this natural order? . . . And is not the creation of justice the institution of a natural order and government of one by another in the parts of the soul, and the creation of injustice the production of a state of things at variance with the natural order? . . . Then virtue is the health and beauty and well-being of the soul, and vice the disease and weakness and deformity of the same? (*Republic* 444D–E, trans. Jowett)

Plato's analogy between the health of the body and the health of the soul is based on the Greek view that health represents a balance or harmony of various elements of the body, such as bodily fluids or material taken into the body. Disease is a disturbance of that balance. Alcmaeon of Croton maintained that health represented a balance of such opposites as the dry, the wet, the hot, the cold, the sweet, the bitter. Other theories were based on the four humors, which were borrowed from Empedocles; or on the four qualities (heat, cold, moisture, and dryness) that produce the four temperaments. When the balance is disturbed or upset, medical treatment consists of restoring them to their right proportions in accordance with nature rather than contrary to it.

The view of health as a condition of the body in which all its elements work together naturally and harmoniously suggested to the Greeks an appropriate analogy with the soul. Moral virtue, after the model of human pathology, was regarded not as adherence to an external code or standard, but as a balance of the elements of the soul. Thus Plato writes:

> The just man will not allow the three elements which make up his inward self to trespass on each other's functions or interfere with each other, but, by keeping all three in tune, like the notes of a scale . . . will in the truest sense set his house to rights, attain self-mastery and order, and live on good terms with himself. When he has bound these elements into a disciplined and harmonious whole, and so become fully one instead of many, he will be ready for action of any kind . . . (*Republic* 443 D–E, trans. Lee).

The elements of the soul spoken of here are neutral, like the elements of the body, but they are capable of producing virtue or vice, depending on whether they are in a disordered or harmonious state. Plato's concept of virtue includes both the virtues of the body and the soul, and hence he makes the virtues of the body (health, beauty, and strength) parallel the virtues of the soul. The body–soul analogy was used not only by Plato but by writers of nearly all philosophical schools. Since health was an integral part of traditional definitions of *arete* and even the greatest of all virtues, it was natural that when Socrates and his successors discussed virtue they should make the care of the body an analogue for the cure of the soul. The philosopher became a physician of the soul.[12] Thus Epictetus, the first-century slave who became a Stoic philosopher, told his students:

> The philosopher's school, sirs, is a physician's consulting-room. You must leave it in pain, not in pleasure; for you come to it in disorder, one with a shoulder put out, another with an ulcer, another with fistula, another with headache. And then you would have me sit here and utter fine little thoughts and phrases, that you may leave me with praise on your lips, and carrying away, one his shoulder, one his head, one his ulcer, or his fistula, exactly in the state he brought them to me. Is it for this you say that young men are to go abroad and leave their parents and friends and kinsmen and property, that they may say, 'Ye gods!' to you when you deliver your phrases? Was this what Socrates did, or Zeno, or Cleanthes? (*Discourses* 3.23.30–32, trans, Matheson; [23], p. 389).

One of the most extensive discussions in classical literature of diseases of the soul is found in Cicero's *Tusculan Disputations*. In this work Cicero deals with the problems of death, pain, and the disorders of the soul. After stating in Book Two that pain must be overcome by virtue and that it is inconsistent with the four cardinal virtues to yield to pain (2.31–32), he goes on in Book Three to praise philosophy as the medicine of the soul. The art of healing the soul has lagged behind the art of healing the body, he says (3.1). "But diseases of the soul are both more dangerous and more numerous than those of the body" (3.5.). There is an art of healing the soul; it is philosophy and we must be our own physicians (3.6). Cicero continues in Book Four his discussion of the disorders that afflict the soul. He uses the traditional analogy of disease in the body resulting from a disordered state to attribute disease of the soul to the "disturbing effect of corrupt beliefs warring against one another" (4.23). These include avarice, ambition, love of women, and other disturbances (4.25–26). After a lengthy discussion of diseases, sicknesses, and defects of the soul, Cicero continues the analogy:

> Moreover, as in evil the analogy of the body extends to the nature of the soul, so it does in good. For the chief blessings of the body are beauty, strength, health, vigour,

agility; so are they of the soul. For as in the body the adjustment of the various parts, of which we are made up, in their fitting relation to one another is health, so health of the soul means a condition when its judgments and beliefs are in harmony, and such health of soul is virtue, which some say is temperance alone, others a condition obedient to the dictates of temperance and following close upon it and without specific difference, but whether it be the one or the other, it exists, they say, in the wise man only (4.30, trans. King).

In Book Five Cicero attempts to demonstrate that virtue alone is sufficient to lead a happy life. It is the disturbances of the soul that produce misery, while tranquillity leads to a happy life. The wise man has control of his emotions and is free from the agitations of the soul. Therefore, the wise man is always happy (5.43).

The concept of health as an analogue of virtue was one of the most fruitful ideas in ancient ethics. Edelstein has suggested that it represents the greatest debt of classical philosophy to medicine ([10], p. 360). From the fourth century before Christ medical terminology was appropriated for the discussion of ethics. Philosophers regularly spoke of the soul as sick or diseased.[13] Emotions were referred to as *pathe*, 'sufferings'. *Sophrosyne*, which originally meant 'soundness of mind', came to mean 'self-control' ([27]), p. 162). *Hygies* ('healthy') was used to describe an idea as sensible or judicious. Lacking a sense of sin, the Greeks conceived of virtue in terms of the harmony of the human body. Hence it has been common to speak of the Greeks as conceiving of virtue in aesthetic terms, with an emphasis on beauty, proportion, and harmony.[14] The human passions were not to be denied or their indulgence regarded as sinful. Moderation (*sophrosyne*) was the key. *Meden agan* ('Nothing in excess') was one of the mottoes inscribed over the temple at Delphi and the theme is repeated often in Greek literature as the key to physical and mental well-being. "Pleasure ought to be roused in moderation, otherwise we lapse into sickness," says the physician Eryximachus in Plato's *Symposium* (187E; [27], p. 162). For Aristotle the 'Golden Mean' became the basis of his whole theory of ethics.

HEALTH AS AN INDICATOR OF VIRTUE

The ethics of Aristotle, like the rest of his philosophy, is teleological. In his *Nichomachean Ethics* he is concerned to find the goal or purpose of man, the end at which he should aim. All men, he says, seek *eudaimonia*, which is often translated as 'happiness', but which means 'well-being', i.e., the state of both acting and faring well. *Eudaimonia* is 'an activity of the soul according to virtue'. Virtue is a state or disposition of character that is acquired by habit through the persistent practice of morally good actions. In contrast to Plato, Aristotle conceives of virtue neither as one, nor innate, nor derived from metaphysical speculation (Book One). It is learned, as arts or crafts are learned, by practice, and it involves making moral choices as rational human beings. We become virtuous by performing virtuous actions. Hence courage is acquired by performing courageous deeds and justice is acquired by acting justly. The virtues are defined by Aristotle in terms of the doctrine of the mean. Aristotle considered all virtues to be means between extremes. The mean is defined not absolutely but 'relatively to us', by which he means the right degree for each individual. The mean is the right proportion that is suitable to every action and circumstance. Hence the behavior of the virtuous man will always be appropriate to every occasion. The virtues are middle points between extremes of excess and defect. Thus courage is the mean between rashness and cowardice; temperance between abstinence and self-indulgence; generosity between meanness and extravagance. The determination of each of these virtues depends on the individual, who is assumed to be a person of good sense. There is no absolute moral standard and hence no one right choice of emotion or action that is independent of the particular circumstances. What is temperate on one occasion might on another be mean and on a third be extravagant (Book Two). In place of a list of absolute rules, Aristotle presents us with a series of moral choices, the solutions of which must be worked out in each particular case. It is our response to these choices that leads to virtuous conduct. We discipline ourselves to act rightly and so develop habits that entitle us to be called virtuous. The result of our conduct will be *eudaimonia*, a state of well-being.

The standard of the mean was an important concept in Greek dietetics. It has been suggested, in fact, that Aristotle's idea of virtue as the mean between the extremes of excess and deficiency was borrowed from Hippocratic medicine.[15] Just as in dietetic medicine treatment must be tailored to individuals, so in moral conduct, in the absence of universal rules, the individual must determine the proper standard for his actions by aiming at what is appropriate to the occasion. Dietetic medicine became in the late fifth century B.C. a method of preserving health by regulating one's regimen.[16] It doubtless began as a means of treating disease by restoring the proper balance to the body. The Greeks had long recognized that health is related to diet and exercise. "Look after your health and be moderate in drink, food, and sports," say the Golden Sayings of Pythagoras (line 32; [27], pp. 161–162). Dietetics soon developed, however, beyond the mere restoration of health; it became a means of insuring continued health by preventing disease. Originally the teacher of gymnastics had specialized in the care of the body. Now physicians came to view physical training as part of the daily routine of a healthy life that required their supervision. The branch of medicine that was concerned with the preservation of health became known as hygiene (*ta hygieina*) and it involved the regulation of both diet and physical exercise. Throughout the rest of classical antiquity the concern with establishing a regimen that would insure good health was as important for physicians as treating the ill. Medicine became "the education of the healthy man" ([13], p. 416). Treatises on hygiene were popular. Perhaps the earliest that we have is the pseudo-Hippocratic *On Regimen in Health*, a short work that was followed by many longer ones like the pseudo-Hippocratic *On Diet*, which gives a detailed description of proper diet and exercise. Another lengthy work that prescribed a daily routine for preserving one's health was written by Diocles of Carystus, who flourished perhaps in the late fourth century before Christ.[17] His treatise on diet is preserved only in fragments quoted by a late author (Oribasius). In it he sets out to describe a daily program based on the standard of the Aristotelian mean. Diocles was influenced by Aristotle's ethics and he applied to medicine concepts and terminology drawn from Aristotle (just as Aristotle had earlier used medical analogies in constructing his theory of ethics). Diocles makes much use of the concept of 'the suitable'. In establishing a daily hygienic program to insure good health a mean must be found that is both suitable and beneficial to the individual and his circumstances. Although he prescribes a regimen that is intended for a man of means who possesses the leisure to devote his entire day to regulating his behavior according to certain standards of exercise and diet, his ideal regimen is one that can be adapted to the needs of those who wish to take care of the body but have less time in which to do it. He also recognizes the need to adapt one's daily regimen to the requirements of different ages and times of the year.

The importance given to dietetics in the classical world reflected the common assumption that health was the greatest virtue. Hence physicians (and others) could write treatises encouraging men to live for the sake of their health. No other society in history before our own placed so much importance on preventive medicine as did that of the Greeks; and no other people so oriented their lives towards physical culture. Much of the background of the Socratic dialogues is the world of the gymnasium and the palaestra and it is not surprising that the discussions of *arete* that we find in them were so deeply indebted to the ideal of physical culture ([14],

Vol. 3, p. 44). But to the Greeks health was not only the greatest of the virtues and an analogue of virtue, it was also an indicator of virtue. Jaeger has described Diocles' attitude to hygiene as "almost an ethical one. His dietetics is, so to speak, the ethics of the body" ([13], p. 417). Given the importance of the body in Greek thought and the belief that physical health provided a paradigm for the harmony of the soul, it was natural to believe that those who are strong and healthy are morally superior to those who are weak and sickly. As a result of the wide acceptance of the parallel between the training of the body and the care of the soul, preventive medicine (i.e., dietietics) came to be viewed as the counterpart of ethics. The care of the body was regarded as a spiritual duty ([14], Vol. 3, p. 45). Plato believed that the inner life of the soul could be improved as the body is trained and educated ([20], p. 353). Eating and drinking were not to be done merely for pleasure but to create and preserve the harmony of the body that was a necessary condition for the harmony of the soul. Since the virtue of the soul and the body were so intimately related, the health of the body mirrored the health of the soul.

According to the principles of dietetics, health can be maintained by right living, while disease is the result of bad habits. A healthy individual reveals himself as one who practices self-control and moderation, Disease and sickness can be avoided and their presence is an indication of the lack of proper regulation of one's life. But overeating or overindulgence of the passions not only leads to bad health, but creates an unhealthy disposition of the soul. Hence medicine and philosophy were thought to complement each other in enabling one to lead a harmonious life, which in turn produced happiness ([10], p. 391). It was in the light of the classical view of health as an indicator of one's own virtue that the virtuous physician was defined. In the Hippocratic Corpus we read that the physician should look healthy and be of suitable weight, "for the common crowd considers those who are not of excellent bodily condition to be unable to take care of others" (*The Physician* 1, trans. Jones). The ordinary Greek would have considered it strange to turn for medical assistance to a physician who could not preserve his own physical *arete*. "Physician, heal thyself!" would have been an appropriate Greek response to such a physician, who ought first to learn to order his own health properly. In ancient medical ethics, as found, for example, in the Hippocratic Corpus, we find much that is motivated by the physician's concern for his reputation: hence the overriding concern with medical etiquette that will create harmonious relations between physicians and patient. Little is said about the 'ideal' or virtuous physician or the moral basis for medical practice. To the Greeks the essential requirement of the physician was competence in the art. Motivation mattered little and was a matter of personal choice. The ancients recognized that physicians might practice from any number of motives. There was no humanitarian or philanthropic impulse that was thought to be necessary for the 'ideal' physician until relatively late in classical antiquity ([5], pp. 1–8). In the so-called Hippocratic Oath, an esoteric document that is not always consonant with the mainstream of Graeco-Roman medical ethics, justice rather than charity is enjoined upon the physician ([10], pp. 6 and 33–37). It is true that one frequently finds the physician depicted in simile and metaphor as an ideal with whom philosophers, legislators, and statesmen are compared. In this sense the physician is characterized as a compassionate and dedicated individual who serves as a model for what the statesman, for example, should be to the state. But from the classical point of view the virtuous physician was regarded in a much more basic sense. He was one who was himself healthy, i.e., who had properly regulated his own body, which was a requisite to the right ordering of his soul, and so was qualified to advise others in matters of health. Hence a physician who was not of suitable weight could not be virtuous. The virtuous physician was not, however, expected to be humanitarian or philanthropic in his practice of medicine. The charitable and humanitarian impulse in medicine was largely derived from Christianity. Here, as elsewhere, in Greek and Roman medicine the difference between the classical and Christian emphases derives from very different ideas of virtue.

THE DEVALUATION OF HEALTH

Beginning in the third century before Christ Greek thought began to undergo a transition from the classical view of health as a necessary virtue and an indicator of virtue to the view that other virtues are superior to health and that happiness may be achieved at the expense of the body. This new view had its origins in the development of the idea that there is a dichotomy between the body and the soul, a belief which E. R. Dodds has said of Greece was "the most far-reaching, and perhaps the most questionable, of all her gifts to human culture" ([9], p. 29). The philosophical sects that arose in the fourth century B.C. – Stoicism, Epicureanism, and Cynicism – did not at first depreciate health; generally they agreed that it was a virtue. But they gained prominence in the period that was marked by the decline of the city-state and traditional religion and the growth of individualism and cosmopolitanism. They aimed at providing practical guidance for indivudual ethics, and their teaching stressed self-sufficiency, independence of all externals, and self-mastery. Hence they were anxious to point out (given the exaggerated importance traditionally accorded to the physical graces and the tendency of dietetic medicine to make the care of the body one's chief concern) that happiness depends less on health than on other factors, particularly on those things that are within one's own control ([10], p. 359).

The dualistic tradition in Greek thought goes back at least to Socrates, who was concerned, by the practice of *askesis* ('training'), to make his soul independent of his body. We find here the beginning of an ascetic tradition that eventually led to the belief that the proper attitude to disease and suffering was one of indifference, while the real concern of an individual should be the care of the soul. The result was a turning inward to the life of individual experience and spiritual health and a consequent disparagement of physical well-being. In the fourth century B.C. this attitude had not fully developed and philosophers like Diogenes the Cynic could appreciate the importance of health and medicine. By the third century B.C., however, the Cynic devaluation of the body had become pronounced ([18], pp. 308–309). Stoicism and Cynicism shared many common beliefs. Stoicism held that virtue is the *summum bonum*, which alone brings happiness. Everything in nature is both rational and good; therefore, a virtuous life is one that is lived in conformity to nature and reason. Since moral virtue is the only real good and moral weakness the only real evil, everything else, including poverty, death, and pain, is an *adiaphoron*, or an indifferent thing. One's happiness is not dependent on whether they are absent or present. As long as one acts in conformity with reason and nature, one is in possession of virtue and therefore independent of change or fortune. Everything that exists, including sickness and disease, is advantageous and intended for our education. No-one can deprive the wise person of virtue and since he possesses the true good he is always happy. Zeno (335–263 B.C.), the founder of Stoicism, hoped to strengthen character and to make men more self-sufficient; and his ethical system comforted many by its teaching that the world had no power over the soul and that one could be master of oneself by disparaging the body and withdrawing into one's soul and there finding peace.

Although Cynicism had a few prominent representatives it was never organized into a school and it gradually declined in the second and first centuries B.C. It enjoyed a revival in the first century after Christ in the form of beggar-philosophers, who wandered throughout the Roman Empire with a stick and knapsack, denouncing all conventions of society and preaching a gospel of simplicity, independence, and a return to nature. Stoicism, on the other hand, became the dominant philosophy of the Hellenistic world. Its strongest opponent was the materialist school of 'the Garden', Epicureanism, which was founded by Epicurus of Samos (341–270 B.C.). The ethics of the Epicureans was based on the belief that the highest good for man is happiness and that happiness is found in the pursuit of pleasure and the avoidance of pain and fear. By this Epicurus did not mean the pursuit of bodily pleasures; rather he meant that happiness is to be found in contemplation and the pursuit of philosophy, which brings tranquillity of soul. He was a quietist not a hedonist. He encouraged withdrawal from the world and taught that the good life consists of the pursuit of quiet pleasures. He recognized that pleasures and pains differ in degrees. While sensual pleasures are merely transitory, those of the mind are more permanent and in order to secure them the wise man will choose suffering and poverty if necessary. But the Epicureans believed that the pleasures of the mind are related to the physical body and valued the health of the body and the removal of pain as a contributing factor to the tranquillity of the soul. But the pleasures of the mind provided resources to be used to overcome sickness and pain when they occurred. Near the end of his life Epicurus wrote to Idomeneus: "On this truly happy day of my life, as I am at the point of death, I write this to you. The diseases in my bladder and stomach are pursuing their course, lacking nothing of their natural severity: but against all this is the joy in my heart at the recollection of my conversations with you" (*ap*. Diogenes Laertius 10.22; trans. Baily; [23], p. 59).

Epicureanism never attracted large numbers of followers and its influence was limited in both the Hellenistic and Roman worlds. Stoicism was introduced into Rome in the second century B.C. and became the dominant philosophy of the Roman aristocracy. With its emphasis on duty, self-discipline, and civic-mindedness, it was well suited to give expression to the ideals of the Roman Empire. Unlike the Cynics and Epicureans, who practiced withdrawal from society, the Stoics encouraged participation in public affairs. They emphasized inner freedom and *autarkeia*, the independence of the virtuous man from external circumstances. Thus Seneca says that one is healthy if one is self-contented and does not depend on external things for personal happiness (*Moral Epistles* 72. 7). Similarly Epictetus taught that we must not let our happiness depend on things that are not in our power. Only our will is in our power and we must keep it pure. He counsels submissiveness to the inexorable and indifference to death, pain, and illness, all of which happen by divine providence and are meant for our good.[18] We are not far removed here from what Edelstein called "the Christian or Romantic glorification of disease" ([10], p. 387). But Stoicism (and the classical world generally) could only enjoin submission and resignation when confronted by pain and suffering, whereas Christianity endowed them with positive value as part of God's discipline and training that were productive of edification and in which a Christian could rejoice (James 1:2–4; Hebrews 12:7–11).

In addition to the tendency in Hellenistic and Roman thought to devalue health and disdain the body in favor of the soul, there was another marked feature of the period that led to a transformation of moral ideas. This was the cosmopolitanism and humanitarianism that began in the Hellenistic era and came to pervade Graeco-Roman thought in the first two centuries of the Christian era. It consisted of an increasing emphasis on the gentler (as opposed to the heroic) qualities and a greater sensitivity to benevolence that spread to all classes within the Roman Empire. This new spirit developed out of a sense of the brotherhood of all mankind that found expression in both popular and philosophical ethics. Both the Stoics and the Cynics preached the common kinship and equality of all men, civilized or barbarian, slave or free, who were regarded as citizens of the world. One finds, particularly in Stoics like Musonius Rufus, Seneca, Epictetus, and Marcus Aurelius, an emphasis on universal kindness and helpfulness, on human behavior and charity, on the forgiveness of those who have wronged us. This spirit influenced medicine, which came to be viewed, at least by some, as a vehicle for compassion – an aspect that it had lacked earlier in the classical world. The concept of *philanthropia*, the 'love of mankind', had been broadened by the new prominence given to the amiable virtues to denote, in the words of Aulus Gellius, "a kind of friendly spirit and good feeling towards all men without distinction" (*Attic Nights* 13. 17. 1). Its Latin equivalent was *humanitas*, a word with many of the same associations as *philanthropia*, with which by the Christian era it had come to be synonymous ([5], pp. 8–9). Galen regarded medicine as 'an especially philanthropic art' because it relieved the sufferings of mankind. He believed that the best physician should also be a philosopher who embodied the qualities of a moral life, a view that he sets forth in a short work entitled *That the Best Physician is Also a Philosopher*. For Galen a physician who was a philosopher would be motivated by *philanthropia*, which he would demonstrate, as Hippocrates did, by treating the poor, advancing the knowledge of medicine, and publishing that knowledge for the benefit of mankind ([5], p. 10). Galen considered philanthropy to be desirable for the physician but not, like competence in the art, essential. Scribonius Largus, a Roman physician of the first century, goes beyond Galen in considering it a requisite quality of the true physician. In a short essay, *Professio medici*, which prefaces his treatise *On Remedies*, Scribonius gives perhaps the highest expression to the ideals of compassion and philanthropy in classical medicine. He writes that medicine should try in every way to give help to the afflicted; she does not regard men's circumstances or character, but aids all alike who seek help. For Scribonius the profession of medicine requires not only competence in the art but sympathy (*misericordia*) and humane feeling (*humanitas*) on the part of the physician ([10], pp. 337–343). Both Galen and Scribonius reflect the influence of Stoicism on the development of a genuinely philanthropic impulse in medicine. This impulse is found elsewhere. It is echoed in a poem composed by a Stoic philosopher of the second century, Serapion, entitled *On the Eternal Duties of the Physician*; and in a progymnasma of Libanius (8) in the fourth century, as a speech made by medicine to a physician who is beginning his practice ([10], pp. 344–345). In both instances the duties of the physician are set forth in terms that speak of compassion, sympathy with the patient, and brotherhood. It is impossible, of course, to know how many ordinary physicians were motivated in their practice of medicine by a disinterested 'love of mankind'. Perhaps it is unrealistic to assume that the number was ever very large. On the other hand, the emphasis on philanthropy and compassion in the philosophical medical ethics of later antiquity will likely have had some influence on medical practice. It clearly represents a transformation of the classical concept of virtue and anticipates the new definition of virtue introduced by Christianity. In this sense it looks both backward and forward and forms a transition between the ideals of classical and Christian medicine.

ACKNOWLEDGEMENT

This publication was supported in part by NIH Grant LM 04108 from the National Library of Medicine.

NOTES

1 See [1], pp. 30–60 and passim; [14], Vol. 1, pp. 3–14; [27], pp. 153–190.

2 The first three lines of this skolion are quoted by Plato, *Gorgias* 415E, while the full quatrain is quoted by the scholiast and by Athenaeus and Stobaeus.

3 See [10], pp. 382–384, and 9–20; and [19], pp. 212–223.

4 See, e.g., Celsus, *De medicina* 5.26. 1C. On the physician's duty to prolong life see [4].

5 As Kudlien has pointed out [17], there developed in fifth-century Greece a concept of relative health that complemented the idea of perfect health. According to this view, chronic disease or physical deformity was not absolutely inimical to the enjoyment of a good life and it was possible even with some physical defect to have 'sufficient' good health to provide relative well-being. Kudlien associates this concept with the movement towards relativism that stressed not perfect but middle ideals. It was propagated by physicians and intellectuals, whereas ordinary people continued to value the traditional concept of perfect health as the highest of all goods.

6 See [20], pp. 217–222.

7 The phrase is Juvenal's (*Satire* 10, line 356). What Juvenal actually says is, "Let us pray that there be a sound mind in a sound body" (*Orandum est ut sit mens sana in corpore sano*). Juvenal's prayer was a common one: cp. Petronius, *Satyricon* 61 and Seneca, *Epistles* 10.4.

8 This view is expressed by Socrates in Xenophon, *Oeconomicus* 4. 2 ([28], p. 17).

9 They are first mentioned in the *Republic* 427E.

10 The phrase, 'cardinal virtues', seems to have been first used by St. Ambrose, *In Lucem*, Book 5.

11 The Platonic doctrine of the tri-partite soul is set forth in Book Four of the *Republic* beginning at 434E.

12 See [22], pp. 17–41; [10], pp. 364–366; [19], pp. 308–309.

13 See, e.g., Plato, *Laws* 862C.

14 See [8], pp. 146–147; [27], p. 163.

15 See [14], Vol. 3, p. 25; and Aristotle, *Nichomachean Ethics* 1104A 13–14 ff.

16 On Greek dietetics see [10], pp. 303–316.

17 Diocles has been traditionally dated to the fourth century B.C. but Werner Jaeger dates him somewhat later (ca 340–260 B.C.): see [13], pp. 409–414; *contra* [10], pp. 145–152.

18 On Epictetus' views of the body and disease see [18], pp. 314–317.

BIBLIOGRAPHY

[1] Adkins, Arthur W. H.: 1960, *Merit and Responsibility: A Study in Greek Values*, The University of Chicago, Chicago.

[2] Alexander, A. B. D.: 1921, 'Seven Virtues', in James Hastings (ed.), *Encyclopaedia of Religion and Ethics*, Vol. XI, Charles Scribner's Sons, New York, pp. 430–432.

[3] Amundsen, D. W.: 1978, 'History of Medical Ethics: Ancient Greece and Rome', in W. T. Reich (ed.), *The Encyclopedia of Bioethics*, Vol. 3, the Free Press, New York, pp. 930–938.

[4] Amundsen, D. W.: 1978, 'The Physician's Obligation to Prolong Life: A Medical Duty without Classical Roots', *Hastings Center Report* 8 (4), 23–30.

[5] Amundsen, D. W. and G. B. Ferngren: 1982, 'Philanthropy in Medicine: Some Historical Perspectives', in E. E. Shelp (ed.), *Beneficence and Health Care*, Reidel, Dordrecht, Holland, pp. 1–31.

[6] Amundsen, D. W. and G. B. Ferngren: 1982, 'Medicine and Religion: Pre-Christian Antiquity', in M. Marty and K. Vaux (eds.), *Health/Medicine and the Faith Traditions*, Fortress Press, Philadelphia, pp. 53–92.

[7] Bowra, C. M.: 1957, *The Greek Experience*, New American Library, New York.
[8] Dickinson, G. Lowes: 1928, *The Greek View of Life*, Doubleday Doran, Garden City, New York.
[9] Dodds, E. R.: 1968, *Pagan and Christian in an Age of Anxiety*, Cambridge University Press, Cambridge.
[10] Edelstein, L.: 1967, in O. Temkin and C. L. Temkin (eds.), *Ancient Medicine: Selected Papers of Ludwig Edelstein*, The Johns Hopkins Press, Baltimore.
[11] Ferguson, J.: 1958, *Moral Values in the Ancient World*, Methuen, London.
[12] Geach, Peter: 1977, *The Virtues*, Cambridge University Press, Cambridge.
[13] Jaeger, Werner: 1934, 1962, *Aristotle: Fundamentals of the History of His Development*, trans. by Richard Robinson, Oxford University Press, London.
[14] Jaeger, Werner: 1939–1945 (1965), *Paideia: The Ideals of Greek Culture*, 3 vols., trans. by Gilbert Highet, Oxford University Press, New York.
[15] Jaeger, Werner: 1957, 'Aristotle's Use of Medicine as a Model of Method in the Ethics', *Journal of Hellenic Studies* 77, 54–61.
[16] Kenny, Anthony: 1978, *The Aristotelian Ethics*, Clarendon Press, Oxford.
[17] Kudlien, F.: 1973, 'The Old Greek Concept of "Relative" Health', *Journal of the History of the Behavioral Sciences* 9, 53–59.
[18] Kudlien, F.: 1974, 'Cynicism and Medicine', *Bulletin of the History of Medicine* 48, 305–319.
[19] Lecky, William E.: 1869, 1902, *History of European Morals: From Augustus to Charlemagne*, 2 vols., Longmans Green, London.
[20] Lodge, R. C.: 1928, *Plato's Theory of Ethics: The Moral Criterion and the Highest Good*, Kegan Paul, London.
[21] MacIntyre, Alasdair: 1981, *After Virtue*, Duckworth, London.
[22] McNeill, J. T.: 1951, *A History of the Cure of Souls*, Harper and Row, New York.
[23] Oates, Whitney J.: 1940, 1957, *The Stoic and Epicurean Philosophers*, Modern Library, New York.
[24] Pence, Gregory E.: 1980, *Ethical Options in Medicine*, Medical Economics Co., Oradell, NJ.
[25] Rist, John M.: 1982, *Human Value: A Study in Ancient Philosophical Ethics,* Brill, Leiden.
[26] Sidgwick, Henry: 1886, 1960, *Outlines of the History of Ethics*, Beacon Press, Boston.
[27] Snell, B.: 1960, *The Discovery of the Mind*, trans. by T. G. Rosenmeyer, Harper and Row, New York.
[28] Strauss, Leo: 1970, *Xenophon's Socratic Discourse: An Interpretation of the Oeconomicus*, Cornell University Press, Ithaca.
[29] Temkin, Owsei: 1973, 'Health and Disease', in Philip P. Wiener (ed.), *Dictionary of the History of Ideas*, Vol. 2, Charles Scribner's Sons, New York, pp. 395–407.

Oregon State University,
Corvallis, Oregon,
U.S.A.

Western Washington University
Bellingham, Washington,
U.S.A.

DARREL W. AMUNDSEN AND GARY B. FERNGREN

VIRTUE AND MEDICINE FROM EARLY CHRISTIANITY THROUGH THE SIXTEENTH CENTURY

In order to examine the relationship of virtue and medicine from early Christianity through the sixteenth century, we shall first discuss the history of concepts of virtue from a variety of perspectives without any anticipation of possible relevance to medicine at this stage. We shall begin with a consideration of virtue in New Testament thought and then examine changing concepts of virtue through the sixteenth century. Our discussion here, though topical, will be essentially chronological, ending with the Protestant Reformation. We shall then look briefly at concepts of virtue in that cultural milieu popularly called the Renaissance which, especially in Italy, occurred during the late Middle Ages and in northern Europe was roughly concurrent with the Reformation. Finally, we shall be in a position to suggest various historical relationships between virtue and medicine.

A point of semantic and conceptual clarification should be made. The words for virtue in Indo-European languages are somewhat ambiguous. They sometimes denote moral virtue. At other times they mean the capacity for action and the power to achieve a particular result, as, for instance, when Galen defines *arete* as the power to accomplish something. When moving from one language to another, the ambiguities increase. For instance, the account in Mark's Gospel (5:25–34) of the healing of the woman with the chronic flow of blood relates that when the ill woman touched him, Jesus felt *dunamis*, 'power', go out from him. The Authorized Version translates this as 'virtue' in accord with contemporary English usage. Renaissance Italian achieved a semantic distinction by calling moral virtue *virtú* and the non-moral sense of virtue, i.e., a capacity or power to achieve a result, *virtù*. In the present paper we shall be limiting our discussion to *virtú*, i.e., moral virtue.

VIRTUE IN NEW TESTAMENT THOUGHT

When Christianity was taking root in the syncretistic soil of the Mediterranean world, the popular Greek word *arete* was pregnant with a wide variety of nuances. It could mean (1) 'excellence', in a general or specific sense; (2) 'manliness' and 'valor'; (3) 'merit'; (4) 'virtue', as a moral concept; (5) 'fame',

Earl E. Shelp (ed.), Virtue and Medicine, 23–61.

'success' and 'praises'; or (6) 'divine self-declaration' and 'divine power'.[1] The Hellenistic Jews who translated the Hebrew Scriptures into Greek (the Septuagint) appear to have studiously avoided using the word to translate any Hebrew word descriptive of human qualities, character, or values, applying it only to God, and that merely six times, all of which come under definitional categories 5 and 6 above.[2] *Arete* is applied to human character, however, in the writings of Hellenistic Judaism, particularly in the wisdom literature and in Maccabees, where it describes "the fidelity of the heroes of faith in life and in death" ([21], vol. 1, p. 459) especially as exemplified in martyrdom. Even in this Jewish literature, which was (to a certain extent) influenced by Hellenism, there is an obvious avoidance of the word *arete* where it might suggest a typically Greek (as well as Roman) anthropocentrism bred of concepts of human merit and innate and potential excellence existing without having its source in, and being dependent upon, God. When *arete* in Hellenistic Judaism (except in the writings of Philo) describes human character, it seems to have reference to the quality conveyed by the Jewish use of the word *dikaiosune* ('righteousness', 'uprightness'), a concept that in Jewish thought was as theocentric as it and *arete* were anthropocentric in Greek thought.

Arete was also very little used by the authors of the New Testament. The near absence there of such a popular word appears quite inexplicable, primarily to those who assume that the authors of the New Testament were influenced by classical sources, which they then sought to blend with certain features of Judaism. If such had been their objective, the concept of *arete* would have been convenient to exploit. But they ignored or perhaps deliberately rejected the word for the most part. It is found nowhere in the synoptic Gospels, in the Johannine literature, in Acts, James or Jude. It is used only once in the Pauline epistles, once in 1 Peter, and twice in 2 Peter. In 1 Peter 2:9 and 2 Peter 1:3 *arete* is used in a manner consistent with the way it is employed in the Septuagint. In the first of these, it describes (in the plural) God's "wonderful deeds"[3] as in definitional category 6 above. In the second reference it is coupled with *doxa* in the phrase "his own glory and excellence," having perhaps the meaning of category 5 or 6 above, although various scholars argue that it here has the force of "the energy of God" ([9], *ad loc.*), leading then to the repetition of the word later in the passage (vs. 5) where it is applied to Christian character perhaps as "the moral energy generated in believers by faith" ([9], *ad loc.*; [26], *ad loc.*). The passage in question reads ". . . make every effort to supplement your faith with virtue, and virtue with knowledge, and knowledge with self-control, and self-control with steadfastness, and steadfastness with godliness, and godliness with brotherly affection, and brotherly affection with love" (2 Pet. 1:5–7). Some scholars see this list of Christian 'virtues' as one of several examples of New Testament authors following the classical, particularly Stoic, precedent of listing cumulative or disparate virtues. It is unlikely, however, that this list of Christian 'virtues', of which 'virtue' is but a part, is modelled on pagan philosophical schemata. Here 'virtue' is but one of an extended list of qualities that must be seen in their distinctly Christian setting. Of course all the words are Greek and were commonly used by classical authors. But *arete* here, coming immediatly after *pistis* (faith), is likely as foreign to its classical counterpart as is 'faith', which in classical usage means merely 'fidelity'. The sequence is: faith, virtue, and knowledge. "Since both faith and knowledge have a specific Biblical thrust, one must conclude that virtue has its own nuance too ...."[4] That nuance here, however, may well be simply 'moral probity', i.e., the most basic and generally-accepted standards of conduct and goodness which the Christian would have in common with all people for whom a minimal moral consciousness may be predicated. This also seems to be the thrust of *arete* in its lone appearance in Paul's epistles.

In his letter to the Philippians Paul writes: "... whatever is true, whatever is honorable, whatever is just, whatever is pure, whatever is lovely, whatever is gracious, if there is any excellence, if there is anything worthy of praise, think about these things" (Phil. 4:8). Here *arete* is translated 'excellence' and is coupled with *epainos* ('that worthy of praise'). Two of the best-known commentators of about a century ago suggest that Paul means, "Whatever value may reside in your old heathen conception of virtue ..." ([23], *ad loc.*) and that "he bids them exercise thought on whatever is rightly called 'virtue', even if not expressly described in the previous words" ([27], *ad loc.*). A more recent commentator maintains that Paul is speaking simply of "normal goodness" ([9], *ad loc.*).

Two matters should be clear at this stage. (1) The authors of the New Testament, by applying the word *arete* only twice to human character, qualities, or values, were not adopting Greek anthropocentric concepts of virtue subsumable under the rubric *arete*. (2) Any effort on our part to focus on the word *arete* in seeking to determine what was regarded as virtue, virtues, or virtuous by the authors of the New Testament would lead to the conclusion that the New Testament possesses and conveys, at the worst, no conception of virtue or, at the best, a pathetically nebulous one. It is safe to assert categorically that while the classical concepts of virtue did not provide the basis for an early Christian formulation and enunciation of virtue, the New Testament is rife with teaching on virtue that is informed by theocentric principles shared with Old Testament and Hellenistic Judaism.

Even a cursory reading of the New Testament reveals that its authors were acutely sensitive to moral values and ethical conduct springing from human character transformed by Christian conversion.[5] Early in his ministry Jesus began to reveal to his disciples the qualities that they should seek to have. He calls those *makarios* (blessed, fortunate, happy) who are poor in spirit, mourn, are meek, hunger and thirst for righteousness, are merciful, pure in heart, are peacemakers, and are persecuted for righteousness' sake (Mt. 5: 3–10). Lowliness, gentleness, purity, and holiness are the substance of these beatitudes, but are only part, albeit a basic part, of the character of those who would follow him. Paul lauds similar qualities, urging the Colossians to "put on compassion, kindness, lowliness, meekness, and patience; forbearing one another and . . . forgiving each other; as the Lord has forgiven you, so you also must forgive. And above all these put on love . . . . And let the peace of Christ rule in your hearts . . . " (Col. 3:12–15). He entreats the Ephesians "to lead a life worthy of the calling to which you have been called, with all lowliness and meekness, with patience, forbearing one another in love . . ." (Eph. 4:1–2). A little later he says, "and be kind to one another, tenderhearted, forgiving one another, as God in Christ forgave you. Therefore be imitators of God . . . and walk in love, as Christ loved us and gave himself up for us . . . " (Eph. 4:31–5:2).

When these qualities are combined with the features described in 2 Peter 1:5–7 (quoted above), we find a broad range of Christian virtues similar to those expressed by Paul in two of his letters. Writing to Timothy he instructs him to "aim at righteousness, godliness, faith, love, steadfastness, gentleness" (1 Tim. 6:11). And in his letter to the Galatians he lists as "the fruit of the Spirit" "love, joy, peace, patience, kindness, goodness, faithfulness, gentleness, self-control" (Gal. 4:22–23). In words reminiscent of 2 Peter 1:5–7, Paul writes to the Roman Christians that "we rejoice in our sufferings, knowing that suffering produces endurance, and endurance produces character, and character produces hope, and hope does not disappoint us, because God's love has been poured into our hearts through the Holy Spirit which has been given to us" (Rom. 5:3–5).

Although we cannot attempt to be exhaustive in culling from the New Testament statements that describe or encourage Christian character, what has been quoted thus far goes a long way in revealing the essential nature of Christian virtue. These are gentle but steadfast qualities, all of which are grounded upon 'love', a word that appears in most of the passages given above. Love itself is described by Paul as having these qualities: "Love is patient and kind; love is not jealous or boastful, it is not arrogant or rude. Love does not insist on its own way; it is not irritable or resentful; it does not rejoice at wrong, but rejoices in the right. Love bears all things, believes all things, hopes all things, endures all things. Love never ends . . . " (1 Cor. 13:4–8). But love is not merely a virtue possessing all those features that are themselves virtues central to Christian character. It goes much deeper than that in New Testament thought. "God is love" (1 John 1:8 and 16; cf. 2 Cor. 13:11) and it is love that motivates God's salvatory relationship with mankind: "In this the love of God was made manifest among us, that God sent his only Son into the world, so that we might live through him. In this is love, not that we loved God but that he loved us and sent his Son to be the expiation for our sins" (1 John 4:9–10; cf. John 3:16). Any response to God is a response to his prevenient love: "We love, because first he loved us" (1 John 4:19). This love is to be both vertical (toward God) and horizontal (toward one's fellow men). Jesus says, "A new commandment I give to you, that you love one another; even as I have loved you, that you also love one another" (John 13:34). But Christian love was not to be extended only to fellow Christians. Shortly after describing Christian character in the beatitudes, Jesus remarked, "You have heard that it was said, 'You shall love your neighbor and hate your enemy.' But I say to you, Love your enemies and pray for those who persecute you . . . . For if you love those who love you, what reward have you? Do not even the tax collectors do the same? And if you salute only your brethren, what more are you doing than others? Do not even the Gentiles do the same?" (Mt. 5:43–47; cp. Lk. 6:27–36). Jesus quoted two passages from the Old Testament which sum up the entirety of God's commandments, both God-ward and man-ward, with the one word 'love': A scribe had asked him which was the great commandment in the law. "And he said to him, 'you shall love the Lord your God with all your heart, and with all your soul, and with all your mind. This is the great and first commandment. And a second is like it, You shall love your neighbor as yourself. On these two commandments depend all the law and the prophets" (Mt. 22: 36–40; cf. Mk. 12:29–33 and Lk. 10:25–27). Love, however, provides only the 'how' and not the 'what' of the law. That is made evident by the plethora of specific injunctions and commands found throughout the New Testament and is emphasized by Paul: "Owe no one anything, except to love one another; for he who loves his neighbor has fulfilled the law." Paul then quotes three of the Ten Commandments and says that these, "and any other commandment, are summed up in this sentence, 'you shall love your neighbor as yourself.' Love does no wrong to a neighbor; therefore love is the fulfilling of the law" (Rom. 13:8–10; cf. Gal. 5:13–14).[6]

The list of Christian 'virtues' given in 2 Peter 1:5–7 begins with faith and ends with love. Faith in the New Testament cannot be separated from love and means much more than the classical idea of fidelity. It means, most basically, trust. God sent his Son for mankind's salvation because of his love. His love is reciprocal and is to be extended by its recipients to all mankind as well. Faith, however, is described as a gift of God: "For by grace you have been saved through faith; and this is not your own doing, it is the gift of God" (Eph. 2:8). It is essential for salvation: "we are justified by faith" (Rom. 1:5). Faith (*pistis*) is derived from the same root as the verb *pisteuo* which means to believe in, or put one's trust in, someone or something. The New Testament never says that God has faith in mankind, or that people are to have faith in each other. What it does state explicitly is that those who have received the gift of salvatory faith are to believe in, or trust in, God. Faith is a virtue in New Testament thought that is as vital and essential to all other virtues as is love. And a functional faith is most basically a condition of humble dependence upon God.

Faith and love are linked with hope by Paul: "So faith, hope, love abide, these three; but the greatest of these is love" (1 Cor. 13:13). Long after this was written, faith, hope, and love were isolated from the other Christian virtues to form the 'theological virtues' to stand beside the Greek 'cardinal virtues'. While this 'triad of Christian virtues' appears only in 1 Cor. 13 in the New Testament, the concept of hope is as foundational to New Testament, concepts of virtue as are love and faith. Hope in the New Testament is considerably different from hope in secular thought. In the latter it is basically a wish or desire that is held in light of the possibility that the opposite, or something considerably less than what is expected, may ensue. But in the New Testament hope, like faith, has its source in and is directed toward God. Hope, when founded upon God and his nature, is a firm and certain expectation of the trustworthiness of his promises, and is thus informed by faith. The author of the letter to the Hebrews defines faith as "the assurance of things hoped for, the conviction of things not seen" (Heb. 11:1). Earlier he wrote of the unchangeable character of God's purposes and the impossibility that he should prove false in his promises which provide the "strong encouragement to seize the hope set before us. We have this as a sure and steadfast anchor of the soul, a hope that enters . . . where Jesus has gone as a forerunner on our behalf . . . " (Heb. 6:17–20). Or, as Paul puts it, "Christ in you, the hope of glory" (Col. 1:27).

While they are in this world, however, Christians are to serve God. This service is to be a manifestation of God's love and the pattern for it is provided by Jesus. Christians are to imitate him. Jesus is recorded as saying that he "came not to be served but to serve" (Mt. 20:28) and he instructed his disciples to follow his example. The example that he left for his followers is two-fold: in service (cf., e.g., John 13:15) and in suffering (cf., e.g., 1 Pet. 2:21). In respect to the latter, the New Testament abounds with affirmation of the necessity and inevitability that those who would follow Jesus' example, who would share his character, or, we could say, those who would have and display the virtues predicated of the Christian in the New Testament, would also share in Christ's sufferings. Jesus had been very blunt about this: "Remember the word that I said to you, 'A servant is not greater than his master.' If they persecuted me, they will persecute you . . . " (John 15:20).

Selfless service and suffering. Such were to be the lot of the Christian in the New Testament. Hardly a happy situation. Yet the New Testament is unambiguous about the joy that is to accompany both these roles: the Christian is to rejoice in all things, including suffering (cf., e.g., Phil. 4:4; Mt. 5:11–12; Rom. 5:3; James 1:2), and give thanks in and for all things as well (cf., e.g., 1 Thess. 5:18; Eph. 5:20). Selfless service and suffering. Hardly an easy lot. Yet Jesus said, "Come to me, all who labor and are heavy laden, and I will give you rest. Take my yoke upon you, and learn from me; for I am gentle and lowly in heart, and you will find rest for your souls. For my yoke is easy, and my burden is light" (Mt. 11:28–30). His description of his own character – "gentle and lowly of heart" – is both what and why his followers are to learn from him as they seek to become like him in character so that they may be like him in service and in suffering. His virtues are to be theirs. They, as we have seen, are to be possessed of holiness and purity – having been the recipients of salvatory faith because of God's love. Their character, anchored in a confident hope, vitalized by a firm faith, and motivated by a selfless love, is to be molded to humility, gentleness, kindness, compassion, self-control, patience, and steadfastness. Those who possess such character transformed by grace, character never perfect but being progressively sanctified, are to serve in the mission of winning souls for Christ and extending his love to the afflicted, the weak, the poor, and the oppressed, by succoring their ills, ministering to their needs, and contributing to their wants. And in the midst of such service, they are to regard, for themselves, the things of this world as of little value, their own lives as only propaedeutic to heaven, and suffering as producing for them "an eternal weight of glory beyond all comparison" (2 Cor. 4:17).

CONCEPTS OF VIRTUE: THE PATRISTIC AGE THROUGH THE SIXTEENTH CENTURY

It is a commonplace to observe that the Greeks had no conception of sin. The closest they came was to speak of moral failure. By contrast Christianity has always been deeply concerned about sin, which is viewed as disobedience and rebellion against God. Since the absence of sin is righteousness and righteousness is often spoken of in terms of virtue (and a righteous person as a virtuous person), much Christian literature, in one way or another, addresses the subject of virtue either explicitly or implicitly. It would require an extensive discussion to do justice to the complexity and diversity of this topic and its relationship with ancillary themes and values. We can only scratch the surface. Three different strands appear in the literature:[7]

(1) A loose and ill-defined use of the word 'virtue' in a popular sense either to mean moral excellence generally or to express a standard of excellence in reference to a specific quality.

(2) A listing of 'virtues' especially valued by the author or considered particularly appropriate to the subject at hand. This at times involves allegory ranging from the elegant and eloquent to the boorish and inane; at other times it reflects a descriptive or popular paraenetic ethics, involving merely enumerations of various virtues or hortatory homilies on the virtuous life. In many instances we see various virtues, in paraenetic lists and in allegory, contrasted with various vices.

(3) Attempts to discuss virtue as a concept particularly with reference to ultimate values and responsibilities. This category was fertile for the treatment of problems in, and the development of concepts of, ethics, free will and providence, and Christian sanctification and growth. It gave rise to extensive discussions of the nature of virtue – i.e., whether innate or infused, natural or supernatural – and is reflected in such developments as asceticism and monasticism.[8]

The first of our three categories warrants no discussion here. Its presence need be noted simply to inform readers of its existence and to alert them to this frequently-encountered but general use of the word (*arete* in Greek, *virtus* in Latin)[9] even in primary sources which may address the subject of virtue or the virtues along the lines of categories 2 or 3.

Category 2 is drawn from a wide and diverse range of sources. From a literary perspective, it is exceedingly rich; from a philosophical or ethical perspective, it is not as fertile. Some early authors singled out certain virtues as most needed. Clement of Rome, for example, in the late first century especially lauded love, repentance, obedience, piety, hospitality, and humility in his Epistle to the Corinthians. About fifty years later the author of the *Shepherd of Hermas*, writing as a visionary, pictured seven virtues as seven women – these were faith, continence, simplicity, knowledge, innocence, reverence and love ("Vision" 3, 8). Later, in the "Similitudes" (9, 5), the author sees twelve virgins and twelve women clothed in black. When he asks who they are, he finds that the virgins are faith, continence, power, patience, simplicity, innocence, purity, cheerfulness, truth, understanding, harmony and love. The women attired in black are unbelief, incontinence, disobedience, deceit, sorrow, wickedness, wantonness, anger, falsehood, folly, backbiting, and hatred. These specific virtues and vices are not paired as exact opposites; nor is the imagery forced to proceed to a conflict between these two groups. But the contrast between them as well as their being presented as personifications soon became a commonly-encountered feature of some types of Christian literature. One element lacking in the *Shepherd*, which soon appears and becomes nearly a constant in allegorical (as well as paraenetic) treatments of virtue, is the theme of conflict.

The theme of conflict is exceedingly important in Christian conceptions of virtue. In Judeo-Christian thought evil is rebellion against God and is essentially, or at least in great part, a matter of the will. This was certainly not the case in classical Greek thought, particularly as examplified by Plato and Aristotle, who thought of evil or vice not in terms of will but reason. Aristotle, especially in his *Nicomachean Ethics* (e.g., 1104 B), maintains that while an effort to be virtuous is an indication of a profligate character, the absence of a need for moral striving is itself an indication of virtue. C. S. Lewis ([22], p. 59) has pointed out how incongruous would the words "Fight the good fight" appear in such a moral system. Not so, however, in pagan (particularly Stoic) philosophers of the early Christian era such as Seneca and Epictetus. "They were vividly aware, as the Greeks had not been, of the divided will, the *bellum intestinum*. The new state of mind can be studied almost equally well in Seneca, in St. Paul, in Epictetus, in Marcus Aurelius, and in Tertullian" (*ibid*., p. 60; cf. [6], p. 63).

Tertullian (second century), to a lesser extent in *De patientia* (15, 4), but very starkly in *De spectaculis* (29), personifies the virtues and vices as two armies fighting for one's soul. In the latter work specific virtues are pitted against the appropriate vices: chastity–lewdness; faithfulness–unfaithfulness; mercy–cruelty; temperance–lasciviousness. Tertullian is the most likely source for Prudentius (fourth-fifth century), whose *Psychomachia* exercised a probably ungaugeable influence on medieval literature and art. In his

Psychomachia, whose very title tells us the thrust of the work, the personified virtues and vices engage in mortal combat. Faith vanquishes idolatry; modesty routs voluptuousness; patience overcomes anger; humility beheads pride; sobriety destroys sensuality; mercy conquers avarice; and finally concord defeats heresy.

The tendency to personify various virtues and vices was very strong in the early Middle Ages and they became, in many minds, as real as angels and demons, having their own distinct personality and iconography. Not content with the uncertainty of *various* virtues and vices, *the* virtues and vices gradually were identified and stabilized, there finally being seven each. The 'seven deadly sins' (pride, avarice, lust, envy, gluttony, anger, and sloth) have a different origin than the very synthetic seven virtues (the four cardinal virtues together with faith, hope, and love, a union that finally occurred in the twelfth century), and they do not provide a true contrast with these seven virtues. Thus when, as so frequently happened, the virtues and vices were contrasted in literature and art, disparate lists of suitably antithetical virtues sprang up, generally called *remedia*, with obvious medico-pathological symbolism. Beginning most markedly with Prudentius and continuing throughout the period under consideration, allegorical works, pitting personified virtues against the 'seven deadly sins', are encountered with varying degrees of frequency, their period of greatest popularity being the twelfth through the fifteenth centuries.[10]

The contrasting of virtues and vices (whether *the* virtues and *the* vices or not) was by no means confined to works in the tradition of the *Psychomachia*. Various genres find room for a presentation of conflict between virtues and vices, whether personified or not. Dante, for example, in his *Divine Comedy* (Purgatorio, cantos 10–26) contrasts various virtues and vices by having the penitents view scenes from history illustrating the particular sin for each terrace and its antithetical virtue. The anonymous *Example of Virtue*, presented to Henry VII, employs various personified virtues and vices as characters in the account of the education of a man from childhood to death. More significantly, from the earliest time, exhortations to virtue were an important aspect of homiletics. Not simply vague calls to the virtuous life, they were usually forthright pleas for the audience to practice various virtues and to flee from and fight against sin in general or distinct vices in particular. Such paraenetic literature, often sermonic in nature, is so well known in Christian tradition that little need be said of it. In the later Middle Ages this literature takes on a peculiar flavor. Speaking of pulpit manuals and treatises from the thirteenth century on, Owst says, "The Vices themselves now strutted upon the scene as well-known types and characters of the tavern or the market-place. The Virtues appeared in the guise of noble women of the times" ([28], p. 87). Mention should also be made of the exceedingly popular and undoubtedly influential morality plays of the late Middle Ages, in which the conflict between virtues and vices is portrayed with varying degrees of sophistication. That concern with the virtues and vices was rampant in the late Middle Ages is aptly illustrated by a recent book that lists the *incipits* (opening words) of untitled Latin manuscripts dealing with the virtues and vices. *Excluding sermons and satirical works*, 6553 different manuscripts dealing with the virtues and vices have been identified from the years 1100–1500 ([7]).

Let us now turn our attention to our third category. Moses Hadas writes of the Greek ideal as expressed by Sophocles, "who had said in the famous first ode of the *Antigone*, 'Wonders are many, but none is more wonderful than man'; and Homer, whose young warriors were enjoined always to strive for excellence and always to surpass all others." Hadas contrasts this with the Christian ideal, " 'Thou shalt worship the Lord thy God, and him only shalt thou serve.' This service required surrender rather than self-assertion, anonymity rather than the fame of the innovator" ([17], pp. 117–118). The values that were the very marrow of Christianity were so different from those of the Greeks as expressed by the authors to whom Hadas makes reference, that the contrast should strike even the most casual reader much more strongly than should the similarities. Yet the earliest Christians who were converted from paganism, whether Greeks, Romans or Hellenized barbarians, however much they were individually transformed through conversion, and their children and grandchildren, however much they were nurtured on Christian values, were members of a syncretistic, pagan culture, whose values, at various levels, were assimilated into their Christianity. In some cases they were oblivious to this; in others they were keenly aware of it, and when aware their reactions varied.

Christians saw the pagans around them as distinctly different from themselves. Yet they sometimes sensed in their own make-up certain affinities and values shared with the uncoverted, that is, in things which were not in themselves *distinctly* Christian. And such things had to be either consonant with their Christian beliefs or discordant with them. That which clashed must be rejected; that which seemed in harmony must be explained and justified. If there was anything of real value in pagan thought, from what source did its value come, from what did its perhaps somewhat obfuscated truth stem? The answer lay in what was regarded as a second *praeparatio evangelica*, the historical groundwork that had set the stage for the Christian gospel. The first *praeparatio evangelica* was Judaism, which had directly prepared the way and continued, through its Scripture, to provide propaedeutic truth. But Judaism was held by many early apologists to have provided, long before, a second, although less reliable, *praeparatio evangelica*; for Hebrew wisdom was regarded as having been the foundation upon which Greek philosophy – indeed all that was good in Greek thought, and through it, Roman thought – had been built. To exploit such truth for its apologetic, evangelistic, and edificatory ends was not compromising the gospel. Rather it was a clear case of "spoiling the Egyptians" to God's glory.

The late Roman Empire was a time of philosophical eclecticism. Seneca's well-known assertion, "Whatever has been well said by anyone is mine," was quoted approvingly by such apologists as Justin Martyr and Clement of Alexandria ([16], p. 109). They took "great pains to argue that the [Christian] revelation was in harmony with philosophy at its best because philosophers had in part been inspired by the Logos" (*ibid*., p. 142). Stoicism and Middle Platonism were oriented to religious concerns ([16], pp. 106–108; [19], pp. 43–44) and the apologists regarded much in Greek philosophy as reinforcing scriptural revelation. Additionally, Christian authors tended to use the language of Greek philosophy in their apologetics. In Clement of Alexandria (ca. 150–ca. 220) "the Christian faith and Greek philosophical tradition became embodied in one and the same individual," thus producing "a highly complex synthesis of Greek and Christian thought" ([19], p. 38). Clement has much to say about virtue and the virtues. And what he says about virtue and the virtues is highly complex and too intricate to warrant untangling here. A scrutiny of his works, however, reveals a concern both with the nature of virtue and the virtues and with their role in the Christian life. In regard to the latter, he made a sharp distinction between 'common Christians', who live by 'simple faith', and 'superior Christians' whose faith is supplemented by *gnosis* (advanced knowledge). These two points then arise in Clement's works and will provide the focus for our discussion of concepts of virtue through the sixteenth century. (1) Virtue and the virtues were viewed and understood in their relation to what we may simply call the created order of things. (2) Virtue and the virtues were seen as an integral part of the Christian life in spiritual growth, moral attainment, sanctification, perfection, and eternal reward. These two areas, which are often entwined in the literature, are sometimes separate. A discussion of the first of these could take an enormous amount of space and become exceedingly complex. We can only present a brief account, touching on some major figures and developments.

An outstanding example is Ambrose of Milan (ca. 339–397) who directly modelled his *De officiis ministrorum* (*On the Duties of the Clergy*) on Cicero's *De officiis*. Ambrose distinguishes three degrees of virtue: fear of God, love of God, and resemblance to God. The virtue that God wishes to be predominant in his children is mercy. The practice of virtue depends upon the will, and the will can be free only when enlightened by revelation and grace. But Ambrose adopts the Greek cardinal virtues. Indeed he is the first to call them cardinal virtues (from *cardo*, "hinge"). He gives to these virtues, however, a distinctly Christian flavor. "Prudence is now the knowledge of God expressing itself in practical piety; justice is transfigured into selfless altruism: courage becomes patience of soul; temperance takes the form of mild and gracious modesty" ([12], p. 530). Virtue is made possible by divine grace, is founded upon faith, and is thus essentially the result and manifestation of a relationship with God. Ambrose's conception of virtue is distinctly Christian; but it was considerably tempered by Stoic principles that appear on nearly every page of his *De officiis ministrorum*. F. H. Dudden neatly captures this strange union: " . . . the good life, as a whole and in all its parts, must exhibit beauty and grace. This conception, of course, was derived through Cicero from Greek philosophy. But it is somewhat startling to find the ascetic Christian bishop exerting all his ingenuity to effect a reconciliation between the austere morality of Christianity and the elegant virtue of the Greeks. It must be noted, however, that the Christian idea of moral beauty is rather different from the pagan. The beautiful character admired by Ambrose is 'gentle' and 'mild'; restraining all violence of passion, abasing itself rather than exalting itself, submitting patiently to guidance, careless of its rights, and when wronged returning good for evil. It displays itself pre-eminently in the modesty which delicately shrinks from everything that savours of impurity, and from all that is coarse and jarring in the life of society. Its beauty is, indeed, the subdued and tender beauty of the 'meek and quiet spirit', which in God's sight is of great price" (ibid.).

Augustine (354–430) had drunk deeply from the well of his classical heritage. When he was converted to Christianity, he saw classical philosophy, particularly Neoplatonism, as having helped to bring him to God. Thus he did not repudiate it but significantly altered it, christianizing it, as it were, as had his mentor Ambrose. Augustine's use of pagan sources and his dependence on their form if not their essence is not as obvious as that of Ambrose. Augustine also adopted the cardinal virtues, which then "were subsumed under love. Virtue consists in nothing else but in loving what is worthy of love; it is prudence to choose this, fortitude to be turned from it by no obstacles, temperance to be enticed by no allurements, justice to be diverted by no pride" ([14], p. 167). For Augustine all that is good is centered upon love; love is the foundation of all good, is the summit of all good. God is the source of love and it is love that unites people to God. Augustine uses three words for love: *dilectio, amor*, and *caritas*. The second and third of these apply to virtue. In the *Catholic and Manichaean Ways of Life* (14), he says that virtue is *amor*. But his favorite definition of virtue is "rightly ordered *caritas*" (e.g., in *City of God*, 15, 22). To one reading Augustine in translation love may appear to be morally neutral. But Augustine employs a fourth word for love, and that is *cupiditas*; and he maintains that God commands nothing but *caritas* and condemns only *cupiditas. Caritas* is the enjoyment of God, self, and others for God's sake; *cupiditas* is the enjoyment of self, others, and material things without reference to God (*On Christian Doctrine*, 3, 10). This distinction is clearly seen in the *City of God* (14, 28): the earthly city is created by *cupiditas*, the city of God by *caritas*. For Augustine every true virtue is founded upon love (*caritas*) of the good and all virtues should lead back to *caritas*.

We should not understand Augustine to be denying that the citizens of the earthly city possess any laudable qualities. But their virtues are penultimate, not ultimate, are based upon false priorities, and thus are incomplete or corrupt. Even these virtues, within the natural order of things, must be seen as gifts of God (*City of God*, 14, 4). It would, however, be wrong to see in Ambrose or Augustine two distinct sets of virtues, i.e., the 'natural' or 'cardinal virtues', shared by pagans and Christians alike, and distinctly Christian virtues (e.g., faith, hope, and love), which are possessed only by God's children as a result of intervening grace. The 'cardinal virtues' or 'natural virtues', as manifested by pagans, are essentially different from virtues of the same name in Christians, for in pagans these virtues are corrupted, incomplete, and imperfect; they are qualities still directed to wrong priorities. It is only when, through grace, these virtues are infused, transformed, and energized by *caritas* that they achieve their potential and become true virtues.

The pervasive influence of Augustine on medieval Christian thought cannot be overstated. It can be clearly seen in medieval concepts of virtue. But there is another strand of medieval thought that also appears which has roots in classical thought as partially christianized by Greek church fathers such as Clement, and passed on by Boethius whose influence on the Middle Ages, although significantly less than Augustine's, was still exceedingly powerful. Boethius (ca. 480–524), who lived in Ostrogothic Italy, is frequently, but with questionable accuracy, called 'the first scholastic', a scholastic being one who attempts to harmonize Christian faith and reason. Little such concern is evident in his most influential work, the *Consolation of Philosophy*, written while he was in prison awaiting the resolution of a charge of treason. This treatise is permeated by optimism, a sweet spirit of resignation and hope, comfort derived from a firm belief in an omnipotent providence (i.e., God in a sense of Greek philosophical monotheism), and in the practice of prayer. If we remove a very few Christian phrases, the work could well have been written by a noble pagan of Neoplatonic or Stoic persuasion, for it is totally silent about Christ and the gospel, ignoring any possibility of seeking or deriving consolation from Christian belief. For in the *Consolation of Philosophy* it is *Philosophia*, personified as a woman, who leads Boethius to 'true contentment' and this true contentment is that which only reason united with virtue can give. And this virtue is as pagan a conception of virtue as might be found in a variety of profound classical thinkers. While its moral flavor is Neoplatonic, its conceptual framework is distinctly Aristotelian.

These two strands of thought concerning virtue, the Augustinian and the Boethian (the latter found particularly in Boethius' commentaries on Aristotle), are seen throughout the Middle Ages, with the Boethian becoming, with the rediscovery of Aristotle in the high Middle Ages, distinctly Aristotelian. While many examples of both could be given, the most outstanding representatives of the Augustinian position are Peter Lombard (ca. 1100–1160), who held that all the virtues were entirely God's work and were themselves identical with grace; and Bonaventura (1221–1274), who saw the natural virtues as imperfect and lacking merit without the intervention of divine grace. Peter Abelard (1079–1142), on the other hand, was Aristotelian in his view, seeing virtue as a *habitus animi optimus*, the fixed disposition of one's soul to choose the just mean in one's actions.

A combination of these strands of thought was most spectacularly developed by Thomas Aquinas (ca. 1225–1274), although in a comparatively superficial way it was anticipated by William of Auxerre (ca. 1150–1231). William, in his *Summa Aurea*, was the first to label the Christian triad of faith, hope, and love the 'theological virtues', distinguishing them from the 'natural' or 'cardinal virtues'. He regarded the former as being infused by God alone, the latter as being caused by one's own actions. But it was Aquinas, that giant of scholasticism, who gave to the Catholic conception of the virtues the form which remained dominant in Catholic thought well beyond the period under consideration.

Aquinas speaks of three categories of virtues: (1) the intellectual virtues – intelligence, knowledge, and wisdom; (2) the moral (i.e., cardinal) virtues; and (3) the theological virtues. He sees the first two as habits or qualities of the mind which are formed by good acts, repetitively performed. Thus *actus* precedes *habitus*. In this he is distinctly Aristotelian. Augustinian thought is seen in Aquinas' view that the theological virtues are infused and depend upon grace. The intellectual and moral virtues, which are natural virtues, achieve their standard by following the mean between excess and defect. There can, however, be no excess or defect of the theological virtues. There is, for Aquinas, a definite relationship between the moral and the theological virtues. He insists that the moral virtues are significantly affected by the theological virtues, especially by *caritas*; further, he distinguishes between *habitus acquisitus* and *habitus infusus*, and his system, of necessity, accommodates both. Aquinas would probably have been horrified by any accusation that he did not see grace as central and essential to this schema of the virtues. Nevertheless, the large section of his *Summa theologica* devoted to the virtues has been interpreted, particulary by those unsympathetic to Aristotelian scholasticism, as placing such an emphasis upon the role that the moral virtues play in contributing to the maturing of the theological virtues, that grace, if not abnegated, is at least depreciated, and the distinction between *habitus acquisitus* and *habitus infusus* becomes garbled and dysfunctional if not downright irrelevant. Thus in Thomistic thought (i.e., late medieval Aristotelian scholasticism), particularly as interpreted by the Protestant Reformers, if virtuous *actus* precedes and produces virtuous *habitus, and* if the moral (i.e., 'natural' or 'cardinal') virtues play a significant role in the growth of the theological virtues in the individual Christian, a natural *actus* can at least affect, if not necessarily effect, a theological (and meritorious) *habitus*. Nevertheless, when grace is carefully identified in the role accorded it by Aquinas, it becomes evident that, at the deepest level, one must be righteous by *habitus infusus* before one can perform a righteous *actus*. The essence of the soul, infused with grace, creates the potentiality of virtuous works. Justification exists then to the extent that grace is manifested in works. Thus grace perfects nature on the one hand, while on the other nature affects the working of grace.

This must all be rejected, according to the Reformers, if the hallmarks of the Reformation, *sola gratia* and *sola fides*, were not to be obscured. Luther especially was adamant in condemning Aristotelian scholasticism as denying grace. Emphasizing the scholastics' insistence that nature has a role in perfecting grace (and perhaps overlooking their equally strong insistence that righteous acts can come only from a nature infused with grace), he asserts that "we do not become righteous by doing what is righteous, but being made righteous we carry out righteous acts."[11]

The Reformers did not deny limited virtues to the unregenerate but attributed them to the distinctly reformed idea of common grace, a doctrine especially dear to Calvin. Calvin frequently asserts in his *Institutes* that mankind's nature is corrupt, but that those who display or strive toward virtue show that there is some purity in their nature. This, however, must be attributed to God's restraining grace (*Institutes*, 2, 3, 3 and 4). All the virtues of the unregenerate are simply images of true virtues and are not praiseworthy in any way, since these too are gifts of God (*ibid*., 3, 14, 2). Even virtuous deeds are sins when exercised without faith since their *telos* is not to serve God (*ibid*., 3, 14, 3). Thus nothing in mankind, however virtuous by the standards of the unregenerate, avails for salvation or earns merit with God.

We turn now to a consideration of our second point, that virtue and the virtues were seen as an integral part of the Christian life in spiritual growth, moral attainment, sanctification, perfection, and eternal reward. A quest for perfection in the Christian life and a recognition of degrees of holiness or sanctification has been a compelling force within the history of Christianity. Sometimes it finds expression in Clementine terms of *gnosis* with two distinct levels of spiritual attainment. It also has given rise to a sharp dichotomy between the clergy and the laity. Further, it has provided an impulse for asceticism and monasticism. Since Christianity teaches denial of self, and since this has often been regarded chiefly as denial of the flesh, asceticism has always been present in Christianity. When asceticism involves withdrawal from society, it is typically called monasticism. Monasticism originated in the late third century in the deserts of Egypt, Syria, and Palestine, the earliest monks being hermits living in individual seclusion. This form of monasticism, called anchoritic monasticism, began to give way in the fourth and fifth centuries to cenobitic monasticism, a form of monasticism involving communities of monks living under a rule of practice. Both forms of monasticism are manifestations of a desire to pursue a high level of Christian virtue by a more rigorous discipline than that of the average Christian, abstaining from various things normally regarded as good, and adding supererogatory requirements and routines.

The four fundamental monastic virtues were chastity, poverty, humility and obedience. The practice of virtue was of two different types. (1) Moral discipline that is directed specifically toward one's own spiritual growth, e.g., regular and often extensive sessions of prayer, meditation, self-examination, and mortification of the flesh by fasting and chastity. (2) Charitable activity for the immediate benefit of those in need, with a residual spiritual capital for oneself. The early (fourth through eighth) centuries of monastic history are marked by the formulation of various rules of practice and a concerted lessening of the excessive and bizarre asceticism that had appeared in some quarters. The ninth through the twelfth centuries were a time of repeated efforts at reform of existing orders and the creation of new, 'uncorrupted orders'. From the thirteenth century through the end of our period monasticism was in great part eclipsed by new movements such as the order of Augustinian canons and the mendicant orders. Especially during the ninth through the eleventh centuries there was a concerted effort to bring ascetic virtues to the laity through what became the sacrament of penance. In the late Middle Ages a tendency toward various ascetic aberrations is seen among some clergy and laity alike. These practices included such matters as the performance of numerous genuflections, *inclusio* (solitary confinement for extended periods in a cell, cave, or hut), and self-flagellation.

The Protestant Reformers of the sixteenth century denied the very idea of a dichotomy between the clergy and the laity or of a spiritual elitism based upon one's position within the ecclesiastical structure. They insisted on the universal priesthood of believers and maintained the view that there is no clear distinction between the sacred and secular in Christian vocation. Further, the reformers heartily rejected sacerdotal celibacy and the belief that any special merit is earned by the display of ascetic practices except for those which they regarded as being the norm of New Testament Christianity expressed in exhortations to a healthy self-denial and spiritual warfare against sin. The Counter-Reformation, particularly in Tridentine deliberations, reacted against the criticism leveled by the Protestants and, while discouraging bizarre and ascetic practices, reaffirmed those principles that most clearly separated Roman Catholicism from the Protestant movements, including a sharp distinction between sacred and secular vocation, monastic principles, and the meritorious feature of the practice of ascetic virtues.

Concluding our discussion of the second of our two points (that virtue and the virtues were seen as an integral part of the Christian life in spiritual growth, moral attainment, sanctification, perfection, and eternal reward), we should mention that underlying much if not all efforts at the practice of Christian virtue, regardless of how eccentric or bizarre some manifestations may seem to us, was the desire to follow Christ and to please God. Often the ardent following of Christ was seen in terms of the *imitatio Christi*. To live a life in imitation of Christ is to seek to model one's thoughts, actions, and character upon his, and thus to adopt his pattern of virtue; to make his virtues one's own, and thus to attain to the ultimate heights of virtue and the virtuous life by Christian standards. The extensive variations in the conception and implementation of an *imitatio Christi* should highlight the vast potential of possible diversity in an understanding of Christian virtue and virtues, historically if not scripturally.

VIRTUE IN THE RENAISSANCE

Traditional appreciation of that historical experience which is usually called the Renaissance, whether in its earlier (Italian) or later (northern) form, and its essential quality, typically labelled humanism, has been governed by two equally erroneous presuppositions. One is wrongly designated the Burckhardtian view, namely that the Renaissance was essentially pagan and anti-Christian. According to this view, with the stirrings of the human imagination stimulated by a re-interest in the classics, light at last had broken in upon the darkness perpetuated by the suppressing force of Christianity in general and medieval scholasticism and ecclesiastical tyranny in particular. Finally some people at last could breathe deeply the fresh air of an atmosphere conducive to the continued development of human dignity and potential, a process that had been interrupted by christianization. A second view is that two disparate strands are evident throughout the Renaissance, one distinctly pagan and the other Christian. Even though these views, especially the former, are still popularly held and perpetuated by some textbook accounts, Renaissance scholars recently have adjusted significantly our understanding of the very spirit of the Renaissance experience.

It is safe to assert that Renaissance thought, although significantly affected by its appreciation and, as it were, re-creation of classical ideas reflected in literature and art, was in its very marrow Christian. When modern enthusiasts assert that a classical Greek or Roman would have felt much more at home in Renaissance Italy than in any environ of the Middle Ages, it should be borne in mind that this hypothetical Greek of Roman would have felt so alien to both that we would be safe only in saying that he or she would have felt very slightly less foreign in the atmosphere of Renaissance Italy than in a medieval environment. And the sense of alienation such a person would have experienced would have been caused by the predominantly Christian presuppositions undergirding nearly all intellectual and artistic expressions. The much-touted anthropocentrism of the Renaissance impulse can only continue to be seriously misunderstood until one sees how much it was informed and qualified by a strong, if not always obviously compatible, Christocentrism. Human dignity – such a hallmark of the Renaissance – was far from any supposed pagan correlate or antecedent, for Renaissance articulation of this theme was based upon a clearly presuppositional acceptance of human dignity based upon creation in the image of God and gracious redemption by Christ. Human self-determination and multipotentiality in Renaissance thought must also be seen within that framework.

There is a discernible tendency on the part of humanists, ranging from and particularly exemplified by Petrarch and Erasmus, to criticize sharply the assertion of pagan philosophers that virtue itself is a sufficient goal for humanity. Humanists generally saw virtue as depending on divine grace. Not that they went as far as the Reformers in denying any merit to natural virtues; but they were, for the most part, closer to the Reformers than to the Aristotelian scholastics on such matters, in spite of obvious differences between, e.g., Erasmus and Luther on the related question of free will.

What could pagan ideas and ideals contribute to Christian understanding? This question had been asked by the church fathers; it was also asked by the Renaissance humanists. Erasmus regarded upright pagan luminaries as having been moved by the spirit of Christ. As strong a tendency to see a compatibility of Greek philosophy with Christianity is evident in many Renaissance humanists as in many Greek church fathers. The motivation was not to paganize Christianity, much less to become pagans themselves; rather it was to profit from the 'divinely bestowed' insights of the great minds of classical antiquity. The incongruities of this seemed to escape them. As one Renaissance scholar writes, they "were unwilling to explore the profound differences between the classical world and Christianity. The existence of a beautiful harmony was above all asserted in the realms of ethics and aesthetics and little attention was paid to the profound incompatibilities that existed for example between the Greek view of immanence and the Judeo-Christian tradition of transcendence or the Greek conviction that knowledge is virtue and the Christian dogma that the intellect is of no avail against a will profoundly corrupted by original sin" ([15]), p. 508).

MEDICINE AND VIRTUE IN A CHRISTIAN CONTEXT: SOME INTRODUCTORY OBSERVATIONS

In the New Testament and in Christian literature generally throughout the entire period under consideration, particularly in genres of a devotional nature and in those which deal with the Christian life, with growth and sanctification, and in works exhorting to an *imitatio Christi*, there is an emphasis on trusting wholly in God and establishing one's earthly priorities in the light of eternal realities. The literature in question heartily agrees with the assertion that the soul is infinitely more important than the body. Emphasis, however, varies considerably when applying this foundational Christian presupposition to a wide variety of questions pertaining to the proper attitude of the Christian toward the material world in general and the human body in particular. Three questions are germane to a consideration of medicine and virtue. (1) Is the created order of things, because of the Fall, inherently and essentially evil and to be depreciated and shunned as much as possible? (2) Is this fleshly body but a physical manifestation and reminder of one's sinful nature, an albatross to be deprecated, even abused, or at the very least endured? (3) Is there a positive spiritual benefit, expiatory as well as edificatory, to be derived from all suffering that the Christian endures, even if self-inflicted? Those within the history of Christianity who would answer these questions with an *unequivocal* 'yes' constitute a relatively small minority. Those whose affirmative answer to the first question is unqualified are often members of heretical movements, e.g., dualistic groups such as the Gnostics, Manichees, and Cathars, or those who display an attitude to which orthodox Christianity (whether Roman Catholic or Protestant) is either overtly hostile or of which it is at least highly suspicious. Those who would give a hearty 'yes' to the last two questions usually demonstrate a severely ascetic disposition which has never been typical of Christianity but has been highly lauded and admired, if not emulated, by the Christian community at various times. It is undoubtedly safe to assert categorically that the vast majority of educated Christians during the period under consideration would have answered our three questions with a 'no'. But this negative answer would have been a highly qualified 'no', qualified by a recognition that each of these questions, although extreme, reflects, even if obliquely, an important point of Christian dogma and practice. Regardless of the answers that Christians of the past would have given to these questions pertaining to the material world, the body, and suffering, it would be a *rara avis*, even on the fringes of Christendom, who would not agree that there is an imperative upon Christians to extend comfort and succor to the destitute and suffering. With these considerations in mind, we turn first to the question, are the practice and use of medicine consistent with Christian virtue? After addressing this question, we shall turn to a discussion of the virtuous Christian in suffering and sickness (the virtuous patient) and the virtuous Christian in caring and curing (the virtuous physician).

ARE THE PRACTICE AND USE OF MEDICINE CONSISTENT WITH CHRISTIAN VIRTUE?

Christianity is a religion of faith in a God who is regarded as the heavenly Father of his children. His love and nurture of his children are central to historic Christian belief. This God is omniscient and omnipotent, causes all things that happen to his children to work out for their good (Rom. 8:28), and instructs them to be anxious about nothing (Phil. 4:6) but to cast all their cares upon him (1 Pet. 5:7). It should not evoke surprise that some Christians, during all periods, have felt that if God permits sickness or injury to befall them it is to him alone that they should turn for healing, while to turn to secular healers shows a lack of faith and dependence. Some would assert that while they would not criticize other Christians for using medicine and physicians, they regard such use as wrong for themselves, at least in the specific circumstances in which they happen to be at that time.

Modern scholarly assessments of the attitudes of the early church to secular medicine are varied. Some find the roots of a condemnation of medicine in various heretical movements. Others see the source of a rejection of human medicine in a severe asceticism which resulted from a non-Christian dualism or a distorted Pauline dualism. Still others maintain that a distrust of secular medicine was very strong in early Christianity, but that as time passed and primitive Christian teachings were tempered and mellowed by an ameliorating Hellenistic influence, medicine was generally accepted as consistent with God's purposes and the occasional voices raised in opposition to it were bizarre and freak occurrences. A recent paper [3] has sought to demonstrate that this last explanation is without merit and that the first two, although partially correct, are inadequate since they account for only a few voices which raised some objection to a Christian's use of medicine and physicians. The paper in question demonstrates that church fathers for the most part had a decidedly positive attitude toward secular medicine and regarded its practice and use as entirely consistent with Christian virtue so long as Christians recognized God as the ultimate and only source of healing whether through his created means and instruments or without them. The attitudes of the fathers are summarized as follows: "Medicines and the skill of physicians are blessings from God. It is not *eo ipso* wrong for a Christian to employ them, but it is sinful to put one's faith in them entirely since, when they are effective, it is only because their efficacy comes from God who can heal without them. Thus to resort to physicians without first placing one's trust in God is both foolish and sinful. Likewise to reject medicine and the medical art entirely is not only not recommended but is disparaged" ([3], p.341). Those fathers who lauded the circumspect use of physicians and medicine, however, rejoiced in instances of miraculous healings, particularly when the efforts of physicians had failed. Some church fathers specifically would have Christians refrain from using medicine in a variety of illnesses (e.g., Basil in his instructions to his monks, *The Long Rule*, 55); others felt that the use of medicine was appropriate for the average Christian, while devout Christians should rely on God directly (e.g., Origen, *Contra Celsum*, 8, 60.).

Throughout the period under consideration very few Christians appear to have expressed unequivocal contempt for secular medicine; only a small number, in other words, would have asserted that the practice and use of medicine are *eo ipso* inconsistent with Christian virtue. Considerable care must be taken when we encounter negative comments made by Christians regarding physicians. Three examples should suffice – Gregory of Tours, Petrarch, and Martin Luther. Gregory of Tours' "enthusiasm for the miracles wrought at the shrine of Saint Martin knew no bounds. He was often delighted to heighten his audience's appreciation of the spectacular nature of a particular healing by describing in detail the sorry attempts of physicians and their abysmal failure and despair in contrast with the efficacy of Saint Martin's miraculous intervention. But when his tales of the failure of physicians, used as a foil to Saint Martin's miracles, are culled from his writings to illustrate early medieval attitudes toward medicine, they do not give a balanced picture even of Gregory's attitude. He himself enlisted the help of physicians and recomended to others that under certain circumstances they do likewise" ([4], p. 119). Petrarch, as an Italian humanist, drew heavily from classical sources for style and subject matter. One relatively popular classical topos was the denunciation of physicians' greed, follies, foibles, and pride. This theme appears to have struck a responsive chord in Petrarch's heart, for he warmed to the occasion with vitriolic vigor when opportunities arose to heap vituperation on physicians. Yet Petrarch, both as a Christian and as a humanist, while despising physicians (probably in great part for literary effect), would have viewed the proper use of the medical art as completely consistent with God's purposes. So also is the case with Martin Luther. Numerous anecdotes can be culled from his writings that, when taken together, would appear to reflect a hostility to and deep-seated suspicion of the medical art, especially of its practitioners. Yet he clearly recognized and acknowledged that God worked through physicians and their medicines. Nevertheless, Luther chuckled with a somewhat perverse glee when he saw

God working around and in spite of physicians' self-assured wisdom and machinations.

These three figures are quite representative of various degrees of hostility to secular medicine. Many more witnesses, however, could be summoned from the sources who highly regarded both physicians and the medical art. All, except for a few nearly bizarre examples, would agree with the attitudes of the church fathers toward the practice and use of medicine as summarized above. We can state categorically that for orthodox Christianity during the period under consideration, the practice and use of medicine were regarded as consistent with Christian virtue as long as one recognized that it was God from whom all blessings come, but that one must rely upon him ultimately, regardless of whether or not one chose to avail oneself of medical means and instruments.

We concluded our discussion of virtue in the New Testament by suggesting that selfless service and suffering were expected to be the lot of Christians. These roles were to be accompanied by joy and growth in Christ-likeness as the character of Christians would be molded to humility, gentleness, kindness, and compassion, self-control, patience, and steadfastness. Such people were to endure suffering and extend Christ's love to the afflicted, the weak, the poor, and the oppressed, by succoring their ills, ministering to their needs, and contributing to their wants, considering the things of this would of little value and this life a training ground for heaven. However much obfuscated by the minutiae of discussion of virtue and the virtues, theses principles are expressed or implied in much of the theological and devotional literature from the patristic period through the sixteenth century.

THE VIRTUOUS CHRISTIAN IN SICKNESS AND SUFFERING

Even a cursory reading of the New Testament should impress upon the reader that suffering is a vitally important theme. The Gospels, the Book of Acts, and the Epistles emphasize the vicarious, representative, and substitutionary nature of Jesus' suffering as a soteriological necessity. He is seen as a participant in various levels of human suffering and is frequently described as being "moved with compassion". Suffering for Christians is expected to be inevitable and to take on the form of persecution coming from unbelievers, or discipline (to include training and punishment) coming from God. The suffering of God's children is to result in their edification and to prepare them for sharing in the glory that is to follow, a glory frequently coupled with a propaedeutic suffering in the New Testament. A fellowship of suffering is to result, both between the believers and their Lord and between believers united by and in their sufferings. This fellowship of suffering prepares the responsive Christian for a ministry of comfort, consolation, and encouragment to fellow Christians. But the ultimate goal of suffering for each individual Christian is a dependence on Christ, and provides a central paradox in New Testament thought: strength comes only through weakness and this strength is Christ's strength that comes only through dependence on him. To New Testament Christians God is a *Person* with whom they may have fellowship in a spiritual depth commensurate with the degree of their commitment to the person of Jesus Christ. It is the intimacy of this relationship, a relationship based upon love and trust, that gives meaning to suffering in the New Testament and enables Paul to say of himself and of those of kindred spirit, "we rejoice in our sufferings" (Rom. 5:3).

The attitude toward suffering in patristic literature is, on the one hand, a curious mixture of popular Stoicism and New Testament teaching. The former is seen much more strongly in the Greek than in the Latin fathers. But even in the Greek fathers the latter emphasis shines through. A general consensus of patristic sentiment is that suffering, including illness, is good only, or at least primarily, because God uses it in the lives of his children, just as he ultimately causes all things to work together for their good. These authors emphasize that Christians must love, trust in, and depend upon their omnipotent Creator who loves and cares for his children, and that they must rely upon him when they are ill, being then aware that any suffering, even illness, can by beneficial when they are open to being edified by it. Such are the sentiments of the ante-Nicene, Nicene and post-Nicene fathers, sentiments on suffering with which the vast majority of sources in Christian literature from the Middle Ages and the Reformation would agree. Suffering benefits Christians when they respond to it in a spirit of dependence on Christ and humble contrition. Most sources that discuss suffering at any length, whether church fathers, medieval homilists, or Protestant reformers, emphasize that suffering is a most salubrious motivation to spiritually beneficial self-examination.

Thus suffering is beneficial to Christians. But when emphasis is placed upon the suffering itself, rather than upon its results, when suffering is glorified for its own sake, some have recoiled and insisted that such an emphasis is contrary to the teaching of Scripture. Excesses have abounded at times and the heroes of bizarre asceticism may be paraded by the historian: the monk Macarius, for example, who exposed his naked body to poisonous flies for six months as a self-imposed punishment for having killed a fly in anger; or Simeon Stylites who lived on a pillar for over thirty years and allowed vermin to eat into his body for some time; or a whole crowd of those who tortured themselves by *inclusio*, or self-flagellation, or by a variety of other attempts to placate a God who was viewed as deriving some propitiatory satisfaction from the sustained suffering of those who sought to earn his favor. Or such did their efforts seem to those who, whether Roman Catholics, Waldensians, or Protestants, claimed that the gospel of God's grace denied to any human suffering (other than Christ's) a propitiatory or expiatory effect.

How then should the virtuous Christian, when ill, view the use of secular medicine? We have seen that some Christians, within the boundaries of orthodoxy, have viewed medicine and physicians with suspicion or hostility. Sometimes this has been more a matter of personal idiosyncrasy than a formulated theological principle. When the latter, it has been very atypical when rigidly applied as a standard by which all Chriatians should live. Others (e.g., Origen and Basil) have taught that one should not have recourse to secular healing under certain circumstances. But the fathers, including Origen and Basil, emphasized that the medical art is one of God's gifts; thus it is perfectly consistent, indeed wise, for Christians to avail themselves of a physician's skill when ill. They should blanket the whole procedure in prayer; resorting to prayer only when the physician's efforts have failed is strongly criticized. Christians owe their bodies many things – nourishment, clothing, shelter, rest, and medicine when ill – but their bodies should never become their master. This is a theme found in some pagan sources: undue concern with one's body keeps one from the pursuit of virtue because it interferes with balance and the mean. Some Greek fathers found this emphasis congenial. But the majority stress that Christians must avoid a consuming solicitation for their bodies. While death should never be sought, life should never be clung to in desperation. Heaven is home and Christians are, in this life, but aliens, strangers in a strange land. Basil expresses succinctly a sentiment with which the majority of sources from the period under consideration would undoubtedly agree, at least in principle. He urges that "whatever requires an undue amount of thought or trouble or involves a large expenditure of effort and causes our whole life to revolve, as it were, around solicitude for the flesh must be avoided."[12]

Not only must virtuous patients not cling to life but they also should refrain from using medicines and physicians for any evil end. The condemnation of abortion in early Christianity is uniquivocal but for most of the centuries under consideration the question becomes obscured by concern with ensouling or quickening. Even here the matter is one of degree of sin. The virtuous patient will not obtain an abortion unless .... The exceptions simply are not resolved with casuistic percision during these centuries, leaving the virtuous patient on the horns of a dilemma. Much more clear is the question of what we call euthanasia. Actively to expedite one's own death is regarded as suicide and suicide is clearly defined as sin during the period under consideration.

It was especially during the period from the fourteenth through the sixteenth centuries that the Catholic Church began to formulate minute definitions of sinful conduct. We find in the confessional literature of this period a much greater emphasis on the responsibilities of a physician to refrain from advising sinful means for restoring health than on the obligation of the patient to refuse to follow such advice. 'Sinful means' that the moral theologians anticipated were fornication, masturbation, incantation, consumption of intoxicating beverages, breaking the church's fasts, and eating of meat on forbidden days ([1], pp. 97–98). Several summists (authors of confessional manuals and works dealing with moral theology) address the question of the obligations of patients.[13] These discussions cannot be considered as inclusive, appearing as they do in sections dealing with the obligations of physicians. A composite includes the following elements. Patients do not sin if they disobey their physicians since physicians have no real authority over them, but can only advise and exhort. It is, however, good to trust one's physician to the extent to which he is expert in the art. If patients knowingly or by gross ignorance consume something deadly, they sin. They do not sin mortally or very gravely, however, if they intentionally consume a substance which they believe is not deadly if by so doing they are attempting to aggravate their illness.

Virtuous patients of the Middle Ages, by the standards of medieval Catholicism, would not obtain medical or surgical services from Jews and Muslims. Medieval canon law is adamant on this point. When we look later at the concerns of the promulgators of medieval canon law with the potential for spiritual harm in medical practice, it should not surprise us that they sought to impose and enforce a policy of restricting medical practice on Christians to Christian physicians.

THE VIRTUOUS CHRISTIAN IN CARING AND CURING

When concepts of Christian virtue are applied to the role of caring and curing, two related but distinct subjects arise: (1) Christian philanthropy;

and (2) Christian medical ethics. We have dealt with the first of these directly, and the second indirectly, in a paper published in a previous volume in the Philosophy and Medicine Series.[14] We shall not, however, simply repeat ourselves here.

There is an unambiguous and unequivocal imperative in the New Testament to extend succor to the needy (e.g., Mt. 25:35–46; James 1:27). In the early church the visitation and care of the poor sick were expected of all Christians. Deacons and deaconesses had a special duty in this regard, a duty eventually overseen by local bishops. The zeal for extending care to the suffering is especially evident during times of epidemics, for instance, during the plague that occurred in the seventh decade of the third century in North Africa. The accounts that survive describe the nearly suicidal enthusiasm of some Christians for extending succor to Christian and pagan victims alike. Beginning in the fourth century, institutions were increasingly established to house and care for the destitute and ill, as well as for orphans and the aged. In late antiquity and the early Middle Ages, monasteries became a refuge for the destitute and the monastic clergy often showed considerable zeal in providing medical care in what may loosely be labelled 'hospitals', but which were more often than not simply hospices. There is, in the sources from the early Middle Ages, a strong emphasis on medical charity as an integral part of the monastic movement. This concern with the extension of free medical care to the destitute was essentially taken over by the mendicant orders beginning in the early thirteenth century. Henry Sigerist believed that Christianity introduced "the most revolutionary and decisive change in the attitude of society toward the sick. Christianity came into the world as the religion of healing, as the joyful Gospel of the Redeemer and of Redemption. It addressed itself to the disinherited, to the sick and afflicted, and promised them healing, a restoration both spiritual and physical .... It became the duty of the Christian to attend to the sick and poor of the community .... The social position of the sick man thus became fundamentally different from what it had been before. He assumed a preferential position which has been his ever since" ([33], pp. 69–70).

The extension of succor to the destitute has remained an obligation throughout the history of Christianity. The extent to which this obligation has been put into practice has varied considerably during the period under consideration. During the Middle Ages the initiative was at some times with the church. At other times it appears to have been more in the hands of individuals or secular organizations, especially in the late Middle Ages when guilds provided many acts of charity, first to the needy within their own ranks, then to the destitute within the community. Hospitals were often founded by guilds or other secular organizations. It has been common to claim that by the time of the Reformation charitable institutions within Catholic states had become nearly moribund and that, by contrast, Protestant efforts in philanthropy far eclipsed those of Catholics. This was supposedly so, it has been argued, because the motivation of Protestants, being free from efforts to earn merit by charitable acts, sprang from a genuine and unselfish concern for the sufferer and thus produced many more channels of charity than did Catholics shackled by the chains of legalism. Not only are such assessments unfair – for probing into the hinterland of the heart to search for purity of motive is a most prejudicial and dangerous basis for scholarly assertions – but also inaccurate, since many Catholic states of the late Middle Ages through the sixteenth century excelled in charitable endeavors.[15]

We have just disallowed an intimate probing into motives. We can, however, examine the literature which reflects or dictates standards of both conduct and motive in medical practice. When we do so, we are dealing with the development of Christian medical ethics, i.e., guidelines for the virtuous physician. While all Christians who wished could endeavor to pursue a life of *imitatio Christi*, the Christian physician who desired to serve and emulate Christ could find his heart warmed and his standards tested by the very highest standard, namely the character of Christ himself who was called *verus medicus*, 'the true physician'. Early Christian authors thus adopted a tradition in classical literature that employed, in simile or metaphor, the idea of the physician as a compassionate, selfless, and philanthropic healer of ills and soother of distress. It was in their care for the destitute and the poor that physicians evinced a Christ-like compassion. Augustine maintains that physicians should always have the care of their patients at heart; otherwise the practice of medicine would be an exercise in cruelty. A highly idealized Hippocrates was adopted by some Christian authors as an exemplar of virtue. Indeed Christ is himself called, "as it were, a spiritual Hippocrates" ([29]), p. 75).

Christ was the physician as an ideal, Hippocrates the ideal physician. If Christian physicians were to look for models of virtue to which they might aspire in their practice, they could hardly aim higher. Unfortunately, we know next to nothing about the effect of such ideals on actual medical practice during the early centuries of Christianity. Especially after Christianity became the official religion of the Empire, Christians who also happened to be physicians may well have felt bound to apply fervently to their practice the philanthropic and moral precepts of their religion. On the other hand, physicians who also happened to be Christians may have found their principles of conduct and responsibility much less tempered by their religion. The extent of physicians' conformity to the Christian ideal may well have been in proportion to their Christian conviction and commitment. Most physicians probably did not act significantly differently from their pagan counterparts of a few generations earlier, except that the religious pressures against abortion and active euthanasia may have deterred many nominal Christians from such practices. While some physicians who were not also members of the clergy were lauded for their compassion and philanthropy, increasingly we find mention of clergy (priests and monks) whose spiritual and medical interests blended into a common concern for the spiritually and physically ill.

Two classes of physicians – two spiritual classes, that is – appear in the sources, just as two distinct classes of Christians generally seem to have been fostered in the climate of late antiquity. This is illustrated by two documents from the pen of Cassiodorus. The first of these was composed when he was in the service of the Ostrogothic king Theodoric in the early sixth century. This document (*Variae*, 6, 19) reinstitutes the office of the *comes archiatrorum*, who appears to have been both the president of the college of civic physicians in Rome and personal physician to the king and the royal household. The text begins with an encomium on the usefulness of the art of medicine, which is labelled 'glorious' because it drives out diseases and restores health. He lauds the nearly uncanny prognostic skill of experienced physicians and praises the art for its being a learned discipline and admonishes physicians to trust its science rather than their own experience. He also chastizes physicians for their bedside bickering, urges them to work together in assisting each other harmoniously, and reminds them that at the beginning of their career they swore an oath to hate iniquity and to love purity. This document could just as easily have been composed centuries earlier by a pagan; little in it is distinctly Christian.

After retiring from the court of the Ostrogothic king, Cassiodorus founded a monastery. In his *Introduction to Divine and Human Readings* he writes to those of his monks who were also physicians: "I salute you . . . who are sad at the sufferings of others, sorrowful for those who are in danger, grieved at the pain of those who are received, and always distressed with personal sorrow at the misfortunes of others, so that, as experience of your art teaches, you help the sick with genuine zeal; you will receive your reward from him by whom eternal rewards may be paid for temporal acts. Learn, therefore, the properties of herbs and perform the compounding of drugs punctiliously; and do not trust health to human counsels. For although the art of medicine is found to be established by the Lord, he who without doubt grants life to men makes them sound. For it is written: 'And whatsoever you do in word or deed, do all in the name of the Lord Jesus, giving thanks to God and the Father by Him' " ([20], 1, 31).

It is worthwhile to compare this exhortation with that which he had directed to the *comes archiatrorum* and, by extension, to physicians in the public medical service. In both he lauds the medical art. Aside from this there is little similarity between the two pieces. While the impression of the ethical basis of Cassiodorus' secular physician is identical to classical descriptions, the peculiar qualities of the monk/physician constitute a picture of the physician as an ideal and the ideal physician of earlier Christian thought. But no longer is this an ideal posited for Christian physicians generally, but for the relatively new group of monastic or clerical physicians.

Some treatises dealing with medical ethics, written before the tenth century and reflecting ideals of monastic or at least clerical medicine, have survived [25]. In these, physicians are encouraged to serve rich and poor alike; if no remuneration is received, a spiritual reward will be. Physicians are urged to be compassionate, feeling the sorrows of those whom they are treating. We know little, however, about the ethics of secular physicians of the early Middle Ages, but by the twelfth century medical literature composed by secular physicians begins to appear. Some of their writings are concerned with ethics. While their tone is very practical, they are not without Christian sentiments. Secular physicians during the late Middle Ages had to fight against a very strong popular prejudice expressed by the adage *Tres medici, duo athei* ("Out of three physicians, two will be atheists"). Especially virulent was the charge that physicians were greedy. In a society in which the imperative to charity was strongly expressed, the question of fees and charitable treatment of the destitute plagued physicians. Their quandary is handily resolved by Henri de Mondeville, who practiced in the late thirteenth and early fourteenth centuries. He advised his fellow practitioners to charge the rich as much as they could in order to be able to extend gratuitous treatment to the poor. The terms in which he expresses this advice reflect an enlightened self-interest as well as an attitude consistent with the theology of his time: "You, then, surgeons, if you operate conscientiously upon the rich for a sufficient fee and upon the poor for charity, you ought not to fear the ravages of fire, nor of rain nor of wind; you need not take holy orders or make pilgrimages nor undertake any work of that kind, because by your science you can save your souls alive, live without poverty, and die in your house" ([18], p. 156).

In popular attitudes there was probably no virtue more desirable in a physician than charity bred of compassion. Much more complex was the Catholic Church's definition of the virtuous physician. We have already mentioned that in confessional literature and in treatises on moral theology written from the fourteenth through the sixteenth centuries attention is given to defining the responsibilities of physicians by specifying what conduct is sinful.[16] The following is composite of their emphases. A physician is not to practice unless competent, even if licensed. To harm a patient owing to ignorance or negligence is a sin. Failure to keep abreast of medical literature and techniques is a sin, as is failure to consult with colleagues when in doubt. Failure to exercise proper diligence and faithfulness in treating one's patients is sinful. Physicians sin if they follow their own fancies rather than the traditions of the art and as a result harm the patient. Similarly, it is a sin to experiment on patients, especially if motivated by the physician's curiosity and resulting in harm to the patient. If in doubt about the effects of a particular medicine or treatment, physicians should leave their patients in God's hands rather than expose them to uncertain dangers. It is sinful to prolong a patient's illness in order to make more money.

It is a sin to demand excessive fees, especially from the poor. Even if there is no hope of recovery, it is a sin to desert one's patient; however, it is also a sin to cause unnecessary expenses or promise recovery when there is little or no basis for hope. Free treatment of the poor is to follow Thomas Aquinas' guideline: Since no one is adequate to bestow a work of mercy on all in need of it, charity should be given first to those with whom one is united in any way. However, if a person is obviously in such a position of need that one cannot see how he or she can be succorred other than by one's own help, then one is obligated to bestow the work of mercy on him. Thus, according to Aquinas, physicians are not obligated always to treat the destitute, or else they would have to put aside all their other patients and occupy themselves entirely with treating the poor (*Summa theologiae*, 2–2.71.1). If it is obvious that an ill pauper will die without medical assistance, the physician sins who does not undertake his or her care. Further, physicians are obligated to treat, even at their own expense, sick people who refuse care, if death will otherwise result. Legal mechanisms were available for collecting one's fee after the fact.

The authors of the literature in question took special interest in the spiritual obligations of physicians, i.e., areas in which a physician's concern with a patient's physical health could conflict with the church's concern with the patient's spiritual health. The Fourth Lateran Council, held in 1215, promulgated a canon (nr. 22) that was included in the official code of canon law (*Decretales* 5.38.13), which states:

> Since bodily infirmity is sometimes caused by sin, the Lord saying to the sick man whom he had healed: 'Go and sin no more, lest some worse thing happen to thee' (John 5:14), we declare in the present decree and strictly command that when physicians of the body are called to the bedside of the sick, before all else they admonish them to call for the physician of souls, so that after spiritual health has been restored to them, the application of bodily medicine may be of greater benefit, for the cause being removed the effect will pass away. We publish this decree for the reason that some, when they are sick and are advised by the physician in the course of the sickness to attend to the salvation of their soul, give up all hope and yield more easily to the danger of death. If any physician shall transgress this decree after it has been published by the bishops, let him be cut off from the church till he has made suitable satisfaction for his transgression. And since the soul is far more precious than the body, we forbid under penalty of anathema that a physician advise a patient to have recourse to sinful means for the recovery of bodily health ([31], p. 236).

Two important matters arise here. (1) The physician's responsibility to ensure that a patient make confession to a priest; and (2) his responsibility to refrain from advising a patient to use sinful means of recovering health. We have already considered the second of these when we discussed the church's concern with the responsibilities of patients. The first of the two points raised by the canon in question is more troubling. This stipulation that physicians must "before all else" admonish their patients to call for a priest created the most frustrating and problematic area in the relationship of physicians to the Catholic Church during the period under consideration. This is especially evident from the tangled discussions in the casuistic literature. Two reasons for the requirement are specified. (1) Since confession has a curative effect it will either make physicians' attendance superfluous or more efficacious. (2) The practice of calling a confessor as a matter of course before undertaking treatment will dispel the notion that physicians only call for a confessor when they have given up hope. It requires little imagination to see the variety of problems this requirement would present in practice and to suspect the negative reaction of many physicians to it. The authors of the confessional literature raised and discussed several questions, such as: (1) Does this apply to each and every case undertaken? (2) Is it the physician's responsibility to ensure compliance? (3) Must the physician withdraw from treatment if the patient refuses to call a confessor? The authorities consulted are in disarray in addressing these questions, leaving the conscientious physician without clear guidelines.

There are very few statements in the medical literature of the late Middle Ages that address the question of calling a confessor for the patient. Before canon law had made confession mandatory, one author of an anonymous treatise on medical ethics, written in the twelfth century, probably in Italy, suggested using the following procedure when taking on a new patient: "When you reach his house and before you see him, ask if he has seen his confessor. If he has not done so, have him either do it or promise to do it. For if he hears mention of this after you have examined him and have considered the signs of the disease, he will begin to despair of recovery, because he will think that you despair of it too."[17] Although this was written before Lateran IV, the action that this anonymous physician recommended and the motive behind it are identical to those of the canon in question promulgated by that council, except that the latter specifies that the act of confession itself may have a curative effect.

There is a deontological treatise attributed to the physician Arnold of Villanova (late thirteenth and early fourteenth centuries) containing advice on confession that strikingly resembles part of that canon and even quotes the same passage from John: "When you come to a house, inquire before you go to the sick whether he has confessed, and if he has not, he should immediately or promise you that he will confess immediately, and this must not be neglected because many illnesses originate on account of sin and are cured by the Supreme Physician after having been purified from squalor by the tears of contrition, according to what is said in the Gospel: 'Go, and sin no more, lest something worse happen to you' " ([32], p. 141). This clearly shows the influence of canon law on a strictly secular treatise on medical etiquette and ethics, as does the following piece which appears in an anonymous plague tractate composed in 1411: "If it is certain from the symptoms that it is actually pestilence that has afflicted the patient, the physician first must advise the patient to set himself right with God by making a will and by making a confession of his sins, as is set forth according to the Decretals: since a corporal illness comes not only from a fault of the body but also from a spiritual failing as the Lord declares in the gospel and the priests also tell us" ([2], p. 416).

In spite of such sentiments from these physicians, by the early sixteenth century there was still no uniformity of formal interpretation or practical application of this stipulation of canon law. To avoid ambiguity and in an attempt to ensure compliance, Pope Pius V in 1566 promulgated his *Super gregem* which required that physicians terminate their treatment of any patient who was unable to produce by the third day a document signed by a confessor certifying that confession had been formally made. Physicians who violated this rule were to lose their license and be expelled from their medical or surgical associations. It should be noted that this tightening of the requirement occurred after the Reformation and thus affected only Roman Catholic physicians and surgeons.

The confessional literature is, by its very nature, negative. It is concerned with identifying, exposing, and correcting sin (sacramentally by repentance and forgiveness), and correcting conduct by the moral education of the penitent. The sections of the confessional literature which deal with physicians are not directly concerned with the virtues of physicians, but rather with their sins, that is, their vices. Early church fathers had raised the question whether virtue is the absence of vice and the answer was a resounding 'no'. For that would be to give virtue a negative value. Generally medieval sources agree. And the purpose of the Sacrament of Penance in the medieval church was not to define virtue but rather to deal with sin.

In a previous paper we wrote: "There is one matter that is conspicuously absent from these casuistic treatments of physicians' moral responsibilities and that is any reference to an obligation *to care*, i.e., any obligation to show personal concern. There are strongly-enunciated imperatives to extend medical care to the destitute and to treat the poor gratuitously, but there is no probing into the hinterland of the heart where genuine compassion and caring are found; there is no attempt to explore and discover the physician's motivation for practicing medicine, no forcing him to examine the 'why' of his practice, but only the 'how'. Had Christian charity – at least the medical charity expected of physicians – become merely the dead letter of the law, devoid of 'the spirit that giveth life?' " ([5], p. 21). Our intention in writing that was not to impugn the intentions of those in the Catholic Church who were attempting to make the confessional act more concise and inclusive and thus more effective. Yet a negative result seems to be an inevitable outcome of a process that is essentially negative in its emphasis if not in its ultimate objectives.

This was also a major objection of the Protestant reformers. But their objections to the Sacrament of Penance went much deeper. It is safe to say that they loathed it; they detested and fought against everything that they saw in it – the power of the keys, the concept of *ex opere operato*, the very theology undergirding it that seemed to them to deny, in the most flagrant terms and in the most manipulative ways, the doctrine of justification by faith alone and the centrality of grace. The Counter-Reformers, on the other hand, defended this sacrament, especially its consolatory function, and denied the criticisms levelled against it. Thomas Tentler observes that for the Reformers "sacramental confession would represent an unchristian instrument of torment and an encouragement to hypocrisy – leaving people proud, licentious, and unrepentant; utterly failing to comfort them; and leading ultimately to eternal damnation." For the Counter-Reformers "it would remain the highest pastoral art, a just and certain discipline; making essential but possible demands on Christian consciences; preserving the divine order in Christ's Church; and assuring, as well as anything in this world could, the consoling gift of eternal salvation" ([34], p. 368).

Some scholars have accused the Reformers of creating a moral vacuum by failing to replace the Sacrament of Penance and its clear and precise definitions of sin with anything of comparable moral utility. Others assert that the Reformers restored pristine Christian liberty and freedom of conscience, making *Scriptura sola* the basis for moral decisions and definitions. Regardless of such value judgments, it is safe to assert that Roman Catholic physicians of the sixteenth century would have had a very clear picture of the expectations of their church simply by an annual interrogation in the confessional. Insofar as they would be able to avoid those sins to which physicians were regarded as most susceptible, they would be virtuous.[18] For Protestant physicians the matter would not have been nearly as clear cut. The highly personal and internal standards of the New Testament, as interpreted and applied by each individual, would have been the only absolute criteria for guidance in virtue.

ACKNOWLEDGEMENT

This publication was supported in part by NIH Grant LM 04108 from the National Library of Medicine.

NOTES

[1] [21], Vol. 1, *s.v.*, ἀρετή.

[2] Is. 42:8 and 12; 43:21; 63:7; Hab. 3:3; Zech. 6:13.

[3] All biblical quotations are from the Revised Standard Version.

[4] [8], p. 699. Cf. [21], p. 460: "Here a notable formal analogy points us to the secular world, to the sphere of 'virtue'. Yet it is almost certain that in this passage ἀρετή is more than the secular parallel ('fidelity') and consequently a similar distinction is likely as in the case of πίστις."

[5] In the following discussion we have deliberately ignored questions of authorship in the New Testament and diversity of theologies (e.g., Johannine, Pauline). Regardless of such questions, one can speak of common New Testament themes that characterize early Christian attitudes.

[6] One should note that James predicates of the "wisdom from above" qualities similar to those expressed in the beatitudes and attributed by Paul to love: " . . . the wisdom from above is first pure, then peaceable, gentle, open to reason, full of mercy and good fruits, without uncertainty or insincerity" (James 3:17).

[7] We have not included a consideration of heroic virtues, as partially christianized with the conversion of, e.g., Germans, Anglo-Saxons, Norwegians. On heroic virtues generally,

see the superlative treatment in [24], pp. 114–122, and also, in the same work, a short discussion of heroic virtues in the early Middle Ages, pp. 154–155.

8 Beginning quite early in Christian history, virginity was singled out as a special, positive virtue. The origins of this idea, and its relationship to virtue in general and other virtues in particular, will not be considered in the present paper.

9 While *arete* was, as we have seen, almost entirely avoided in the Septuagint and by the authors of the New Testament, the Greek Fathers used it often without the same caution. The Latin Fathers appear to have used *virtus* in about the same way. Jerome's translation of the Bible into Latin employs *virtus* at least 100 times in the Old Testament, nearly 150 times in the Apocryphal books, and almost 130 times in the New Testament. However, if the Hebrew Scriptures had been translated into Latin by devout Jews at the time the Septuagint was produced, and if the New Testament had been originally written in Latin, it is likely that *virtus* would have been as studiously avoided then as was *arete* because of its equally anthropocentric force in classical thought.

10 Some noteworthy examples of this genre are the *Aenigmata* of Boniface, the Anglo-Saxon missionary to the Germans; Alcuin's *Liber de virtutibus* and *De animae ratione*; and *The Assembly of Gods*, wrongly attributed to John Lydgate.

11 As quoted by [13], p. 90.

12 As quoted by [3], p. 341.

13 This matter is not discussed in [1]. The following works have a section devoted to this subject:

Bartholomaeus de Sancto Concordio, *Summa casuum* (c. 1338; Venice, 1473), copy at University of Pennsylvania; generally cited as *Pisanella*.

Antoninus of Florence, *Summa theologica* (or *Summa moralis*) (1477; 1740; reprint, Graz: Akademische Druck- und Verlagsanstalt, 1959).

Baptista Trovamala de Salis, *Summa de casibus conscientiae* (c. 1480; Venice, 1495), copy at College of Physicians of Philadelphia; generally cited as *Baptistina*.

Angelus Carletus de Clavasio, *Summa Angelica de casibus conscientiae* (c. 1486; Lyons 1494), copy at Free Library of Philadelphia; generally cited as *Angelica*.

14 [5], especially pp. 12–24.

15 See, e.g., the discussion in [30] which extends, in its implications, well beyond Renaissance Venice.

16 See [1] for a detailed discussion.

17 Latin text is in [11], vol. 2, p. 74.

18 Restitution to anyone harmed by one's sin was a *sine qua non* for forgiveness. "There was a strong tradition in the church, from Augustine through Gratian, in canon law and conciliar legislation, that whatever was gained through dishonest practices had to be restored in full before remission of sin could be given, before *ego te absolvo* could be pronounced efficaciously. Restitution was an absolute condition of forgiveness and the confessor had considerable freedom of discretion in determining situations in which restitution must be made. The principle of restitution was applied to much besides ill-gotten gain, indeed to damages generally" ([1], p. 111).

BIBLIOGRAPHY

[1] Amundsen, D. W.: 1977, 'Medical Deontology and Pestilential Disease in the Late Middle Ages', *Journal of the History of Medicine and Allied Sciences* **32**, 403–421.

[2] Amundsen, D. W.: 1982, 'Casuistry and Professional Obligations: The Regulation of Physicians by the Court of Conscience in the Late Middle Ages', *Transactions and Studies of the College of Physicians of Philadelphia* (N.S.) **3**, 22–39, 93–112.

[3] Amundsen, D. W.: 1982, 'Medicine and Faith in Early Christianity', *Bulletin of the History of Medicine* **56**, 326–350.

[4] Amundsen, D. W. and G. B. Ferngren: 1982, 'Medicine and Religion: Early Christianity through the Middle Ages', in M. E. Marty and K. L. Vaux (eds.), *Health/Medicine and the Faith Tradition: An Inquiry into Religion and Medicine*, Fortress, Philadelphia, pp. 93–131.

[5] Amundsen, D. W. and G. B. Ferngren: 1982, 'Philanthropy in Medicine: Some Historical Perspectives', in E. E. Shelp (ed.), *Beneficence and Health Care*, D. Reidel, Dordrecht, pp. 1–31.

[6] Bloomfield, M. W., W. R. Bowie, P. Scherer, J. Knox, S. Terrien, N. B. Harmon: 1952, *The Seven Deadly Sins: An Introduction to the History of a Religious Concept, with Special Reference to Medieval English Literature*, Michigan State University Press, Ann Arbor.

[7] Bloomfield, M. W., *et al.*: 1979, *Incipits of Latin Works on the Virtues and Vices, 1100–1500 A.D.*, The Mediaeval Academy of America, Cambridge, Mass.

[8] Bromiley, G. W.: 1973, 'Virtue, Virtues', in *Baker's Dictionary of Christian Ethics*, Baker, Grand Rapids.

[9] Buttrick, G. A. (ed.): 1951–1957, *The Interpreter's Bible*, Abingdon, New York and Nashville.

[10] Cayré, F.: 1935 and 1940, *Manual of Patrology and History of Theology*, trans. H. Howitt, Desclée, Paris.

[11] De Renzi, S.: 1852–57, *Collectio Salernitana*, Naples.

[12] Dudden, F. H.: 1935, *The Life and Times of St. Ambrose*, Clarendon Press, Oxford.

[13] Ebeling, G.: 1970: *Luther: An Introduction to His Thought*, trans. R. A. Wilson, Fortress, Philadelphia.

[14] Forell, G. W.: 1979, *History of Christian Ethics*, Vol. 1: *From the New Testament to Augustine*, Augsburg, Minneapolis.

[15] Gilmore, M. P.: 1974, 'Erasmus' Godly Feast', in C. H. Trinkaus (ed.), *The Pursuit of Holiness in Late Medieval and Renaissance Religion*, E. J. Brill, Leiden, pp. 505–509.

[16] Grant, R. M.: 1970, *Augustus to Constantine: The Thrust of the Christian Movement into the Roman World*, Harper and Row, New York.

[17] Hadas, M.: 1960, *The Greek Ideal and Its Survival*, Harper and Row, New York.

[18] Hammond, E. A.: 1960, 'Incomes of Medieval English Doctors', *Journal of the History of Medicine and Allied Sciences* **15**, 154–169.

[19] Jaeger, W.: 1961, *Early Christianity and Greek Paideia*, Belknap, Cambridge, Mass.

[20] Jones, L. W. (trans.): 1946, Cassiodorus, *An Introduction to Divine and Human Readings*, Columbia University Press, New York.

[21] Kittel, G. and G. Friedrich (eds.): 1964–1976, *Theological Dictionary of the New Testament*, trans. G. W. Bromiley, Eerdmans, Grand Rapids.

[22] Lewis, C. S.: 1936, *The Allegory of Love: A Study in Medieval Tradition*, London, Oxford University Press.

[23] Lightfoot, J. B.: 1913, *St. Paul's Epistle to the Philippians*, Macmillan, London.
[24] MacIntyre, A.: 1981, *After Virtue: A Study in Moral Theory*, University of Notre Dame Press, Notre Dame, Indiana.
[25] MacKinney, L. C.: 1952, 'Medical Ethics and Etiquette in the Early Middle Ages: The Persistence of Hippocratic Ideals', *Bulletin of the History of Medicine*, **26**, 1–31.
[26] Mayor, J. B.: 1907, *The Epistles of Jude and II Peter*, Macmillan, London.
[27] Moule, H. C. G.: 1893, *The Epistle to the Philippians*, Cambridge University Press, Cambridge, Mass.
[28] Owst, G. R.: 1961, *Literature and Pulpit in Medieval England*, 2nd ed., Blackwell, Oxford.
[29] Pease, A. S.: 1914, 'Medical Allusions in the Works of St. Jerome', *Harvard Studies in Classical Philology* **25**, 73–86.
[30] Pullan, B.: 1971, *Rich and Poor in Renaissance Venice: The Social Institutions of A Catholic State, to 1620*, Harvard University Press, Cambridge, Mass.
[31] Schroeder, H. J. (trans.): 1937, *Disciplinary Decrees of the General Councils*, Herder Book Co., St. Louis.
[32] Sigerist, H. E.: 1946, 'Bedside Manners in the Middle Ages: The Treatise *De Cautelis Medicorum* Attributed to Arnald of Villanova', *Quarterly Bulletin of the Northwestern University Medical School* **20**, 136–143.
[33] Sigerist, H. E.: 1943, *Civilization and Disease*, Cornell University Press, Ithaca.
[34] Tentler, T. N.: 1977, *Sin and Confession on the Eve of the Reformation*, Princeton University Press, Princeton, New Jersey.

Western Washington University,
Bellingham,
Washington, U.S.A.

Oregon State University,
Corvallis,
Oregon, U.S.A.

DIETRICH VON ENGELHARDT

VIRTUE AND MEDICINE DURING THE ENLIGHTENMENT IN GERMANY

BACKGROUND AND INTERRELATIONSHIPS

Medicine in the Enlightenment was influenced by the philosophical, pedagogical, and theological trends of the period [10, 12, 16, 17, 19]. This was also the case for the views of virtues and their significance for medicine, for the sick person, the physician, and society. The health care practitioners of the 18th century remained at a distance from scholasticism and the philosophy of Descartes, and increasingly from the philosophy of Leibniz. They considered metaphysical systems to be suspect or superfluous, recommended a return to the thought of Hippocrates, ascribed more relevance to sensualism and popular philosophy, hardly recognized materialism, and held the progress of empirical and psychological knowledge over human behavior to be of central importance. The turn of medicine away from metaphysics found an exemplary expression in the controversy between Leibniz and the physician Stahl. An echo of popular philosophy in medicine was to be found in Unzer's magazine 'The Physician'. In 1798, relying on Kant's formula, Osterhausen defined medical enlightenment as "the exit of a person from his minority into things dealing with his physical well-being" ([20], 1, p. 8). In the tradition of Locke and Hume, Weikard could see only a 'fanciful hunting' and convulsive illness in the transcendental philosophy of Kant. Then in the time of the Romantics health care practitioners were once again concerned with metaphysical bases of medicine due to the influence of Schelling.

The views of philosophy, pedagogy, and theology in the 18th century were not in agreement concerning virtues and duties. Through all the diversity and change the traditional virtues retained their importance: the four platonic virtues of wisdom (*prudentia*), moderation (*temperantia*), courage (*fortitudo*), and justice (*justitia*), and the three Christian cardinal virtues of faith, hope, and charity. Changes were evident in the world of the Baroque as middle-class virtues expanded in opposition the aristocratic virtues of generosity and leisure. Geulincx placed in the foreground the virtues of industry, obedience, justice, and modesty. The pedagogues Basedow, Salzmann, Campe, and Pestalozzi emphasized virtues like order, frugality, cleanliness, and industry; Campe even elevated order to the 'mother of virtues'. Kant understood virtue

Earl E. Shelp (ed.), Virtue and Medicine, 63–79.

as a moral strength to fulfill one's duties, not in the sense of a habit, but rather in a completely new achieved consciousness and affirmation. At the end of the 18th century virtue in general possessed a high degree of attraction; the 'virtue league' around 1800 was a characteristic example. The formulations of German idealism marked a completion and conquest of the Enlightenment. According to Hegel, duties correspond to a world of established morality, while virtues belong to states of emergency in which confirmation is possible and which single out the individual.

The Platonic and Christian cardinal virtues also appear in medicine; medical writings are full of quotations from the Bible and from Roman authors. One can recall the double meaning of *arete* as technical and moral excellence, the Aristotelian definition of virtue as the middle way between the two extremes, the resonance of Cicero's words about the honorable and sympathetic lie, Augustine's explanation of sickness as expiation and trial, or the Stoic concept of the inner connection and the naturalness of virtues. Later concepts of virtue are also worthy of examination. Medical deliberations about virtues and duties during the 18th century manifest an increasing independence from philosophy and theology; psychological-constitutional presuppositions are stressed, medically immanent changes take effect. The progress of medicine gives rise to new problems, which in no way always were followed by moral progress. During the 18th century, it remained a worthwhile endeavor to present and to analyze in some detail the discussions of virtue in theology, philosophy, and pedagogy with the medical ethical debates in their agreements and disagreements.

In a great number of medical dissertations, papers, and monographs of the 18th century, the virtues, duties, and rights of the sick person, the physician, and society are discussed, sometimes ironically and sometimes seriously, and, as was customary, often in Latin. My concentration on the German Enlightenment should not obscure the international and historical connections; Bartholin, Boerhaave, Gaub, Tissot, Rush, and Gregory are continually mentioned together with many other foreign physicians who have written about ethical questions. The tradition remains current, quotations from the Hippocratic corpus are cited, as well as the works of Galen, Avicenna, and Paracelsus. Major authors from the beginning of the epoch are Stahl, Storch, and Hoffman; the transition to the 19th century was accomplished with Vogel, Mai, and Hufeland.

Medical ethics is composed of an inner structure and outer conditions. The inner structure consists of three centers: the patient, the physician, and the state, as well as the nine relations arising between these centers, since each center also has a relation to itself, the patient to the illness and to other patients, the physician to medicine and his colleagues, the state to the social lower classes and to other states. The outer conditions refer to the areas of society and culture which operate on the sick person, the physician, and the state in its theoretical-practical association with sickness. The meaning and function of virtues, duties, and rights in medicine are described and interpreted as suitable for the Enlightenment according to this structure of medical ethics.

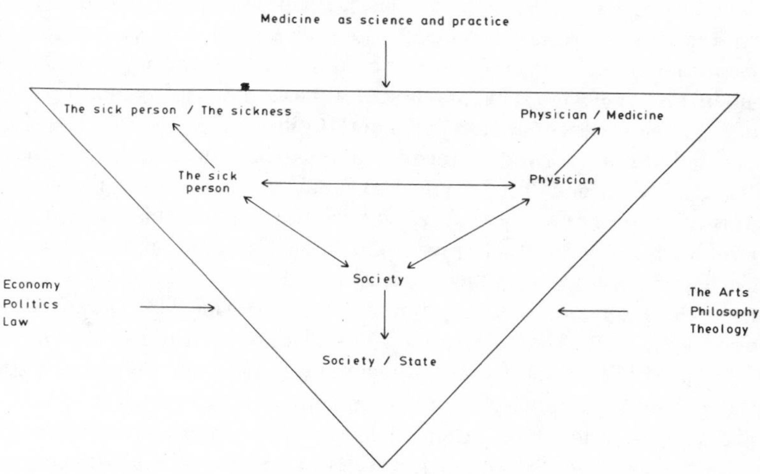

Fig. 1. The Structure of medical ethics.

THE PHYSICIAN

There is hardly any profession that should be so dependent on virtues, so constructed upon duties, as medicine, according to the belief of the physicians of the Enlightenment. In no situation outside of the military is the mortality so high as it is with physicians. The basis of the medical profession is seen in a philosophical and religious bearing. From this basis true knowledge is shown to be a connection between experience and reason, as well as true action to be a connection between mercy and justice.

Stahl and Storch include, at the beginning of the century, the virtues of piety (*pietas*), erudition (*eruditio*), preparedness (*alacritas, promptitudo*), industry (*diligentia*), courtesy (*humanitas*), patience (*patientia*), courage (*animositas*), wisdom (*prudentia*), truth (*veritas*). A little later Hoffman characterizes humility, generosity, friendliness, and discretion as the essential virtues of the physician. Ploucquet, towards the end of the century, considers patience, philanthropy, sympathy, mildness, modesty, courtesy, industry, and the love of order to be the chief virtues of the physician. They had to be present in the proper measure, but in any case their lack would have worse consequences than their exaggeration. Their foundation is the love of God and neighbour; nevertheless, their psychological basis is not infrequently ascribed to a sanguine-melancholic temperament. Collisions between the different virtues should not be avoided. Intrinsic and extrinsic qualities and virtues are differentiated, divided into physical and spiritual; according to Stark the beautiful figure of the physician is extrinsically physical, his cultural education is extrinsically spiritual ([26], p. 198). Definitions of virtues are also always definitions of vices, which have been dealt with in essays by physicians. General characteristics of the physician were described by, among others, Stahl and Storch (1728), Hoffman (1738), Unzer (1759), Rübel (1758 and 1766), Mai (1777), Stark (1748), Hildebrandt (1795), and Ploucquet (1797).

The physician's knowledge should include the natural sciences and the clinical subjects, should be extended to philosophy, particularly logic, and even the belles artes; the command of Latin and Greek and some modern languages is necessary, as well as the possession of historical information. Self-knowledge is also essential, and finally a broadening of the consciousness through travel. Reason, knowledge, observation, memory, and discernment are all necessary for the physician. Zimmermann's article "Von der Erfahrung in der Arzneykunst" (1763/64) is the convincing representation of these theoretical qualities. In principle the physician should not overestimate his knowledge and not consider himself above other men: "A physician should be neither courtly nor proud, but rather modest" ([24], p. 7). Erudition should also have an essential fundament in religious belief, without constantly endangering the virtues of the physician.

The quarrel with superstition and charlatanism/chicanery occupies much space, because they are dangerous for the physician as well as for patients; they are the reason medicine is scorned; they are the focus for the battle of physicians from the Enlightenment to the transition into the scientific 19th century. The path of pure experience is dangerous and deceptive, it is chosen by the most varying swindlers, by shepherds, farriers, executioners, pharmacists, surgeons, barber-surgeons, monks, old women, and mid-wives: "the world succumbs to the burden of kind-hearted murderers of life" ([26], p. 149), as Stark complains, who considers no science and art so susceptible to bungling as medical science is. In contrast, the right path is achieved by means of a scientific education at a university and through one's own personal experience; only this will lead to a practice based on reason and make Rousseau's criticism of medicine superfluous. Uncertainty and limitations in medicine must also be recognized; the physician cannot set himself up against time and nature, and he displays his moderation (*temperantia*) when he respects these limits.

The physician also displays the virtues of moderation and modesty in his politeness and self-control, in his refined speech, restrained gestures and mimicry, and his decent clothing. He should avoid gluttony, drunkenness, gambling, and scandalous speech, but also not fall into the habits of greed, pettiness, and taciturnity. The generosity of the 16th and 17th centuries collides with the stronger stress on modesty in the 18th century. The appearance of the physician is also rewritten as 'gallant'; the physician should not be a misanthrope or eccentric, but rather as a social being be in command of the forms of society and possess conduite (the virtue of management and discretion). The physician should be serious about his profession; care, industry, and work receive in medicine the highest esteem: "Intelligence demands that we be hard-working and industrious and abhor indolence" ([26], p. 10). In his practice the physician may not be too hesitant nor too foolhardy; he must adhere to the correct middle way here also, which enables him to act but restrains him from dangerous experiments.

The physician is guided by humanity and helpfulness. As a person he also confronts the person in the sick individual and not only the diseased organs or a defective machine. The author Longolius wrote in 1727 that the physician will "have a steadfast attention to both body and soul in all bodily accidents of a living person" ([13], p. 43). The physician wants to please the patient and tries to win his trust. The sick person must be convinced not only of the knowledge but also of the moral character of the physician. Eminent physicians are said to have healed through trust alone. Faith and hope grow with trust. The physician, however, should defend himself in contrast against, as Ploucquet expressed it, "irritable, impolite behavior of the sick person and his family" ([22], p. 138), if he doesn't want to lower himself to the level of the fawning charlatan. He should not return insults for insults but rather, with Wedekind, recognize the background and show understanding: "thus our passion diminishes, and sympathy with the erring brother takes its place" ([32]. p. 20). What the physician has to suffer in annoyances and resistance is treated in a comprehensive manner in the 'Medicus afflictus' of 1715.

The moral education of the sick person is expected from the physician, he is supposed to influence the sick person, to bring him to a correct attitude concerning the sickness, and to guide him to a sensible manner of living. The physician is responsible for the dietetics of the sick person as well as the healthy person. In the 18th century dietetics still meant the area of the ancient *sex res non naturales* (the six non naturals), the association of air and light, movement and rest, waking and sleeping, eating and drinking, retentions and excretions, passions and emotions. Beginning with the suggestions of Stahl and Hoffman, the range of dietetic writings extends from the impulses of Unzer to Mai's 'medical lenten sermons' (1793/94) and Hufeland's 'art of lengthening human life' (1796). For physicians looking at the plethora of examples, it appears indubitable that virtues lengthen life but vices shorten it: "Virtues promote health and restore it. Vices are evil and unhealthy" ([29], 1, p. 692).

The physician must take care that the patient follows the therapy and also be ready to help when problems arise, as when the patient cannot swallow the medicine because of nausea. The physician's assistance will always be improved by means of psychological knowledge about the power of the passions and their place between physiology and ethics. Vogel relates the *sçavoir faire* of the physician to the association with patients and colleagues; the psychological perspective is dominant, factual knowledge and morality come into play in "the best talents and virtues" ([31], p. 86). The readiness of the physician to help has its basis in love of God and man and basically knows no boundaries. The physician according to Rübel will assist "both day and night his suffering neighbor with help and advice, without regard for class or wealth or person" ([24], p. 13). Views about the attitude towards terminal illnesses differ. Because of the dangers for the physician, Hoffman demands that he "refrain from visiting during the time the patient is dying" ([7], p. 183). The Hippocratic warning is often refused, as it is with Ploucquet: "one still dedicates his strivings even to the dying, should that take place, in order to diminish the sum of his suffering" ([22], p. 114). The physician may not cause the death, whether by his own action or by giving the sick person adequate means. According to the opinion of the physicians active euthanasia contradicts the explanation of the highest and holiest duty of medicine, to preserve life. Hufeland fears the greatest dangers for the future if this duty is violated: "If the line should be crossed, then there are *no longer any boundaries*" ([9], p. 20). Faith and morals forbid any active euthanasia, they also demand the highest care with experiments on humans, which should properly be conducted by the physician on himself, according to Hufeland.

The practice of medicine is bound by the duty of confidentiality; Rübel calls discretion the "virtue of not saying the truth when it avails nothing but instead brings certain ill" ([24], p. 12); discretion refers to the relationship to the sick person as well as to the environment. Physician and priest are bound by the same constraints concerning the knowledge of the secrets of their fellowmen. What the sick person trusts to the physicians, according to Wedekind should be a "holy deposit" ([32], p. 23) in his breast and may not be related to other persons. The physician should be both discrete and open at the same time; the duty of explanation as well as the duty of silence leads him. Conflicts and compromises cannot be avoided. The obligation of confidentiality ends where other persons come into danger. Prostitution and abortion also place limits on the duty of confidentiality. The physician should explain but at the same time recognize the freedom of the patient, which can lead to a rejection of the truth and even to self-destruction. Illusion and deception touch the physician's understanding of explanation. The physician must conscientiously test to see what the sick person can bear. Impending death should be concealed from the sick person. According to Stark the patient "often becomes unhappy in body and soul" ([26], p. 349), if this possibility is withheld from him. If the sick person cannot bear this information, the family members at least should be told, even if in many cases restraint should be employed. Other physicians support the belief that patients should be left in ignorance. Hoffman suggests in the case of dangerous diseases, offering a judgment "from which questioners will not know what they should conclude" ([7], p. 188). In general, the physician is reminded that the prognosis of death cannot be given with an absolute certainty regarding time, but a high risk can be indicated.

The physician may expect payment for his services to the sick person. At the same time the payment for him occupies a position of tension relating to the high, nearly holy meaning of medicine, which should not be expressed in financial terms. A Christian physician should practice his profession out of compassion and humanity, not because of the monetary recompense. Therefore, it will also be expected and demanded of the physician that he accept no fee from the poverty-stricken [7, 13, 22, 32]. The physician should not increase his income by means of unnecessary visits; he should base his fee on the ability of the patient to pay.

The physician of the 18th century had to recognize the class differences. Poor and genteel, worldly and spiritual people want to be treated differently. Particularly important are circumspection, discretion, versatility, and stability. The upper class is known for its sensitivity. "Pains are much more suitable for common people" ([24], p. 72), as Rübel reports from his practice and considers this opinion to be correct; people from the lower class simply have a "strong, thick, and less sensitive skin" ([24], p. 73). Stark recommends reserve with genteel persons, which is not the case with simple people, who are themselves so shy and deficient in vocabulary ([26], p. 13). The duty of confidentiality in respect to the genteel can also be suspended when inferiors contract syphilis; according to Stahl and Storch their superiors should be informed of the disease, since they would experience a greater aversion and disgust than normal people and could even become ill from unexpectedly seeing this disease [28].

Women and children also demand special attention. They serve as pitfalls for every physician: "There are two difficult things in medicine, namely, to cure children and pregnant women" ([7], p. 169), as Hoffman says. The physician should always stand fast and let himself be led by his knowledge and his humanity – unperturbed by fashion and pity, recompense, flattery and force, never going in contradiction to his conscience. Behind all superficial differences he should always see the person who needs his sympathy and his aid, or in the words of Stark: "He deports himself as a person among other persons, treats them as persons, loves them as his brothers, serves them willingly as friend, tries to ease their fates" ([26], p. 236). The physician in this sense recognizes no differences of class, no differences of religion, sex, or education; for him all men are the same.

During the 18th century the relationship with one's colleagues is repeatedly the theme of discussions concerning virtue and duty. Contentiousness and rivalry among physicians were not without reason a beloved object of the literature and satire of that period. Politeness and respect, camaraderie and aid should characterize the relationships among physicians. Particular attention is paid to the relationship to the surgeon, the apothecary, and the midwife. According to Stieglitz, art and the sick person should compel the physician "to keep the honor of the consultation intact" ([27], p. 72). At the same time the physician is compelled to point out fraudulent colleagues; truth comes before collegiality. If an accusation cannot be proved against a colleague, then he should not be openly scolded or criticized in front of a patient. A discussion of the case at the sick-bed is definitely improper. The physician should seek the advice of his colleagues and always be ready to acknowledge his uncertainty and mistakes. A consultation protects one from incorrect diagnoses and doubtful therapies, but it serves not only the patient, but also the progress of medicine. Contact with surgeons is necessary on both sides. The physician should not prescribe more medication than is really necessary, so as not to increase the earnings of the apothecary. The prescriptions should conform to the norm, prescribing by apothecaries should, as far as possible, be circumscribed, since according to Rübel they do not possess "the necessary conditions to practice prudently" ([24], p. 25).

The qualities and capabilities of the physician influence his relationship to the environment of the patient, to other groups in society, and to the state. The duties of silence and explanation reach out from the patient into the life of other people. If the prognosis cannot be related to the patient, the physician must decide whether to tell family or friends. In the case of contagious diseases there is a collision of the duty of confidentiality with the general duty of the physician to protect healthy persons from contagion. According to Rübel, the physician must act cautiously and deliberately, "because not only the welfare of one person but often the health and happiness of an entire republic are dependent on it" ([24], p. 31). There are many points of contact when one considers the clergy. An agreement concerning the care of the soul and particularly the sacrament must be reached. The consequences of the sacrament can in no way be seen as negative only, physicians can even observe emotional calming, which is said to lead to healing. But discretion and circumspection are also repeatedly demanded from priests and ministers, since they have to recognize on principle their lack of medical knowledge. But physicians consider the care of the soul to be important, since sickness and health are always bound with psychic aspects. Dietetics cannot be separated from religion [21]. Finally, the physician must produce expert opinion for the state. On the one hand, careful observation of the facts and rational conclusions are demanded of the physician; on the other hand he must recognize the limits of medicine and relinquish the final decision; according to Hoffman, he must relinquish "the reasons for decision to the scholars of jurisprudence" ([7], p. 85). The physician must continually be careful when instructing the public, so that individual honor will be offended as little as possible and society can feel secure as far as possible and not be unnecessarily alarmed.

THE PATIENT

A sick person has a relationship to his disease, his physician, and his society. According to Unzer, patients have "to observe certain duties concerning themselves as well as concerning other people, but especially concerning physicians" ([29], 1, p. 13). Like the physician, the sick person should also have a philosophical and religious attitude and let himself be guided in his dependence on help by politeness and respect in his relation to his environment; right conduct can also be expected from the sick person. Longolius wants to have a 'philosophical' and 'gallant' patient. The sick person must find an appropriate attitude to his illness; he should neither pretend to be sick when he is well nor pretend to be healthy when he is suffering. A person feigns health, according to Longolius, because he fears that in the condition of being ill, "his security, his pleasure, or his honor will suffer" ([13], p. 13). Concealing an illness hinders the physician who is trying to recognize the cause and takes from the environment the possibility of assistance – "and therefore that which is most necessary to convalescence" ([13], p. 18). Just as lacking in virtue are those persons who try to draw attention to themselves with feigned diseases and speak of their sufferings at every opportunity. These 'patient-masques' risk not being believed in an instance of real disease. Fashion dictates the choice of simulated diseases. In the 17th and 18th centuries podagra, venereal diseases, and hypochondria were chosen – podagra as the suffering of the rich and the mighty, venereal diseases as the result of a depraved life, hypochondria as the characteristic of scholars: "Because one commonly finds this disease among scholars, the most stupid persons usually maintain that they are hypochondriacal" ([29], 2, p. 407). However, the sick person should also not accept his disease, if it should not be permitted to conquer.

Christian and Stoic virtues prove their value. In overcoming pain and suffering the sick person can attain a higher condition than the healthy person. Patience and hope should serve the sick person. Association with sickness and death reveals the moral character of a person: "In need and misery, in sorrow and sickness, it will be decided who has a noble soul" ([24], 1, p. 626), as Unzer describes it. Just as reason, belief is a presupposition of patience, *patientia* becomes the central virtue of the sick person – patience in the face of pain, patience in the face of unsuccessful therapy, patience in the face of unsympathetic physicians, patience in the face of chronic illnesses.

If disease appears, a physician should be sought. The sick person has a duty to describe his true condition to the physician; he should also be convinced that the physician can help him. Above all, the diseases of the lower classes are often said to be caused by ignorance and lack of money. The success of therapy presupposes early contact with a physician. The sick person can by means of his choice increase the number of genuine or false physicians. Physicians complain again and again about the incapacity of people to form an appropriate judgment about physicians and to choose the right physicians. Among rich and poor, the former transgress "usually in excess, the latter in defect in the choice of physicians", according to Sklerander ([25], p. 16); only too often sick persons would choose flattering instead of honest physicians.

The sick person must inform the physician of the kind and origin of his disease, and of his life-style; according to Unzer he should truthfully advise the physician of his observations "of his nature, his arrangement of his life, and his perceptions of the first onset and during the progress of the disease" ([29], 1, p. 125). The physician's duty of confidentiality will enable the patient to impart his observations. According to Henning, this duty is only too often violated: "The voluntary practice of confidential communication in this sense has been hindered by the bad conduct of so many garrulous and flippant physicians" ([5], p. 50). The sick person must be ready to let himself be thoroughly examined, which was perceived to be unpleasant in the 18th century, especially by women. The anecdote of one woman is characteristic. She allowed her pulse to be taken only when a cloth covered her wrist, whereby a physician covered his hand in the same way and is said to have commented: "A linen-covered pulse demands a medication of cloth" ([13], p. 59).

A further duty of the sick person in relation to the physician lies in the observance of the therapeutic instructions. The physician is responsible for the therapy, the patient is bound to give obedience. He should not fail to take instructions from the physician. According to Henning, it may never be forgotten that the physician can lose interest in an uncooperative patient ([5], p. 61). The patient must inform the physician immediately of any problems with the medication, so that a different method can be considered. According to Longolius, patients can be observed again and again "who know how to deal with the physician as well as with the medication" ([13], p. 51). The sick person has to test himself critically to see whether his aversion is not exaggerated, since "imagination and affection can easily be hidden under a natural aversion" ([13], p. 52). Vanity plays a role whenever expensive and rare medications are considered more efficacious than inexpensive and common ones. Certain worries naturally produce surgical interventions. The sick person also should not demand too much from the physician, since the physician must respect time and nature. Also, "the most clever physician can do no more than to watch carefully during this time and duly to ask nature for help" ([13], p. 57).

Sickness demands a particular regimen, which is always oriented to a general dietetic schema. Traditional attitudes are bound with an esteem for order and cleanliness, as is characteristic for the Enlightenment. Housekeeping becomes the primary metaphor for the relation of the inner with the outer world, the health of the individual with the health of the state. The sick person should avoid arguments and violent emotions, since he is still much susceptible than a healthy person.

Finally, the sick person is required to be thankful to the physician; this refers both to a friendly demeanor as well as to payment for the physician's efforts. From the patient, according to Longolius, the physician can expect that he "not only especially likes or esteems the physician as a particular friend of the human race, but also rewards his labor according to his means" ([13], p. 58; [25], p. 80). In contact with the physician, the sick person should adjust his behavior according to the physician, and even orient himself to the character and inclinations of the physician in their conversations, which will always begin with the purely medical. It is the duty of the sick person to take notice of the physician's health, not to call him more often than necessary, and to greet him in clean clothes and clean surroundings. The concept of regimen also forms the relationship between patient and physician. Every physician, according to Henning, will not be allowed to forget that he cannot expect from the sick person that cheerfulness "that recommends him as a companion in the days of blooming health" ([5], p. 45).

To his environment, the sick person has not only the right but also the duty to demonstrate virtue to the persons caring for him, his relations, and the state. Healthy persons have the right, which the sick person must also respect, to expect understanding and moderation from him. The sick person should not be either too worried or too carefree; he should accept help thankfully and not be too impatient if the help does not come subtly, quickly, or continuously enough. A sick person can even be a model for ministers and, according to Osthoff, lead them back "to the path of virtue and duty" by his behavior ([22], p. 301). Also, in relation to visitors who come for different reasons, the sick person may not lose his self-control; even the sick person is a social being.

SOCIETY AND THE STATE

Medical ethics is not limited to the relationship between patient and physician; society and the state reach into this relationship: the state regulates prevention, therapy, and rehabilitation, and social groups govern various forms of support and immediate help. The sick person and the physician possess a quite specific relation to the state and to society.

The healthy person should show sympathy to the sick; compassion and faithfulness are his essential virtues as concerns the suffering person. In his visits and reports about everyday life in contrast to the sick-world, the healthy person should show moderation. The need of the sick person for quiet and yet for company departs especially in refined circles from the established rules; visitors can become a form of martyrdom for the sick person ([29], 1, p. 675). The healthy person has a duty and in fulfilling it can demonstrate the virtues of tact and steadfastness and persuade the sick person to go to the physician and avoid quacks. In relation to the physician he can be candid and keep the physician's injunctions to himself. Finally, it will also depend on him, to what extent the sick person can bear his suffering with strength and patience. Above all, parents have this duty in regard to their children. Besides the family, other groups of society are also affected by the illness and, in their deportment to the sick person and the physician, should adopt a convincing demeanor or withdraw.

However, the healthy person should perceive himself as responsible for his own health and guard against disease in his life-style. He should not immediately consider every affliction to be a disease. The importance of regimen is demonstrated again and again. Hoffman, who has written widely on regimen, considers as absurd and destructive the idea that only the physician is responsible for health; every person must be 'his own physician' and can observe himself and recognize what is good or bad for him. According to Unzer, health is partly a natural gift, partly a result of life-style, condition of the environment. Far too often have persons neglected the rules for ordering their lives, far too often is health considered a matter of luck, "as one can have gratuitously" ([29], 2, p. 5). The emotions, as the sixth sphere of hygiene, display an immediate union between morality and disease, insofar as passions are brought into an alliance with virtues. Burdens risk the health of the individual and the well-being of the community; according to Mai, it is "only too true that the burdens of individual men are avenged not only on the health but also more often on the welfare of the state and civil society" ([14], 1, p. 407). Activity and movement receive a new accentuation with the re-evaluation of work and industry. However, work should be balanced with necessity and not be stressed onesidedly; inactivity is necessary for recuperation. Unzer declares proper moderation to be a main virtue, "whereby we are able to extend our lives even more than the usual goal of finite persons" ([29], 1, p. 213). Other values are more important for the romantics. Friedrich Schlegel sees in industry and usefulness "the angel of death with his firey sword, who prevents man from returning to paradise" (Lucinde). Eichendorff titles a novel: *From the Life of a Do-Nothing.* The healthy person should not only approach and accept the physician in the medical perspective, he should wish to discuss only disease and therapy. The manner in which the healthy person conducts himself with the physician determines the form of subsequent care, according to Sklerander: "Greet your physician cordially when you are healthy, so that he will visit you in a timely fashion when you are sick and free you from your pains, insofar as it is possible for him to do so" ([25], p. 37).

The state has charge of medical affairs. It opposes man's finitude, epidemics, suicide, it watches over hospitals and prisons, it is responsible for the frequency of diseases. Frank recognizes a 'mother of diseases' in national misery [4]. Medical regulations determine the education and activity of physicians; even the actions of pharmacists and mid-wives are regulated by the state. In these regulations can be found assertions concerning the duties of physicians in respect to the sick, to colleagues, and to the state. The duty of confidentiality can lead to conflicts with the duty to testify. The views of physicians on this point are not uniform. Wedekind takes the position that the physician cannot be forced to give testimony in a court of law, if the sick person does not agree, except for certain crimes ([32], p. 23). The physician must disregard the duty of confidentiality when the reporting of contagious diseases is concerned.

In the time of the Enlightenment, the state was required to declare itself for the spread and observance of virtues and duties on the part of physicians as well as of the sick. Ignorant as well as inhumane physicians should be expelled; according to Unzer, the government must "be certain that their physicians possess an upright, mature, amiable, compassionate character" ([29], p. 13). Johann Peter Frank's *System of a Complete Medical Police* (1779–1819) is the theoretically systematic compilation of various initiatives, which had their beginning in the Enlightenment, which were directed to making health care a central concern of the state. To the reproach that the freedom of the individual is endangered by this development, Frank replies that social life is always based on limits; whoever wants to follow Rousseau may allow men to go back to the forests and their animal-like half-brothers.

CONCLUSION

Virtues and duties were intensively discussed by physicians in the time of the Enlightenment, particularly with reference to the sick, the physician, and society. Philosophical-theoretical concepts have consequences, the traditional schema of the seven virtues is also alive in medicine. Alteration exists next to permanence; the transition from an aristocratic to a bourgeois world has consequences; virtues of the Enlightenment such as order, cleanliness, and industry come to the fore. Class structure has weight, social forms play a large role, the physician has not yet attained the social position that has been self-evident for him since the 19th century. Virtues and duties in medicine are at the same time dependent on changes in internal medicine, on advances in theory and practice, which are, after all, an expression of the growing emancipation from philosophy and theology. Virtues have a certain affinity to the individual professions. Justice is still the central virtue of the judge, love the central virtue of the physician, and the central virtue of the sick person is patience. All virtues have their essential basis in love of God and love of fellow-man. One cannot avoid collisions of virtues; solutions should be able to be found less in the area of judicial laws than in the spirit of humanity and the power of belief.

This essay dealt with the views of the physician in the 18th century. Many allusions illustrate the conduct of the state; the attitudes of the sick person himself were never expressed, nor were the contributions of philosophers, theologians, and pedagogues, Leibniz and Wolff, Knigge, Lavater, and Kant concerning virtues and duties in medicine. How much reality conformed to these views and claims of the physicians is a further and exceptionally important question, which could not be answered here. What is obvious requires no judicial or moral standardization. If charlatanry and superstition were attacked, if there were complaints about ungrateful patients, if conscienceless and greedy physicians were condemned, then without doubt one would encounter the general features of reality. The satirical, ironic, and sarcastic tone of many writings of this century, such as Swift, Sterne, Voltaire, and Lichtenberg, is also significant. The 'medical Macchiavelli' became a slogan of the time, physicians concurred with the criticism of Molière. In 1750 Ernst Gottfried Kurella published his Maxims on *Becoming a Rich and Famous Physician Without Loss of Time and Energy*: above all, the physician can pull profit even from poor patients, who would allow therapeutic experiments to be performed without many restrictions, to show "which therapy would be best in this or that case" ([11], p. 82). The center of such an ethics is always the problem or the task of making money.

The Enlightenment meant freedom and reason, education of the individual and of humanity. The boundaries of these ideals have been manifest, new methods and new orientations had to be found. The ethical views of the physicians of the Enlightenment are a continuation of the past and point to the future – the future of a scientific medicine establishing itself, whose impressive progress conquers disease and misery and at the same time has produced intolerable forms of technologizing and anonymity. The current interest in philosophical questions and the ethical aspects of medicine also turns attention towards the past and obtains suggestions and perspectives from previous descriptions and interpretations of being sick, of therapy and the relation between patient and physician. The period of the Enlightenment possesses in this respect a special significance.

BIBLIOGRAPHY

[1] Baldinger E. G.: 1793, 'Anfrage wegen Zweyer Casuum conscientiae für den Arzt', *Neues Magazin für Aerzte* **15**, 138–144.

[2] Engelhardt, D. von: 1982, 'Betrachtungen zur Grundstruktur der medizinischen Ethik', in A. J. Buch und J. Splett (eds.), *Wissenschaft, Technik, Humanität*, Frankfurt a.M., pp. 99–119.

[3] Frank, J. P.: 1779–1819, *System einer vollständigen medicinischen Polizey*, vols. 1ff, Mannheim.

[4] Frank, J. P.: 1794, *Akademische Rede vom Volkselend als der Mutter der Krankheiten*, Marburg, Nachdruck Leipzig 1960.

[5] Henning, Fr.: 1791, *Von den Pflichten der Kranken gegen die Ärzte*, Leipzig.

[6] Hildebrandt, Fr.: 1795, *Ueber die Arzneikunde*, Erlangen.

[7] Hoffman, Fr.: 1738, *Medicus politicus*, Leyden, 2nd edn. 1746, deutsch Leipzig 1753.

[8] Hufeland, C. W.: 1796, *Die Kunst, das menschliche Leben zu verlängern*, Jena.

[9] Hufeland, C. W.: 1823, 'Von dem Rechte des Arztes über Leben und Tod', *Journal der practischen Heilkunde* **56**, 1, 3–28.

[10] King, L. S.: 1978, *The Philosophy of Medicine. The Early Eighteenth Century*, Cambridge, Mass., and London.

[11] Kurella, E. G.: 1750, *Entdeckung der Maximen ohne Zeitverlust und Mühe ein berühmter und reicher Arzt zu werden*, Berlin and Potsdam.

[12] Lesky, E.: 1966, 'Medizin im Zeitalter der Aufklärung', in *Lessing und die Zeit der Aufklärung*. Veröffentlichung der Joachim Jungius-Gesellschaft der Wissenschaften Hamburg, Göttingen, pp. 77–99.

[13] Longolius, J. D.: 1727, *Galanter Patiente*, Budissin.

[14] Mai, Fr. A.: 1777–1807, *Stolpertus*, Vols. 1–5.

[15] Mai, Fr. A.: 1793/94, *Medicinische Fastenpredigten*, Mannheim.

[16] Mann, G.: 1966, 'Medizinische Aufklärung. Begriff und Abgrenzung', *Medizinhistorisches Journal* **1**, 63–74.

[17] Maurer, H. J.: 1972, 'Medizin der Aufklärung', in H. J. Schoeps (ed.), *Zeitgeist der Aufklärung*, Paderborn, pp. 173–197.
[18] *Medicus Afflictus, Oder: der wohl-geplagte Medicus*: 1718, Budissin.
[19] Nussbaumer, A.: 1965, *Die medizinische Berufsethik bei Johann Storch (1732) und seinen Zeitgenossen*, Zürich.
[20] Osterhausen, J. K.: 1798, *Ueber medicinische Aufklärung*, Zürich.
[21] Osthoff, H. C. A.: 1810, *Briefe über Aerzte und ärztliches Wesen*, Sulzbach.
[22] Ploucquet, W. G.: 1797, *Der Arzt, oder über die Ausbildung, die Studien, Pflichten, Sitten und die Klugheit des Arztes*, Tübingen.
[23] Rübel, J. Fr.: 1758, *Der Character oder die Eigenschaften eines Medici*, Frankfurt and Leipzig.
[24] Rübel, J. Fr.: 1766, *Das wahre Portrait eines geschikten Arztes*, Frankfurt and Leipzig.
[25] Sklerander: 1733, *Der einfältige Patiente*, Görlitz.
[26] Stark, J. C.: 1784, *Versuch einer wahren und falschen Politik der Aerzte*, Jena.
[27] Stieglitz, J.: 1798, *Ueber das Zusammenseyn der Aerzte am Krankenbett*, Hannover.
[28] Storch, J.: 1728, *Praxis Stahliana*, Leipzig, 2nd edition 1732.
[29] Unzer, A. J.: 1769, *Der Arzt, Eine medicinische Wochenschrift*, 2nd ed., Hamburg.
[30] Vogel, S. G.: 1795, 'Einige allgemeine Bemerkungen über das Sçavoir in der medicinischen Praxis', *Journal der practischen Heilkunde* 1, 3, 295–324.
[31] Vogel, S. G.: 1799, 'Noch einige Bemerkungen über das Sçavoir faire in der medicinischen Praxis', *Journal der practischen Heilkunde* 8, 3, 85–96.
[32] Wedekind, G. Chr. G.: 1791, *Vom Zutrauen*, Mainz.
[33] Weikard, M. A.: 1799, *Philosophische Arzneykunst*, Frankfurt a.M.
[34] Zimmermann, J. G.: 1763/64, *Von der Erfahrung in der Arzneykunst*, Vols. 1–2, Zürich.

Institut für Medizin- und Wissenschaftsgeschichte, Lübeck, Germany

LAURENCE B. McCULLOUGH

VIRTUES, ETIQUETTE, AND ANGLO-AMERICAN MEDICAL ETHICS IN THE EIGHTEENTH AND NINETEENTH CENTURIES

In medical ethics the place of the virtues, habits or traits of character that dispose one to act on one's duties, is uncertain. In part this is owed to the attention given to the obligations of the physician in such matters as confidentiality, informed consent, and the care of the dying patient. Because in much of ethical theory moral principles are taken to be the primary source of obligations, it is no surprise that principle-based medical ethics has emerged to preeminence [2]. As a consequence of this focus on moral obligations and principles, moral virtues have received less attention in medical ethics. Medical etiquette, prescriptions about the particulars of professional behavior, has been all but ignored.

This was not always the case. In the eighteenth and nineteenth centuries one finds texts in Anglo-American medical ethics that address the physician's obligations and virtues in patient care. Some of these texts serve as precedents for contemporary efforts to bring virtue ethics into medical ethics. They do so by first establishing the physician's duties and then asking what attitudes and habits are required to make fulfilling one's duties less of a moral struggle and more a matter of routine. The virtues are not an independent source of obligations. Instead, they complement them and complete an account of the physician's moral responsibilities.

During this period, etiquette, which included attention to everything from interprofessional relationships to dress style, seems to have been subsumed under the virtues. This essay will show that admonitions about etiquette need not necessarily be conceived in terms of mere add-ons to medical ethics and thus be devoid of moral content. These two themes are addressed in the first section of this essay.

Finally, I shall advance – though not exhaustively defend – a thesis about the development of Anglo-American medical ethics from the eighteenth to the late nineteenth centuries: Medical virtues and medical etiquette gradually lost their foundations in a comprehensive account of the physician's moral responsibilities. It should therefore come as small surprise that contemporary medical ethics struggles to find the proper place for the virtues and etiquette in the moral life of medicine. I address these themes in the second section of the essay.

Earl E. Shelp (ed.), Virtue and Medicine, 81–92.

VIRTUES AND ETIQUETTE IN EIGHTEENTH CENTURY MEDICAL ETHICS

The most important example during the period in question of a comprehensive or complete medical ethics – one that addresses principles, obligations, virtues, and etiquette – is John Gregory's (1724–1773) *Lectures on the Duties and Qualifications of a Physician* [5]. Gregory was Professor of Physick at the Univeristy Edinburgh and published these lectures after his students had earlier published a version of them from their own transcripts.

Gregory bases his account of the physician's moral responsibility on sympathy. He uses this term in the way that his contemporaries in moral philosophy used it, namely, as a natural disposition to experience the feelings of another and, based on the pleasure or pain accompanying those feelings, judge the moral status of an action that arouses those feelings. David Hume provides a nice example:

> Were I present at any of the more terrible operations of surgery, the preparation of the instruments, the laying of the bandages in order, the heating of the irons, with all the signs of anxiety and concern in the patient and assistants, wou'd have a great effect upon my mind, and excite the strongest sentiments of pity and terror ([6], p. 576).

Gregory speaks to the 'humanity' or 'sympathy' of the physician as the basis of moral responsibility in medicine: "that sensibility of heart which makes us feel for the distresses of our fellow creatures, and which, of consequence, incites in us the most powerful manner to relieve them. Sympathy produces an anxious attention to a thousand little circumstances that may tend to relieve the patient . . . " ([5], p. 22).

On this basis, Gregory addresses the physician's obligations as well as the physician's virtues, the "genius, understanding, and temper that naturally fit a man for being a physician," "the moral qualities to be expected from him, the exercise of his profession," and "the decorums and attentions peculiarly incumbent upon him as a physician . . . " ([5], p. 15).

Consider his treatment of how the physician should respond when a patient has a suggestion regarding his own care. Consistent with his understanding of the basic goals of medicine – "the art of preserving health, of prolonging life, and of curing diseases" ([5], p. 6) – Gregory argues that the patient's safety is the primary consideration. Nonetheless, he argues, patients have certain rights in these matters.

> Sometimes a patient himself, sometimes one of his friends, will propose to the physician a remedy, which, they believe, may do him service. Their proposal may be a good one.

It may even suggest to the ablest physician, what, perhaps, till then, might not have occurred to him. It is undoubtedly, therefore, his duty to adopt it. Yet there are some of the faculty, who, from a pretended regard to the dignity of the profession, but in reality from mean and selfish views, refuse to apply any remedy proposed in this manner, without regard to its merit. But this behavior can never be vindicated. Every man has a right to speak where his life or his health is concerned: and every man may suggest what he thinks may tend to save the life of his friend ([5], pp. 34–35).

Gregory goes on to address the virtues that should be cultivated in these sorts of encounters.

It becomes them [the patient or the patient's friends] to interpose with politeness, and a deference to the judgment of the physician. It becomes him to hear what they have to say with attention, and to examine it with candour. If he really approve, he should frankly own it, and act accordingly. If he disapprove, he should declare his disapprobation in such a manner, as shews it proceeds from conviction, and not from pique or obstinacy. If a patient be determined to try an improper or dangerous medicine, a physician should refuse his sanction: but he has no right to complain of his advice not being followed ([5], pp. 35–36).

In short, Gregory understands that knowing one's duties in the practice of medicine is not enough. The physician must also develop an appropriate moral character, one for which acting on one's duties will not be so much a matter of continuous moral struggle as a matter of routine. Moral struggle in the fulfillment of duty emerges when the physician is tempted to act on his own self-interests rather than the interests of the patient. In the present instance, the physician's pride and desire to control patient care – elements of the physician's authority – may interfere with obligations to act in a way consistent with the patient's best interests, as the patient or the patient's friends have come to see them.

This account of the physician's virtues is based in a principle of beneficence founded on sympathy. Sympathy directs the physician to the best interests of the patient and an assessment of them in terms of the goods of medicine. The physician's duties or obligations are to seek the greater balance of good over harm that defines those interests. But acting on one's duties requires the moral agent to blunt self-interest through the cultivation of certain attitudes, habits, or traits of character that direct him or her to the interests of others – in short, the virtues. In the example above, the physician is obligated both to protect the patient from the harm of ineffective or dangerous remedies and to acknowledge that the patient or the patient's friends may bring the physician new knowledge about effective and safe remedies. To complement these duties, Gregory says, the physician should cultivate virtues of attentiveness and candor. The latter is a disposition "which makes him open to conviction, and ready to acknowledge and rectify his mistakes." ([5], p. 31)

Gregory's accounts of confidentiality and truthfulness to the dying display this same systematic character. Confidentiality is especially important, he says, in treating patients, where "discretion, secrecy, and honour of the physician" ([5], p. 29) are essential elements of the physician's responsibility to maintain privacy. Similarly, telling patients that they are seriously ill "is one of the most disagreeable duties in the profession: but it is indispensible. The manner of doing it, requires equal prudence and humanity" ([5], p. 37). Prudence directs the physician to the timing of disclosures and humanity to a sensitivity to the patient's reactions to those disclosures. These virtues shape the manner of discharging the physician's duties in such a way that the patient is benefitted and not unduly frightened or harmed.

Matters of etiquette are addressed in a similar way, namely, in terms of preserving goods essential to an effective relationship with patients. Of interest in this respect are Gregory's admonitions regarding the physician's dress styles.

> In some cases, there is a great impropriety in a physician's having any distinguishing formality in his dress or manners. I do not hint merely at the disagreeable impression, not to say terror, with which this sometimes affects the minds of children. Even among people who possess the greatest vigour and firmness of mind when in health, there is often a feebleness and depression of spirits attendent on sickness, that renders the sight of any stranger whatever very painful. In such a state of mind, the visit of a physician, even when wished for, is often particularly dreaded; as it naturally wakens the apprehensions of danger; apprehensions, which a formal dress, and a solemn behavior, are ill-calculated to dispel. Surely, if there be, at any time, a propriety in an easy, cheerful, soothing behavior, it must be on such an occasion, where it is so necessary to forget the physician in the friend ([5], pp. 55–56).

Thus medical etiquette is firmly grounded in a more general account of the physician's obligations to patients. Gregory's remarks about interprofessional disagreements are treated in the same way: the patient's, not the physician's, best interests come first ([5], pp. 40–41). (I return to this particular subject below.)

Gregory also addresses the general virtues that should be cultivated as essential to the *role* – and not simply particular obligations – of the physician. The physician should not be 'insensible', lacking in sympathy for the patient, since sympathy is "one of the most natural and powerful incitements to exert himself for the relief of his patient" ([5], pp. 12–13). Sympathy provides the moral ground for the application of the principle of beneficence to the physician's role. At the same time, overidentification with the patient's plight is equally dangerous. "On the other hand, a physician of too much sensibility may be rendered incapable of doing his duty from anxiety, and excess of sympathy, which cloud his understanding, depress his spirit, and prevent him from acting with that steadiness and vigour, upon which perhaps the life of his patient in a great measure depends" ([5], p. 13). Those concerned with the dehumanization of medicine might gain considerable insight from Gregory's views in these matters.

There is another trap that the physician can avoid through the cultivation of virtues, namely, allowing his sometimes negative reactions to difficult patient(s) to become the basis of his relationship to them. This can occur, for example, when the physician encounters the patient in an emergency situation or must respond when the patient's condition takes an unexpected turn for the worse. For patients, Gregory says, such situations "are apt to flutter the spirits of a man of lively parts and of a warm temper" ([5], p. 21), which can lead, in turn, to stress in the relationship with a patient. Anticipating by more than a century Sir William Osler's remarks on *aequanimitas* [7] Gregory writes about how the physician should respond.

> Accidents of this kind may affect his judgment in such a manner as to unfit him for discerning what is proper to be done, or, if he do perceive it, may nevertheless, render him irresolute. Yet such occasions call for the quickest discernment and the steadiest and most resolute conduct; and the more, as the sick so readily take the alarm, when they discover any diffidence in their physician. The weakness too and bad behavior of patients, and a number of little difficulties and contradictions which every physician must encounter in his practice, are apt to ruffle his temper, and consequently to cloud his judgment, and make him forget propriety and decency of behavior. Hence appears the advantage of a physician's possessing presence of mind, composure, steadiness, and an appearance of resolution, even in cases where, in his own judgment, he is fully sensible of the difficulty ([5], pp. 21–22).

We have then, in Gregory's *Lectures*, an example of what might be called a complete or comprehensive medical ethics, in the sense that it provides a ground or foundation for both the obligations and the virtues (including the etiquette) of the physician. Under the virtues fall particular or special virtues like secrecy, honor, prudence, and avoidance of "little peculiarities" of manners ([5], p. 57), as well as general virtues like steadiness and candor. The latter direct the physician to the best interests of the patient by creating the conditions for sympathy to play its crucial role in Gregory's medical ethics (and its basis in the moral psychology of his time). The former direct the physician to the best interests of the patient by systematically creating conditions for fulfilling duties grounded in sympathy. These conditions are traits of character and habits that also embrace etiquette, which is thus made part and parcel of a complete medical ethics. The history of Anglo-American medical ethics in the century following Gregory is the story of the dissolution of his comprehensive approach.

VIRTUE AND ETIQUETTE IN NINETEENTH CENTURY MEDICAL ETHICS

That dissolution begins with Thomas Percival's famous and influential *Medical Ethics* [8]. This book was prompted by a dispute among the various medical practitioners in the Manchester Infirmary. Percival had been called in to attempt to allay the problems that led to the dispute and his book is the result of his efforts in this regard.

This interest in settling disputes seems to have had a profound influence on Percival's book. One does not have to read very far to discover that he is preoccupied with interprofessional matters. His treatment of consultations is a clear example. Patients should first be presented by the most junior physician or surgeon present and then on up through the most senior. The disposition of the patient's case should be determined by a majority of those present. The order of presentation changes when there is a 'mixed case' ([8], p. 81), one in which both surgeons and physicians are involved. In these circumstances the surgeons present first, junior to senior, followed by the physicians. (Seniority is to be determined by the date of one's appointment in the hospital.) It is an interesting lacuna that Percival does not prescribe the same voting mechanism to determine the patient's course of treatment in these 'mixed cases'. He does note that the concerns of the physicians and/or surgeons should be kept from the patient "and every just precaution used to guard him from anxiety or alarm" ([8], p. 82).

Note the contrast to Gregory's discussion of consulation. The focus is far less on the mechanics of interprofessional relations and more on the best interests of the patient.

> There are often unhappy jealousies and animosities among those of the profession, by which their patients may suffer. A physician, however, who has any sense of justice or humanity, will never involve his patient in the consequences of private quarrels, in which he has no concern, Physicians in consultation, whatever may be their private resentments, or opinions of one another, should divest themselves of all partialities, and think of nothing but what will most effectually contribute to the relief of those under their care ([5], pp. 38–39).

The difference is stark. Gregory's beginning and ending point is the well-being of the patient. The virtues and behavior appropriate to consultation have their ground in that well-being. Indeed, the physician should seek consultations, Gregory later adds, in those cases where he is uncertain about the balance of good over harm that might be achieved by a particular intervention ([5], p. 39).

Percival, on the other hand, introduces his discussion of consultation from quite another perspective.

> The harmonious intercourse, which has been recommended to the gentlemen of the faculty, will naturally produce *frequent consultations*, viz., of the physicians on medical cases, of the surgeons on chirurgical cases, and of both united in cases of a compound nature, which falling under the department of each, may admit of elucidation by the reciprocal aid of the two professions ([8], p. 80).

Percival's goal, like Gregory's, is harmonious relations among the professions, but the moral justification for that goal in terms of the patient's best interests is largely absent in Percival. (The patient's best interests are mentioned in passing later in the text [(8), p. 97].) The result is that an account of the physician's virtues is separated from an account of the physician's duties in terms of moral principles. Indeed, at times Percival's medical ethics, itself, seems devoid of any clear moral foundation, as is plain in the opening paragraph of *Medical Ethics*.

> Hospital physicians and surgeons should minister to the sick, with due impressions of the importance of their office, reflecting that the ease, the health, and the lives of those committed to their charge depend on their skill, attention, and fidelity. They should study, also, in their deportment, so to unite *tenderness* with *steadiness*, and *condescension* with *authority*, as to inspire the minds of their patients with gratitude, respect, and confidence ([8], p. 71).

This passage, which appears nearly verbatim at the beginning of the first "Code of Ethics" of the American Medical Association in 1847 [1] has three distinguishing features – each marking a sharp break in the history of medical ethics after Gregory.

The first concept emphasized is the importance of the office or standing of the physician or surgeon in the hierarchy of the hospital. Obviously, for Percival their place is preeminent. This is owed, he later seems to indicate, to their superior education and training ([8], pp. 96–97). In short, the *authority* and *power* of physicians vis-a-vis other professionals – and inevitably patients – moves to center stage, from which Gregory had firmly excluded it. This signals a radical shift from Gregory's approach, and begins to decouple an account of the physician's virtues from an account of his or her moral obligations.

In second place in this opening statement appears a reference to the physician's duties. The implication is that they find their justification in the physician's authority. This becomes plain in the opening sentence of the next paragraph:

The *choice* of a *physician* or *surgeon* cannot be allowed to hospital patients, consistently with the regular and established succession of medical attendance. Yet personal confidence is not less important to the comfort and relief of the sick-poor, than of the rich under similar circumstances. And it would be equally just and humane to inquire into and to indulge their partialities, by occasionally calling into consultation the favored practitioner ([8], p. 71).

When we contrast this view with that of Gregory on the patient's role in choosing treatment interventions, it is plain that Percival's book signals a major shift in kind, not just degree, in this history of Anglo-American medical ethics. The duties of the physician are founded in authority and no justification is given for the legitimacy of that authority. Gregory, by contrast, grounds the whole enterprise of medical ethics in the patient's best interests, including matters of authority.

The consequences of this shift for the third part of the opening paragraph of *Medical Ethics*, Percival's remarks on the basic, general virtues of the physician and surgeon are far-reaching. These are the virtues of the authoritarian paternalist, not the benign, sympathetic paternalist of Gregory's medical ethics. This is plain in the very meaning of 'condescension', The physician is to put aside his own sense of lofty station of power and authority and descend to the lower standing and station of the relative powerlessness of the patient. But he is to do so only for a time, it seems. 'Authority' requires the physician to direct the patient's care, including denying the patient his choice of physician or treatment. In this context, reference to tenderness and steadiness is somewhat ineffectual. Indeed, this reference seems little more than a borrowing from Gregory, to whom Percival had made passing reference in his preface ([8], p. 68).

Other apparent borrowings from Gregory appear later in the text, when Percival addresses virtues that are part of what he calls the "moral rules of conduct" ([8], p. 90). Whereas Gregory had addressed these virtues at some length and provided for them a substantial moral context and grounding, Percival merely enumerates them.

> The *moral rules of conduct*, prescribed toward hospital patients, should be fully adopted in private or general practice. Every case, committed to the care of a physician or surgeon, should be treated with attention, steadiness, and humanity. Reasonable indulgence should be granted to the mental imbecility and caprices of the sick. Secrecy and delicacy, when required by peculiar circumstances, should be strictly observed. And the familiar and confidential intercourse, to which the faculty are admitted in their professional visits, should be used with discretion and with the most scrupulous regard to fidelity and honour ([8], p. 90).

The result of Percival's approach to medical ethics and to the place of the virtues in an account of moral responsibiltiy in medicine is to break the systematic connection between duties and virtues that characterizes Gregory's approach. Moreover – and, in the long run, perhaps more important – Percival shifts the terms of medical ethics away from a moral grounding in the patient's best interests to claims to authority and power.

Given the interest of the American Medical Association some fifty years later, it is not surprising that they found this approach attractive. As Jeffrey Berlant has shown, a principal concern of these initial attempts to organize medicine was to gain monopoly power over the practice of medicine [3]. This interest appears in the introduction of the first "Code of Medical Ethics" of the American Medical Association, where it is claimed that the obligations of the physician imply certain rights of the physician, including rights to non-interference by society ([1], pp. 26–27). Thus, the theme of authority and power that seems to have begun in Percival receives a powerful and influential formulation in the first professional code of ethics of organized medicine in the United States. The *moral* groundings that the authors of this code mention are anemic indeed: Medical ethics, as a branch of general ethics, "must rest on the basis of religion and morality. They comprise not only the duties, but also, the rights of a physician . . . " ([1], p. 26). Nothing more is said that in any way substantially elaborates on this claim.

By the middle of the nineteenth century, then, modern Anglo-American medical ethics was at a far remove from Gregory's. It had been removed from any attempt to ground it in a moral theory of the patient–physician relationship, or even a moral psychology. Instead, the first section of the Code appeals to the 'greatness' of the physician's mission and to his 'conscience' ([1], p. 29). Contrast this language with Gregory's on the 'appearance of resolution' when dealing with *difficult* patients (not *all* patients). Gregory emphasizes resolution so that the unnecessary anxieties of some patients may be alleviated whereas the AMA code emphasizes uniform greatness of the physician's role, implying power and authority to control all aspects of the relationship with the patient.

There then follows in the Code a passage taken almost *in toto* from the first paragraph of Percival's *Medical Ethics*. The dissolution of medical ethics as Gregory understood it, begun by Percival, reaches its culmination in the American Medical Association's first code of ethics.

The consequences for its and later attempts to provide an account of the physician's virtues and of medical etiquette were considerable. There is no sense in the Code of where the virtues of the physician come from, as there was in Gregory. This is because there is no *moral* ground provided for the physician's duties – essential to an account of the physician's special virtues – or for the relationship of physicians to patients – essential to an account of the physician's general virtues.

In a commentary on the Code thirty-five years after its adoption, Dr. Austin Flint provides a glimmer of understanding of the place of the virtues in medical ethics.

> The sentiments so admirably expressed in the foregoing first paragraph of the code need no arguments for their support, nor any comments to increase their force. They antagonize undue influences arising from self-conceit, and irritable temper, indolence, devotion to pleasure or to occupations which divert from professional duties, and all mercenary considerations. At the same time, they do not contravene self-respect and a proper regard for personal interest ([4], p. 8).

Flint captures part of the point of the virtues: They blunt the physician's self-interest and direct him to the best interests of the patient. But an account of why the physician should be primarily committed to the patient's best interests in the first place seems absent from Flint's essay.

This brief review provides at least *prima facie* evidence for a major thesis of this essay: In the course of the nineteenth century the virtues and, as a subcategory of them, medical etiquette, gradually lost the foundation in a comprehensive medical ethics that John Gregory had provided for them. The reasons for this change are two-fold. First, the ground of medical ethics changed, with Percival, from a moral account of the patient-physician relationship and the physician's role to an account shaped by concerns of authority and power in that role. Second, the consequence of this change for the virtues of the physician is that accounts of virtue are decoupled from accounts of the physician's duties. Hence, accounts of virtue and etiquette in this period take on an anemic character and come down to us separated from the central concerns of medical ethics. It is, then, no surprise that, with the recrudescence of medical ethics in the past two decades, we struggle to regain for the virtues and medical etiquette their rightful place in an account of the moral responsibilities of physicians.

CONCLUSIONS

In this essay I have attempted to provide an account of the historical project of a comprehensive or complete medical ethics that John Gregory advanced in the latter half of the eighteenth century. In such a medical ethics, the physician's virtues, much less medical etiquette, do not present a problem. Indeed, they are indispensible features of that medical ethics. These matters have come to be problematic for contemporary medical ethics because of the significant changes introduced into the history of Anglo-American medical ethics by Percival's *Medical Ethics* and the first "Code of Ethics" of the American Medical Association. These two works, respectively, initiated and completed the dissolution of medical ethics, thus setting the terms for the inquiry undertaken in much of the rest of this book. If we reach back past the nineteenth century, to the earlier efforts of John Gregory – and, no doubt, his forebears in the history of medical ethics – we may well discover important precedents for that inquiry.

BIBLIOGRAPHY

[1] American Medical Association: 'Code of Medical Ethics', *Proceedings of the National Medical Convention 1847–1848*, American Medical Association, Chicago. Reprinted in S. Reiser *et al.* (ed.): 1977, *Ethics in Medicine*, MIT Press, Cambridge, Mass., pp. 26–34.

[2] Beauchamp, T. L. and J. Childress: 1983, *Principles of Biomedical Ethics*, Oxford University Press, New York, 2nd ed.

[3] Berlant, J.:1975, *Profession and Monopoly: A Study of Medicine in the United States and Great Britain*, University of California Press, Berkeley.

[4] Flint, A.: 1883, *Medical Ethics and Etiquette*, Appleton and Company, New York.

[5] Gregory, J.: 1772, *Lectures of the Duties and Qualifications of a Physician*, Strahan, London. All quotations from 1817 edition, M. Carey & Son, Philadelphia.

[6] Hume, D.: 1978, *A Treatise of Human Nature*, in L. A. Selby-Biggs (ed.), 2nd ed. rev. P. H. Nidditch, Oxford University Press, Oxford.

[7] Osler, W.: 1932, *Aequanimitas: With Other Addresses to Medical Students, Nurses and Practitioners of Medicine*, Blakiston, Philadelphia, 3rd ed.

[8] Percival, T.: 1803, *Medical Ethics, or a Code of Institutes and Precepts Adapted*

to the Professional Conduct of Physicians and Surgeons, S. Russell, Manchester. Reprinted as *Percival's Medical Ethics*, ed. by C. Leake, R. E. Kreiger Publishing Co., Huntington, N.Y., 1975.

Georgetown University,
School of Medicine,
Washington, D.C.,
U.S.A.

SECTION II

THEORIES OF VIRTUE

BERNARD GERT

VIRTUE AND VICE

"Dispositions when they are so strengthened by habit that they beget their actions with ease and with reason unresisting, are called *manners*. Moreover, manners, if they be good, are called *virtues*, if evil, *vices*" ([4], p. 68). Relationships cannot be virtues because they are clearly not dispositions at all. Thus if one holds, as MacIntyre seems to do, that for Aristotle friendship is "a type of social and political relationship" ([7], p. 147), one cannot at the same time hold that for Aristotle, "friendship is itself a virtue" ([7], p. 146), and still be using virtue in the sense that I, and I think most writers on the subject of virtues, use that concept.

There are two types of virtues and vices that I shall be concerned with in this paper; I shall call them personal and moral virtues and vices. Personal virtues are those traits of character that all rational persons want for themselves; personal vices are those traits that no rational person wants. Moral virtues are those traits of character that all impartial rational persons want everyone to have; moral vices are those traits that all impartial rational persons want no one to have, unless they have some special reason.

Prudence, temperance, and courage are the primary personal virtues; imprudence, intemperance, and cowardice are the primary personal vices. Kindness, truthfulness, trustworthiness, and fairness are characteristic moral virtues; cruelty, callousness, and untrustworthiness are characteristic moral vices. (These virtues and vices are sometimes called by different names, e.g., trustworthiness is called by Hobbes 'justice' and by others 'fidelity',) What I shall try to show in this paper is that when all of these terms are adequately explained there is virtually complete agreement on the nature of the virtues and vices.

Part of the problem in discussing the virtues and vices is that the terms are not always understood in the same way by everyone. This not only leads to disagreement over whether a particular character trait is or is not a virtue, for example, patience, but also to pseudo-agreement, everyone agrees that courage is a virtue, but they may disagree on the analysis of courage. What I should like to do is to provide a clear enough account of the virtues and vices that both pseudo-agreement and disagreement are virtually eliminated.

The main outlines of the theory that I am presenting are those presented

Earl E. Shelp (ed.), Virtue and Medicine, 95–109.

by Hobbes. Though I pick Hobbes because I think that his theory is closest to being adequate, there is the added benefit that Hobbes's place in the history of philosophy shows that the view being presented has a long philosophical history. This is especially true since Hobbes, as Strauss has shown, was well versed in classical philosophy and was strongly influenced by Aristotle's *Rhetoric* ([9], pp. 35–43).

Hobbes's theory is very simple. All rational persons want to avoid death, pain and disability for themselves. The virtues are those traits of character that help one to avoid these evils. The *moral* virtues are those that "appertain to the preservation of ourselves against those dangers which arise from discord" ([4], p. 152). They are those virtues that are "suitable for entering into civil society; and . . . those whereby what was entered upon can be best preserved" ([4], p. 70). As Hobbes says in *Leviathan* "they come to be praised as the means of peaceable, sociable, and comfortable living" ([5], p. 104). These virtues include such standard virtues as "*justice, gratitude, modesty, equity,* [*and*] *mercy*" ([5], p. 104). Hobbes is primarily concerned with the moral virtues and vices, so much so that he sometimes talks as if all the virtues and vices were moral virtues and vices. However, he is aware that there are some vices that are not necessarily related to discord among men but that tend "to the destruction of particular men" ([5], p. 103).

Hobbes's discussion of the personal virtues and vices in both *De Cive* and *Leviathan* is very brief, only two sentences in the former and one sentence in the latter. In *De Cive*, after finishing his discussion of the moral virtues, he says: "But there are other precepts of *rational* nature, from whence spring other virtues; for temperance, also, is a precept of reason, because intemperance tends to sickness and death. And so fortitude too, that is, that same faculty of resisting stoutly in present dangers, and which are more hardly declined than overcome; because it is a means tending to the preservation of him that resists" ([4], p. 152). And that is all he says. In *Leviathan* he says essentially the same thing about temperance, citing drunkenness, but leaving out any mention of fortitude ([5], p. 103).

It is in *De Homine* that he has somewhat more to say, but not much more. After pointing out that all moral virtue is contained in justice and charity, he continues, "However, the other three virtues, (except for justice) that are called cardinal – *courage*, *prudence*, and *temperance* – are not virtues of citizens as citizens, but as men, for these virtues are useful not so much to the state as they are to the individual men who have them. For just as the state is not preserved save by the courage, prudence, and temperance of good citizens, so it is not destroyed save by the courage, prudence, and temperance of its enemies" ([4], p. 69). This distinction between the moral and the personal virtues is taken up, perhaps surprisingly, by Kant in the very first paragraph of Chapter 1 of *Groundwork of the Metaphysic of Morals* where he distinguishes the good will from "courage, resolution, and constancy of purpose, as qualities of *temperament*" ([6], p. 61).

When philosophers talk about certain pairs of concepts such as rational and irrational, moral and immoral, virtue and vice, it is usually the first member of the pair, the positive member that is thought of as fundamental. However, with all of these pairs, it is in fact the second member, the negative one that is basic. An adequate account of rational depends upon a prior account of what is irrational; acting morally is best understood as simply not acting immorally, though, of course, being morally good involves more than simply being moral. Similarly, it is necessary to understand the vices in order to understand the virtues even though to have a particular virtue requires more than not having the associated vices. It is not necessarily the case that everyone either has the virtue or the vice and that no one has neither. Many people are neither courageous nor cowardly, neither truthful nor deceitful.

In order to explain this fact the analysis of virtue and vice must take into account what it would be reasonable to expect a person to do. If one is in a situation in which it would be reasonable to expect a person to be so affected by danger or fear that he would act irrationally, then if one does so act, we do not count the action as showing the person to be cowardly, but we do count it as showing that he is not courageous. Being courageous involves acting rationally even in those situations in which it would be reasonable to expect a person to act irrationally. One counts as having courage to the extent that one acts rationally in situations where it would be reasonable to expect a person to act irrationally; the greater the expectation that a person will act irrationally, the more courage is shown when one acts rationally. Of course, this analysis also allows for persons to be courageous when faced with some kinds of dangers or fears but not with others. Thus, we might distinguish physical courage from other kinds of courage if it turned out that some people regularly acted rationally when faced with physical dangers that made it reasonable to expect one would act irrationally, but did not act rationally when faced with public disapproval or economic loss.

Cowardice is shown when one acts irrationally when it is reasonable to expect a person to act rationally in that situation. If one does act rationally in this kind of situation, then this does not show courage, but is necessary for courage. Though courage is shown by acting rationally when this is not what it is reasonable to expect, it also requires acting rationally when it is reasonable to expect this kind of behavior. Thus courage necessarily excludes cowardice, though it is not merely lack of cowardice. Extreme cowardice, that is, acting irrationally when faced with a very low risk of evil, or a risk of a very small evil, may cease to be a vice and become a pathological condition, such as a phobia. There is no sharp line between a pathological state and that of a personal vice. This is true not only of cowardice, but also of foolhardiness, intemperance and imprudence, indeed of all personal vices. We are often not sure whether we are dealing with a genuine lack of volitional ability ([2], pp. 109–125), or only a weakness of will, even though intemperance is sometimes simply described as weakness of will. This close connection between the personal vices and pathological states may explain our attitude toward those who exhibit the personal vices; we are not sure whether to condemn or to pity.

This is not true of the moral vices, unless that vice seems to be the result of a personal vice. Only if someone is deceitful because he is cowardly, or intemperate, do we sometime pity rather than censure him. Most often, moral vices do not depend upon personal vices, rather they are simply the result of people not acting impartially, usually not caring as much for the people they harm as for the people they benefit. To have a moral vice, e.g., deceitfulness, involves acting in ways that no rational person would publicly advocate ([3], p. 89), when there are no non-moral reasons that make it reasonable to expect a person to act in this way. It does not count as an exhibition of deceitfulness if one deceives when one would publicly advocate deception, indeed this does not even count against one having the virtue of truthfulness. However, though it does not count as having the vice of deceitfulness if one deceives when there are coercive non-moral reasons for deceiving, this does count against one having the virtue of truthfulness. Truthfulness, taken as a moral virtue, requires never deceiving except when one could publicly advocate such deception, and one counts as truthful to the extent that one resists non-moral reasons for deception.

No moral virtue ever requires acting immorally. This seems to me beyond dispute. It also seems clear that acting morally never counts as showing that one has a moral vice. What is not yet clearly determined is whether acting morally can ever count against one having a moral virtue. If there are times when it is morally justifiable to deceive or not to deceive, then though deceiving does not count as showing that one is deceitful, does it count against one being truthful? There are problems with answering either yes or no. If we say that deceiving, when it is morally allowable to do so, counts against one having the moral virtue of truthfulness, then there will not be agreement among impartial rational persons that truthfulness is a completely desirable trait. If we say that deceiving, when it is not morally required to do so, does not count against one's having the virtue of truthfulness, then we will be allowing a truthful person to deceive when it is morally allowable for him to tell the truth. This same problem arises with regard to trustworthiness, fairness, honesty, kindness, and indeed all of the moral virtues.

These two problems do not seem to me to be on a par. I think one can accept that truthfulness, trustworthiness, etc., do not require acting on the associated moral rule whenever this is morally allowed, without a drastic change in one's concept of the virtues. It is already clear that these virtues not only do not require following the rule when this is morally unjustified, they prohibit such action. However, if we take away from the virtues the characteristic that they are completely desirable traits, that all rational persons want to have all of the personal virtues themselves, and publicly advocate that everyone have all of the moral virtues, we will be unable to arrive at any coherent account of the virtues.

These points can be understood more clearly if we take some examples of the various virtues. Let us first consider the personal virtues of courage, prudence, and temperance. All rational persons would want to be courageous, prudent, and temperate, at least they would, when the proper analysis of these virtues is provided, because they benefit from having these virtues and increase their risk of suffering harm if they have the corresponding vices. But rational persons may differ on whether or not they want others to be courageous, prudent, and temperate, because they may have different attitudes toward others. It is most likely that some would want their friends to have these virtues and their enemies not to have them. Even moral persons might not want all others to have the personal virtues. They might want only those who were moral, who had a good will, to have the personal virtues. They might believe that if those who were immoral had the personal virtues that would very likely increase the amount of evil that would be caused.

Let us now consider some moral virtues, e.g., trustworthiness and kindness. Every rational person would want others to be trustworthy and kind, for their being so increase his own chances of being helped or benefitted and decreases his chances of being harmed. If others are trustworthy, then I can take their promises seriously and make my plans accordingly. If others are kind, then it is more likely that I will be helped when I need help. I benefit when others have the moral virtues; I do not necessarily benefit when I have them myself. I might be better off simply appearing to have these virtues while actually not having them.

It is one of the premises of this paper that all rational persons want to have all of the personal virtues, e.g., courage, prudence, and temperance, themselves. No account of these virtues or of rationality can conflict with this premise. If it does, I take that to be conclusive evidence of its inadequacy. This means that having one of these virtues cannot require having some other personal vice, for if it did then rational persons could not seek to have all of the personal virtues. This provides us with some clue as to the proper interpretation of the personal virtues and vices.

Prudence and courage must be given an interpretation that does not result in their requiring cowardice and imprudence respectively; the same is true of any two sets of personal virtues and vices. Given this relationship between the virtues and the vices, it becomes clear that the virtues and vices cannot be regarded primarily as involving distinctive ways of acting or reacting to the same situation; at least, not if one of the virtuous ways of reacting requires acting in what is correctly described as exemplifying a personal vice. If this were the case, then it would be impossible for a rational person to seek all the personal virtues and, as noted above, this is not an allowable conclusion.

Let us see how this helps us deal with some possible interpretations of courage and prudence. Suppose that one took courage to be the trait of acting so as to overcome the present danger regardless of the possible harmful consequences to oneself and that one took prudence to be the trait of acting so as to minimize the possible harmful consequences to oneself. Both of these accounts have some plausibility, but given what we have said in the previous paragraph we can see that they cannot both be correct. For if they were, one sometimes would, when faced with danger, be forced to choose between being either courageous and imprudent or prudent and cowardly; one could not be both courageous and prudent. Many accounts of the personal virtues and vices lead to this same difficulty: one can discover a situation that requires one to exemplify one vice when exercising another virtue. This difficulty, which will be discussed in more detail later, is partly a result of not distinguishing clearly enough between the virtues and vices, i.e., character traits, on the one hand and personality traits on the other.

Following Aristotle, we all realize that courage can be considered a mean between the extremes of rashness and cowardice. (I am not putting forward Aristotle's more general account of virtue as the correct one.) However, all too often we simply contrast courage with cowardice and forget about rashness altogether. This results in our tending to equate courage with fearlessness and cowardice with fearfulness. But fearlessness can lead to rash action as easily as, if not more easily than, it leads to courageous action. Thus it is a serious mistake to equate courage with fearlessness, for courage is a virtue, a trait of character all rational persons want for themselves, but fearlessness is not necessarily a trait that all rational persons want.

Fearlessness is a personality trait, something that children are sometimes born with. There are some infants who seem to be born fearless, others who seem to be born fearful. This may not be true; it may be that very early childhood training is at least partly responsible for a child's fearlessness or fearfulness. But it is quite clear that children can be fearful and fearless at ages far below those at which it is appropriate to ascribe any character traits, that is, virtues or vices, to them at all ([3], pp. 153, 154).

It is commonly believed, and it may even be true, that being fearless is more likely to be related to being courageous and being fearful more likely to be related to being cowardly. But this is at most a contingent connection and does not justify equating courage with fearlessness and cowardice with fearfulness. For a fearful person can be a courageous person and it is not at all uncommon for a fearless person to be rash or foolhardy rather than courageous. It is not how a person feels that determines whether or not he is courageous, it is how he acts.

If we are to guarantee that courage be a trait of character that all rational persons want for themselves, then it is necessary that courageous action must always be rational. If courageous action were ever irrational then it would not be the case that all rational persons would want to be courageous. How can one guarantee that courageous action will always be rational? The simplest and most direct way to do this is simply to include as part of the definition of courage that it involves acting rationally. But, obviously, this is not a sufficient account of courage, for what we have said of courage also applies to prudence and temperance; both of these virtues must also include in their definition that they involve acting rationally.

But if courage, prudence, and temperance all involve acting rationally, what is it that distinguishes them from each other? This seems fairly straightforward; what distinguishes the virtues from one another is the situation or circumstances in which their exercise is appropriate. Thus courage is the trait of character that involves not allowing danger or fear to make one act irrationally. Prudence is the trait of character that involves not allowing present concerns to make one act irrationally by neglecting significant consequences with regard to oneself or those for whom one is concerned. Temperance is the trait of character that involves not allowing strong emotions or desires to make one act irrationally.

Interestingly, this account of the personal virtues naturally achieves the unity of the virtues that Plato and others have sought. The personal virtues are distinguished from one another by the situation that calls for the virtue. It is danger or fear that calls for courage; it is lust or strong desires that call for temperance. On this account, the virtues are consistent with a very wide range of personality traits, desires, emotions, etc. They require only that one act rationally when faced with certain kinds of situations.

On this account of the virtues, it does not make any difference how many separate personal virtues that one invents or discovers, all of them will be consistent with each of the others, for they will differ from one another only in the situation that calls for them. Thus fortitude will be the character trait of not allowing continuing difficult circumstances to make one act irrationally. Perhaps patience can be defined as the character trait of not allowing long delay to make one act irrationally. This account of the personal virtues shows quite clearly that the vices are more fundamental than the virtues, for each of the vices can be defined by simply leaving out the 'not' in the definition of the corresponding virtue, e.g., intemperance is allowing emotions to make one act irrationally.

I am sure that others far more skilled in depicting situations will be able to give a more adequate account of each of the particular personal virtues and vices. But if they apply the schema that I have outlined, then it will be quite clear that it will be possible for someone to have all of the personal virtues or to have all of the personal vices or to have some of one and some of the other. That is, the account I have given of the personal virtues and vices allows both for their complete coexistence and also for their existing independently of each other.

A person may have no trouble acting rationally when confronted with danger or fear, but find that he cannot consistently act rationally when he has a strong desire for sex or drink. It is, of course, partly due to one's personality traits that one acts rationally in some kinds of situations and not in others, thus I am in no way minimizing the importance of personality traits. However, if one is to come up with an account of the personal virtues that makes them such that all rational persons want to have all of the personal virtues themselves, then it is essential to distinguish sharply between personality and character traits.

Though one may agree with all that I have said so far about the personal virtues and vices, that agreement will not mean much unless there is also agreement about what it is to act rationally. There are many accounts of rationality, but I shall criticize only one of these before I present my own account. This account is the one accepted by almost all social scientists, and most of those philosophers who deal with social issues ([8], p. 143).

On this account, acting rationally involves acting so as to maximize the satisfaction of one's desires. This view has many modifications: the agent must have full information, be in a cool moment, etc. ([1], pp. 10, 11). But it is not necessary for our purposes to examine them.

This view seems to make very good sense of the account of the virtues that I have presented. We can decide if a person is courageous by seeing if he acts rationally, i.e., acts so as to maximize the satisfaction of his desires, when confronted by danger or fear. This account of courage makes no judgment on the content of the desires, they can be concerned with oneself alone, or with people in general. All that is required for courage is consistent action in order to maximize satisfaction of one's desires in those situations when faced with danger or fear. It certainly seems as if we would all want to have courage, given this definition of rationality.

This account of rationality also results in very plausible interpretations of temperance and prudence. Temperance involves consistently acting so as to maximize the satisfaction of one's desires, especially those one has in a cool moment, in those situations when one is in the grip of some strong positive desire or emotion. Prudence becomes almost synonymous with rationality, simply involving consistently acting so as to maximize the satisfaction of one's desires, i.e., not ignoring more important future desires for less important present ones. It is not necessary to go through each of the personal virtues and show how this account of rationality results in a plausible interpretation of the virtue. There may be some who do not like these accounts of the personal virtues simply because they are accounts of personal, not moral virtues, but granting that courage, temperance, and prudence are personal rather than moral virtues, it seems to me that these interpretations are extremely plausible.

Nonetheless, this account of rationality does not seem to me to be adequate, It is inadequate because there is no limit on the content of the desires whose satisfaction one is seeking to maximize. This is true even with modifications demanding full information and a cool moment, etc. For unless these modifications entail that there are certain desires whose satisfaction is not rational, they do not limit the content of desire in an adequate way. If these modifications do entail that certain desires are ruled out, then the essential feature of the maximum satisfaction of desires view of rationality has been lost, namely its being completely person-relative, and hence completely tolerant and non-dogmatic. I think that this tolerance and non-dogmatic feature of the standard account of rationality is precisely what is wrong with it. We do regard some desires as irrational in themselves independent of their relationship to the maximum satisfaction of our other desires.

If someone has an unmotivated desire for death, pain, or disability, that is, if he simply desires to have these things, not for any ulterior purpose, but simply for themselves, we regard such desires as irrational regardless of their relationship to the maximum satisfaction of desires. The same is true for unmotivated desires for loss of freedom, opportunity, or pleasure. I call all of these objects of irrational desires, evils. Thus death, pain, disability, loss of freedom, opportunity, and pleasure are all evils. It is not irrational to want one of these evils if one wants it in order to avoid one of the other evils, or even to obtain some good, goods being abilities, freedom opportunity, and pleasure. The avoidance of evils and the obtaining of goods count as reasons whether they are for oneself or someone else. Rationality is not the same as self-interest; it is also rational to be concerned with others. But an unmotivated desire for the evils is irrational. Even if someone were to have this kind of desire to such a degree that satisfying it would provide the maximum satisfaction of desires, he would still be regarded as acting irrationally if he acted on it ([3], pp. 20–59).

If we regard the maximum satisfaction of desires view as limiting itself to rational desires, then I think that we can regard it, for practical if not for theoretical purposes, as an adequate view. This means that we do not regard someone as courageous who undergoes significant danger for no reason other than that he wants to undergo that danger. Similarly, when we add to the list of irrational desires the desires for loss of freedom opportunity, or pleasure, we see that we cannot regard as temperate someone who refuses to satisfy his strong present desire simply in order to deprive himself of pleasure. But with these very plausible limitations, we can simply adopt the rest of the maximum satisfaction of desires view of rationality and the result seems to me to be an acceptable account of the personal virtues.

On this account of the personal virtues, there are no rules to be followed: the personal virtues are in no way rule dependent. This will not be the case when we discuss the moral virtues. Some moral virtues are going to be rule dependent: to have the virtue involves following the appropriate moral rule ([3], p. 154). But as we shall see, even in these cases the virtue does not involve a blind following of the rule, but acting on the rule in the way that an impartial rational person would advocate. This means that it is not possible to think of the virtues as a simple following of a rule, for in some cases impartial rational persons will advocate breaking the rule, and this will not count against one's having the virtue.

Indeed, it may even be possible to eliminate any mention of the rules and to describe the moral virtues as those virtues that an impartial rational person would want both for himself and for all others. From what we have said so far, the only difference between the moral virtues and the personal ones is that the personal virtues are wanted by all rational persons for themselves, whereas moral virtues are wanted only by impartial rational persons, but they want them not only for themselves but for everyone.

We have already noted that all rational persons want to avoid death, pain, disability, and loss of freedom opportunity, and pleasure for themselves unless they have some reason. When the rational person is also impartial, he will want everyone to avoid these evils; he will want to minimize the overall suffering of evils. Thus he will want everyone, himself included, to have those traits of character which are most likely to result in the least amount of suffering of evil overall.

One of these traits of character will be what I call kindness. This virtue involves acting so as to prevent or relieve the suffering of evils by others. But like all of the moral virtues, truly having the virtue of kindness involves having the appropriate personal virtues as well. Someone does not have the moral virtue of kindness if his compassion, a personality trait that involves suffering because of the suffering of others, leads him to relieve someone's present suffering when he knows or should know that this will result in greater suffering at a future time.

Kindness, when considered to be a moral virtue, involves acting as an impartial rational person when confronted with the suffering of others. The combination of rationality and impartiality requires that one try to relieve or prevent the suffering of others, but not when this will result in equal or greater suffering by either the person himself or others. This understanding of kindness distinguishes it from indifference or callousness on the one side and paternalism on the other. It also makes clear that kindness, far from being incompatible with the personal virtues of prudence, temperance, and courage, requires these personal virtues for its fullest expression.

Kindness, like the personal virtues, does not involve following rules, though, of course, it involves not unjustifiably breaking any moral rules, not only those prohibiting the causing of evil, such as those against killing and causing pain, but also those prohibiting deception, breaking promises, and cheating. But, whereas kindness seems directly related to the aim of morality, the minimization of suffering, the moral virtues of truthfulness, trustworthiness, and fairness seem more directly connected to rules and less directly connected to achieving that end. It is these moral virtues that provoke the pointless disputes between those who call themselves deontologists and those who call themselves consequentialists.

There may have been, and there still may be, some philosophers who are pure deontologists or pure consequentialists, but any plausible moral theory involves elements of both. That is, all plausible moral theories will involve both rules and consequences. Any statable moral rule will have exceptions and it will sometimes be necessary to consider the consequences of violating the rule in order to determine what exceptions are justifiable. Further, even the justification of the rules themselves will involve consequences, either directly or indirectly. However, consequences alone will not be sufficient, for without rules there will be no way for that kind of impartiality which is recognized by everyone to be an essential feature of morality, to be adequately incorporated into the theory. Both of these points will become clearer when we discuss some of these moral virtues in greater detail.

Truthfulness clearly involves not unjustifiably violating the moral rule prohibiting deception. But just as clearly truthfulness as a moral virtue does not involve following this rule when this would be unjustifiable. Truthfulness involves avoiding deception except when one could publicly advocate it, that is, when as an impartial rational person one could advocate deceiving. Kindness involves not causing suffering except when one could publicly advocate it. Using this understanding of the virtues, let us see if we can deal adequately with the situation where these two moral virtues seem to come into conflict with one another.

It sometimes arises, especially in the practice of medicine, that one has painful news to tell. There are times when not telling this news, a grim diagnosis or prognosis, counts as deception as clearly as making a false statement. However, telling the news will clearly cause suffering. Here it seems as if being truthful requires that one be unkind, whereas kindness requires that one be deceitful. If truthfulness demands telling and kindness demands not telling, there does seem to be a conflict between the virtues. However, it may be that one can tell, but tell in such a way that minimizes the suffering of the patient, thus seeming to satisfy the demands of both truthfulness and kindness. But this is too easy a way out. Though one can and should minimize the suffering of the patient by telling in certain ways, it will still often be the case that significantly more suffering will occur if you tell, no matter how, than if you deceive either by not telling or by making a false statement.

This situation requires more precision in the account of the virtues than I have provided so far. Up till now I have characterized the personal virtues as if they were character traits that involved acting consistently in ways that rational persons would want to act, and have characterized the moral virtues as character traits that involved acting consistently in ways that impartial rational persons would want to act. I have claimed that what distinguishes one personal virtue from another is the situation in which the exercise of the virute is called for, and have made the same claim with regard to moral virtues.

This claim strongly suggests that the virtues cannot conflict with one another, because they are called for in different situations. However, this suggestion neglects the possibility that the same situation can call for the exercise of two personal or moral virtues and yet it be the case that if one of these virtues is exercised the other will not be. This seems to be the situation we have been discussing where both truthfulness and kindness are called for, yet it does not seem possible to exercise both. A similar situation can occur with the personal virtues where one may be in a dangerous situation that calls for both courage and prudence and yet if one acts in one way it will be appropriately described as courageous, but not as prudent, and if one acts in the other it will be appropriately described as prudent, but not as courageous.

Does this show that courage and prudence must sometimes come into conflict so that courage sometimes requires imprudence and prudence, cowardice? Is the same true for truthfulness and kindness? If this is so, then it cannot be correct that all rational persons seek all of the personal virtues and seek to avoid all of the personal vices or that all impartial rational persons seek all of the moral virtues and seek to avoid the moral vices. But I think that these claims are correct. I think that we can reconcile these claims with the situations described above in the following way.

To seek the virtue is not necessarily to seek to exemplify it in every situation. One may feel that in some situations it would be preferable to exemplify a different virtue. Thus, a situation may allow many different choices to be made, among them one that exemplifies courage and another one that exemplifies prudence. In choosing to act in the courageous way one is not thereby committed to the view that acting in this way is imprudent, only that prudence is not the virtue that one's action is intended to exemplify. Similarly, acting in the way that would be called prudent does not require that one regard the action as cowardly, only that courage was not exemplified by the action. Thus situations which seem to call for one or the other of two virtues to the exclusion of the other should not be seen as showing the incompatibility of the virtues, but only the difficulty of performing an action which examplifies both of them. The same point can be made about the seeming conflict between truthfulness and kindness.

However, the view that exercising one moral virtue must sometimes require exemplifying another moral vice may arise from the fact that impartial rational persons can disagree about what is the morally best way to act. If two people do disagree, one holding that a given choice would count as kind, e.g., not telling the patient the truth, and another person holding that it would be deceitful not to tell, then it seems as if there is a real conflict between the virtues. But this is a mistake: to the persons involved there is no conflict, the one who thinks it kind not to tell does not regard it as deceitful, and the one who thinks it deceitful not to tell does not think it kind. Rather, one thinks it is justified to deceive and the other does not; thus they have different views about what is the morally acceptable way to act. They disagree about whether not telling counts as an act of kindness or not and this disagreement is a disagreement about whether or not this prevention of suffering counts as a morally adequate reason for not telling this patient the truth.

This account does not have any place for misplaced kindness. But can't one be kind when it would have been better if one had not been? Here, I would like to mention the fallacy of ignoring the modifier. Misplaced kindness is not kindness anymore than a false friend is a friend or a rubber duck is a duck. If one knows that it would be wrong not to tell the patient and yet one cannot bring oneself to tell him; this is not kindness, at least not kindness if it is to be regarded as a moral virtue. Rather, it is a manifestation of the personality trait of compassion or pity. It is, of course, true that someone with compassion is more likely to be kind than someone without compassion, but confusing kindness with compassion is as much a mistake as confusing courage with fearlessness. It is very important to distinguish between the virtues and the personality traits that are usually related to them in such an intimate way. It is only by recognizing that all impartial rational persons favor all of the moral virtues that we can show that misplaced kindness is not really kindness at all.

Similarly, someone who claims that one is not being truthful if one refuses to tell, when all impartial rational persons would advocate not telling, is confusing the virtue of truthfulness with compulsive truth telling. A truthful person never deceives when it is morally unjustifiable to deceive. But when it would be morally wrong not to deceive, a truthful person deceives. Though this sounds paradoxical, it is clear that if truthfulness is to be a moral virtue it must never require doing what is morally wrong. Someone who tells the truth regardless of consequences is not truthful but tactless. The names of the virtues are sometimes misleading. The view of the virtues as necessarily desirable is essential for coming up with an adequate analysis of them.

NOTE

Preparation of this paper was aided by a NEH–NSF Sustained Development Award, R11–8018088 A03. The views represented herein do not necessarily reflect the views of these agencies.

BIBLIOGRAPHY

[1] Brandt, R. B.: 1979, *A Theory of the Good and the Right*, Oxford University Press, Oxford.
[2] Culver, C. M. and B. Gert: 1982, *Philosophy in Medicine: Conceptual and Ethical Issues in Medicine and Psychiatry*, Oxford University Press, New York.
[3] Gert, B.: 1970, *The Moral Rules: A New Rational Foundation for Morality*, Harper and Row, New York.
[4] Gert, B. (ed.): 1972, *Man and Citizen*, Doubleday Anchor, Garden City, New York.
[5] Hobbes, T.: 1651, *Leviathan*, Oakeshott, M. (ed.), Basil Blackwell, Oxford.
[6] Kant, I.: 1964, *Groundwork of the Metaphysic of Morals*, H. J. Paton, (trans.), Harper and Row, New York.
[7] MacIntyre, A.: 1981, *After Virtue*, University of Notre Dame Press, Notre Dame, Indiana.
[8] Rawls, J.: 1971, *A Theory of Justice*, Harvard University Press, Cambridge, Massachusetts.
[9] Strauss, L.: 1952, *The Political Philosophy of Hobbes: Its Basis and Its Genesis*, The University of Chicago Press, Chicago.

Dartmouth College,
Hanover, New Hampshire,
U.S.A.

EDMUND L. PINCOFFS

TWO CHEERS FOR MENO: THE DEFINITION OF THE VIRTUES[1]

Socrates: What do you yourself say virtue is?
Meno: First of all, if it is manly virtue you are after, it is easy to see that the virtue of a man consists in managing the city's affairs capably, and so that he will help his friends and injure his foes while taking care to come to no harm himself. Or if you want a woman's virtue, that is easily described. She must be a good housewife, careful with her stores and obedient to her husband. Then there is another virtue for a child, male or female, and another for an old man, free or slave as you like; and a great many more kinds of virtue.
Socrates: I seem to be in luck. I wanted one virtue and I find that you have a whole swarm of virtues to offer.

Plato, *Meno*

Meno must surely rank as one of the easiest to push over of Socratic respondents, but, if he had just pursued it, he could have salvaged two or three points from his weak beginning. One point, overlooked by nearly every contemporary philosopher who attempts a definition, is that, if we are to understand what virtue is, it might be well to begin with a list of qualities that we intuitively recognize as virtues and vices. A second point is that it may, for some qualities, depend on the context (city or household, for example) whether they are virtues or vices or neither. A third point is that by distinguishing different sorts of virtue we will be in a better position to evaluate answers to such large questions as whether virtue can be taught, whether virtue is one, and whether virtue is knowledge. In this paper, I will offer a preliminary survey that could be useful in arriving at nuanced answers to these large questions.

Meno starts badly, not because he tries to distinguish sorts of virtue, but because he begins with an inadequate sorting scheme. Socrates maintains that justice is justice whether predicated of a man or a woman, and that to speak meaningfully of justice in the city or in the household we must have a notion of what justice is wherever it is exhibited. So the range to be covered by an adequate definition of virtue is indicated by terms like justice and temperance rather than by terms like female virtue and male virtue, slave virtue and freeman's virtue. But if this is the range, and we can add an indefinitely long list of terms that intuitively belong within it, then why should

Earl E. Shelp (ed.), Virtue and Medicine, 111–131.

we suppose that there is a Yes or a No answer to such questions as whether virtue can be taught? Far from presupposing that there is one answer to such questions, we might even use them as means of differentiating sorts of virtue: virtues that can and cannot be taught, virtues that do and do not consist in, or require, knowing something, and so forth.

But if Meno rates two cheers, Socrates rates two and one-half, for Socrates recognizes the difficulty of defining the virtues, as Meno does not. Socrates arrives at no definition but only a hypothesis concerning the origin of virtue, irrelevant, as he recognizes, to the question what virtue is. This negative conclusion, that no satisfactory definition of virtue has been found, is the note on which the *Meno* ends; but Socratic modesty is not a characteristic of recent writing on the concept of virtue. At the end of this paper, I will mention some contemporary definitions of virtue that have, I believe, been too confidently put forward.

THE FIELD TO BE EXAMINED

Well short of definition, some preliminary things can and need be said about where virtues and vices (VV's) belong. They are qualities of persons; and the first problem is to set them off from other sorts of qualities that persons may have. Some qualities or properties may be of persons but not personal qualities. For example, being attracted by gravity is a property of every person (with the possible exception of angels), but we would not want to call that property a personal one. It does not enable us to distinguish between persons; and I will take it that that is a necessary characteristic of a personal property. On the other hand, some qualities that do enable us to distinguish between persons do not yet seem personal in the appropriate sense, the sense in which they are relevant to the kind of distinction we would like to make. Thus, some persons maintain a body temperature of 98.1° farenheit and others maintain a temperature of 97.9°, some have black hair and some have red. It tells us something about Rocky as a person that he has red hair only if we associate having red hair with properties of a different sort, like being pugnacious or aggressive.

As a first approximation, we can say that VV's are a subclass of those properties referred to in answer to the question, 'What kind of person is *A*?' Not all answers to that question consist of a list of properties. One sort of answer is to tell a revealing story, another is to smile approvingly or to frown and shake one's head. It is also clear that not just any list of properties like 'born in Chicago' or 'over 150 pounds' will do. Another sort of answer, closer to the mark, will not do either. It does not tell us, except by inference from the speaker's values. what sort of person *A* is if the speaker replies that he is good, bad, odd, praiseworthy, wonderful, or despicable. To understand these remarks we must find out what sort of person the speaker would count as good, etc. He has not told us so much what kind of person *A* is as what his opinion of *A* is. The terms he uses are descriptively empty, and we want description, or at least some descriptive content, but descriptive content of a certain sort. A third kind of answer won't do either. We can't say what kind of person *A* is by mentioning merely transient or ephemeral qualities. It won't do to say that *A* is bored or that he is angry, unless what is meant is that he is typically a bored person or an 'angry man'.

As a second move, let's spread before us the names of some qualities that can serve as answers to the question what kind of person *A* is. Appendix I is a list of such names.

These properties do have some common features. They are not static ones, like specifying weight or place of birth, but are dynamic, having to do with tendencies, dispositions. They concern the way *A* typically or always moves through life – his reactions and attitudes as well as his actions. They also seem at least potentially to be grounds for preference or avoidance. But the preference or avoidance is of a different sort from that in which we would, for example, prefer a person over 150 pounds for an experiment in nutrition, or would avoid a person raised in Chicago if we wanted a good example of a Southern accent. Furthermore, although these properties are (for short) dispositional, the dispositions in question are determinable as opposed to determinate ones. A disposition to blush upon hearing a certain word, to tremble, to mispronounce 'nuclear', to blink rapidly is determinate. The term for the disposition is descriptive in a straightforward physical way of just what the person does. We could teach a child what 'tremble' or 'blink rapidly' means by showing him people who tremble and blink rapidly. We could construct dolls for him that act and react in these determinate ways, but it is open to conceptual question whether ingenious technicians could construct dolls for him that act and react in the ways appropriate to the properties on our list. These properties are determinable. You cannot teach a child what charitability is by showing him someone putting money in the hands of another person. There are indefinitely many ways, not just one way in which charitability can be evinced. The terms on our list vary in their determinability, but none of them is determined in the sense indicated.

WHAT KINDS OF QUALITIES ARE VV'S?

Some sorts of persons we prefer, others we avoid. The properties on our list can serve as reasons for preference or avoidance. If we are asked to pick out the virtues on our list, and the vices, then, I suggest, we will be looking for grounds for preference or avoidance. We might think that cruel people are nearly always to be avoided, and that cheerful ones are nearly always to be preferred; but we may think that it depends much more heavily on the circumstances whether ambitious persons are to be preferred or complacent ones avoided. I suggest that the natural home of the language of virtue and vice is in that region of our lives in which we must choose, not acts, lines of action, or policies, but persons. Persons, unlike acts or policies, are not right or wrong, beneficient or disastrous, although policies or acts, like persons, can be cruel or just. I will not take up here the question of the relations between the virtues of persons and of acts, policies, or practices.

'Choosing persons' is, of course, a very global sort of expression. Sometimes we must, quite directly, choose *A* or *B* or *C* where all are candidates for the same job, public office, or scholarship. Sometimes we must choose not between persons but whether, and with what reservations, to enter into some relationship with a particular person: as landlord, tenant, lawyer, contractor, wife, friend, confidant, or guest. These choices, once made, can lead to further and yet further choices at many levels and of many degrees. We must not only choose relationships but continually adjust to them in a variety of ways depending upon our assessment of other people and of ourselves. We can think of the language of the VV's as providing the set of categories in terms of which we can justify our person-choices.

Suppose no choice could be made between persons. This could be so if it were impossible to distinguish between 'persons', or if 'choice' were made for us. If we lived in a world of identical creatures, then the determination who should be president and who should be hod-carrier would be an arbitrary one; and if the 'choice' were made *for* us between non-identicals, we would have no occasion for the justification of choice, for the adducement of grounds.

OBJECTIONS TO A DISPOSITIONAL ANALYSIS

I am not sure whether any philosophers would deny that VV's are dispositional properties that provide grounds for preference or avoidance of persons. There are at least three possible objections to a dispositional analysis of VV's. The first might arise on the part of a reader who has been accustomed to think of character traits as matters of habit rather than disposition. So he will not be ready to concede the laurels to dispositions as a matter of definition. The crux of the matter seems to be that to say of a person that he has a habit of doing anything is to imply that at some time or other he learned to do it and that he might have learned otherwise, that it would have been possible for him to do so had he chosen. It always makes sense, given that Otto has a habit of doing something, to ask when he learned (or picked up, or acquired) the habit. The implication, correctly or incorrectly, is that if he learned it, he might have learned some other habit instead, so that he is *responsible* for what he does as a result of the habit. It is not an excuse, to put the matter another way, that what Otto did was a result of habit, since, if what he did was wrong, Otto should not have such habits. To begin by making personality traits and hence character traits, dispositions, is, so the objection might go, to settle the fundamental question of responsibility by definitional fiat.

Two answers can be given in justification of a preference for dispositions over habits. The first is that 'habit' is a term of very narrow scope. There are a great many traits that are clearly neither habits nor the result of habit. It would be stretching the term unmercifully to refer to cleverness, stupidity, or cupidity as habits. There would be no answers to the questions when we acquired the habits in question, what we did to encourage them, or whether we might not now break them. Not only this but there are a great many habits that are evidence for dispositions. For example, the habit of twitching one's foot is evidence for the proposition that a person is nervous. The second answer is that speaking of dispositions allows us to capture more easily the language in which we assess character. Dispositions are determinables; habits are typically determinate. When we say that Otto has a habit of tipping his hat to ladies, we say exactly what Otto does in certain cirsumstances; but when we say that Otto is polite we do not. Courage is not the habit of leading instead of following one's troops, plus other habits. In fact, to the extent that action can be shown to be 'merely' habitual, then Otto's doing that thing is not evidence of Otto's courage.

I suspect that the preference for habit over disposition arises partly through confusion. The inculcation of habits may be the best way to develop dispositions. That is why we can give children 'moral training'. But moral training is at best imparting habits that are likely to bring about appropriate dispositions. Otto can quite easily be taught the habit of tipping his hat; but it does not follow that he can easily be taught to be polite, unless by politeness is meant merely the habitual performance of such rituals. No less a question than whether virtue can be taught is at issue here; but to speak of personality traits and hence of character traits as habits would offer us an all-too-facile answer. Of course habits can be taught; but we cannot resolve so difficult a problem by so simple a move.

The second source of discomfort with a definition of personality traits as dispositions may arise from a behavioral interpretation of 'disposition', as in the paradigmatic dispositional term 'brittle'. A kind of material is brittle if when struck it shatters, similarly 'disposition' may be thought to refer always to publicly observable behavior. But the objector rightly wants to reserve a place for the 'inner' dispositions or tendencies to feel, to experience emotion, to sense the feelings of others, to react privately. These too must be taken into account in the assessment of character; and if by definition they are excluded, then the definition must be refashioned. The answer to this objection is simply that I agree and that it is not my intention to define disposition behaviorally. Dispositions as I understand the term are not just tendencies to act in certain ways but also to feel, think, and react, to experience 'passions'. I do not need to take a position on the confused question whether these feelings, etc., can somehow be 'reduced' to external behavior. Whether or not they can, I wish to include them in the data on which we should rely in the attribution of traits of personality, and hence of character traits.

Finally, it might be objected that, leaving aside the question of 'internal' vs. 'external' dispositions, it is a mistake to define personality or character traits in terms of dispositions at all, since a person's character can be established by one incident in his lifetime. For example, the objector might continue, no matter what else he might have done in his lifetime, Ivan Karamasov's nobleman, who set his hunting pack upon a serf child who had injured his hunting dog's paw, was a cruel man. Nothing he could do afterwards would make this an unjust or unwarranted accusation. The one incident alone is enough to establish his character. But the objector must answer the question why he feels there is nothing Karamasov's nobleman can do to erase the cruelty of his character. I suggest that it is because he feels that his carefully planned non-accidental deed *reveals* the nobleman's character in a way that less dramatically cruel deeds might not. He believes that a man who could do such a thing could not be otherwise than cruel. His nature has been exhibited and it cannot henceforth be successfully concealed. But what is the objector saying if not that the deed reveals a covert disposition? I would not wish to deny that there are dispositions that can be more or less successfully concealed. It is undeniable that men do sometimes reveal themselves by a single dramatic action, and that, unless he was insane or otherwise ineligible for judgment, our nobleman *was* indeed a cruel man. This is a place to pay attention to tenses. I would be most hesitant to say that henceforward throughout his lifetime he *is* a cruel man. That would be to take the dubious moral and psychological stand that no transformation of character is possible for a man who has descended to such depths of cruelty. I am not at all sure that this is true. I do not say that he should not be ashamed of his action for the rest of his life or that he has not done something inexcusably wrong. I do not want to argue that the transformation of character somehow releases a person from the opprobrium he has rightly incurred for his past action. In any case, the objection presupposes a dispositional interpretation of personality traits, the one deed supposedly revealing a covert disposition. Since I am willing to grant that dispositions may sometimes be successfully concealed, there is no argument.

THE TENDENCY TO OFFER REDUCTIVE DEFINITIONS

If we understand VV's as dispositional properties that provide grounds for preference or avoidance of persons, then the list of VV's will be indefinitely long, and it will be functionally various. There can be *many* different sorts of reason for preference, some of which we will survey in a moment; and a definition of virtue that picks out one or a very few dispositions as constituting the whole of virtue must shoulder the burden of showing that it is not unjustifiably reductive. Thus, for Kant, conscientiousness is *the* virtue; for Mill, *the* virtue is benevolence; for some ancient philosophers there were four cardinal virtues; and the church fathers, accepting these, were willing to add only three theological ones. Yet unless there is some prior reason to believe that there is but one, or that there are but very few bases for preference of persons, the reductive mode of analysis may be misconceived from the beginning. Why should we suppose that only conscientiousness matters in preferring one person to another, or only benevolence, or only the cardinal virtues? A dozen neglected dispositions come to mind. What reason do we have to suppose that they can all be neatly tucked under the favored qualities? Why should we say of a person who is careful, cautious, charitable, cheerful, clever, civil, co-operative, and courteous that the only reason for preferring him to others is, really, that he is conscientious, or that he is just? To say this is to imply, I think, that either these other apparent grounds for preference and choice are not such, or that the dispositions in question are really the favored dispositions in disguise, or are derivable in some way from them. But that would then remain to be shown; and I do not know of any place that it is shown.

A more promising way of encompassing virtue than nominating some one or a few virtues as the only real ones, is to look more carefully at preference and choice of persons, to ask of the philospher who would restrict the virtues so radically what sort of choice he has in mind. What are the most general grounds of choice that must be made of persons? That is not a small question; but it would seem that until we have attained some clarity on it we will be unable to show what dispositions provide grounds for choice. This might make it seem as if we should start again with Meno and say that there is one set of grounds for choosing or preferring slaves, another for freemen, one set for men and another for women, and so on. In slaves we will want obedience, patience, and vigorousness, say. But since there are as many categories of persons as there are bees, such an approach will yield an unordered 'swarm of virtues'. For there are old and young slaves, field slaves and house slaves, and on and on. Besides, if my suggestion is correct, VV's are best understood as answers to the question what sort of person *A* is, not just what sort of slave, freeman, athlete, poet, etc., he or she is. (Of course we can speak, not of virtue proper, but of the virtues of slaves, etc. This is not what Socrates wanted, and not what is wanted for moral philosophy.) If I am correct, if *V* is a virtue, then to say that *A* has *V* is to provide a reason why *A* should be preferred to other persons, whatever *A*'s status, even though for a particular status one sort of virtue may be more relevant to choice than another. The first task, then, is to survey the most general grounds for preference of persons. Such a survey is, I think, the best antidote to reductive definitions.

A SCHEMA FOR SORTING THE VIRTUES[2]

A way of beginning to distinguish the grounds of preference for, or choice of, persons, and, hence, of categorizing the virtues (for simplicity I ignore the vices), is to set off considerations having to do with the aptness or appropriateness of the person for the accomplishment or achievement of goals or objectives, from considerations that do not have to do with these teleological intentions. Let me begin, then, by speaking of *instrumental* vs. *non-instrumental* virtues. Instrumental virtues in a person are those that, in relatively direct fashion, make it more probable that he will successfully pursue goals, ends, objectives.

If we were to pursue this distinction in more detail, it would be necessary, with a bow to Ryle, to distinguish these qualities from another set with which they might be confused: the virtues that concern the doing of tasks well, as opposed to the present set, those that concern pursuing goals successfully. There are task virtues like neatness, thoroughness, carefulness, and sensitivity that fall in the first class; virtues like persistence, alertness, and courage fall more naturally in the second, instrumental, class. The instrumental virtues are at home in talk about winning wars, finding the sources of rivers, finishing novels, reducing deficits, and crossing oceans. The task virtues are at home in talk about bricklaying, painting, gardening and typing where we can and do distinguish the question how successful the overall undertaking was from the question how well it was done. A good bricklayer may be either a person who gets the job done quickly (and passably well), or one who is unusually neat, or artistic.

The easiest and most obvious side of this overall classification is the instrumental one. Persons, as individuals, are more likely to succeed in undertakings of any difficulty if they are persistent enough not to be easily discouraged, courageous enough to face daunting challenges, alert enough to perceive pitfalls and opportunities, careful enough not to make needless errors, resourceful enough to devise alternative strategies, prudent enough to plan ahead for eventualities likely to be encountered, energetic and strong enough to carry through what they have planned, cool-headed enough to meet emergencies without panicking, and confident and determined enough not to give way to evanescent feelings and desires that would lead them away from their tasks. In addition, there are qualities that make persons worthy of preference for participation in joint or communal undertakings, such as cooperativeness, the virtues of leaders and followers, and the kind of practical wisdom about the best means to ends that comes from experience in group endeavors, the sort of wisdom that has to do with the division and apportionment of tasks, and the choice of appropriate social structures and instruments for the undertaking at hand.

The non-instrumental VV's are a more varied lot, and a good deal more intriguing candidates for analysis. They may make the attainment of individual or group goals more or less likely, but if so their contribution is not a direct one; and they are not typically valued for that reason. Some non-instrumental VV's have no easily detectable relation to success in individual or joint undertakings. I will speak of three general sorts of non-instrumental VV's: aesthetic, meliorating, and moral.

Aesthetic VV's are qualities that are farthest away from being instrumentally valued or depreciated. Aesthetic virtues are appreciated for what they are, for the vision of themselves; we are grateful for their presence; they are examplars of what humans can be; their absence is regretted as impoverishing life. There are at least two general sorts of aesthetic VV's: the noble and the charming. Noble virtues include dignity, virility, magnanimity, serenity, and, of course, nobility itself. The similarity to the Stoic list of virtues is not accidental. The Stoic's objectives seem to have been twofold: to achieve serenity – a certain level or tone in life, and to make of oneself a kind of model of what humans might aspire to, a model of nobility, of the sort of person who lives and dies in a high and admirable way.

Charming virtues attract, not by their altitude, but by their beauty. We appreciate gracefulness in a person: gracefulness in posture, in movement, in expression, in meeting the ordinary exigencies of life. Wittiness can be charming, as can liveliness, imaginativeness, and whimsicality. People who have such qualities attract, simply because life is better with such people than without them. The corresponding vices make for dull and unattractive common life: ungracefulness, lack of wit, and so on.

Meliorating virtues occupy a middle range on the instrumental–non-instrumental continuum. I have placed them on the non-instrumental side because their contribution to common life is not the direct one of making success in individual or common endeavor more likely; it is rather the making of common life, whether or not structured for common endeavor, more tolerable. This dark description may perhaps be lightened by the sorts of examples I will give. In general, there seem to me to be at least three sorts of meliorating virtues: mediating, temperamental, and formal.

Given that, in the ordinary course of life, differences, disputes, and quarrels are likely to arise, peacemakers, negotiators, appeasers are needed; and the qualities appropriate to these roles are generally valued: tolerance of views and attitudes different from one's own, reasonableness in assessing opposing points of view, tactfulness in not arousing unnecessary hostilities. These, then, are *mediating* virtues; they will be more or less emphasized as the historical, sociological, and particular and personal circumstances demand; but it is hard to conceive of a human situation in which they will not have value.

Like other meliorating virtues, the mediating virtues have value in a less-than-perfect world. In a world in which, like the Distant Earth in Dostoyevsky's *Dream of a Ridiculous Man*, everyone truly loved one another, and there were never never invidious comparisons, frictions, or disputes, there would be no place for mediating virtues; and the next sorts of qualities I want to discuss, the temperamental and formal ones, would be simply taken for granted as to be expected in everyone, and, hence, would not count as virtues at all.

Temperamental virtues form a large class. Their general characteristic is simply that the persons who have them are easier and more pleasant to live with, that the avoidance of the corresponding vices constitutes a gain in communal life. Let me mention a few: gentleness, humorousness, amiability, cheerfulness, warmth, appreciativeness, openness, even-temperedness, non-complainingness, and non-vindictiveness. The most direct road to appreciation of the temperamental virtues is to focus on the corresponding vices, to think what life is like in the presence of vindictiveness, continual complaints, frequent outbursts of temper, and so on.

There is a special problem about the temperamental virtues signalled by the label I have given them. If they are matters of temperament, then how can they be virtues at all? If virtues are qualities for which the bearer is to some degree responsible, and different people, like different breeds of horses or dogs may just be the happy or unhappy possessors of a particular temperament, given by birth, how can there be temperamental virtues? I have not, however, wanted to confine the field of virtue and vice to qualities for which the agent is responsible. It is, I think, merely a superstition of some philosophers that virtue need be so confined. We can and do prefer and avoid people on account of qualities they have when we have not the least idea whether they may justly be held responsible for those qualities. Some people may well inherit qualities or acquire them so early that they cannot be said to have any control over their acquisition, and these qualities may nonetheless provide substantial grounds for preference or avoidance. Preference and avoidance must not be thought of as instruments for moral training by means of which we induce people to develop certain qualities. Sometimes, often, we may prefer or avoid a person for qualities we do not praise or blame him for having.

This is not to say that there are no degrees of responsibility for what I have labelled temperamental virtues. Some of them may be, to some extent, under the control of the agent. If he tries to change himself, he can be more cheerful or open. Perhaps the least tractable of the virtues I have mentioned is humorousness. If a person has no sense of humor, then he can hardly be blamed for being humorless. One can't give oneself a sense of humor in the way that one can, to some extent, give oneself cheerfulness. A sense of humor depends on a certain capacity to catch the point of a joke or to see the funny side of things; cheerfulness can be forced, and if forced enough, and in the right way, can become second nature. But even if a person cannot be blamed for humorlessness, we would prefer the person with a sense of humor to him; and we will always have a residual suspicion that if it is (to use the language of Samuel Butler) not a person's fault that he has no sense of humor, it is nevertheless a fault in him. He is too narrowly focussed on the literal and the everyday, too little concerned with side-perspectives, with the leavening of the flat and the ordinary. Humorlessness may in fact be symptomatic of a whole syndrome of qualities, some of which *are* changeable by the agent's efforts.

However that may be, there is a third group of meliorating virtues that we must now notice: the *formal* virtues. These include civility, politeness, decency, and hospitableness. Their common characteristic is that they meliorate by adhering to customs, common understandings, or practices, that themselves meliorate. The practices that fall under the general heading of courtesy simply make common life more livable, easier, pleasant, and less strained by confrontation and push-and-shove. It is not easy or pleasant to live with persons who are incivil, impolite, have no decency, or are habitually inhospitable.

'MORAL' VIRTUES AND VICES

There is a particular sort of non-instrumental VV's that deserves separate treatment. Persons who exhibit the virtues and vices in question are regarded in a special sort of way, a way that can be distinguished from the way in which those persons are held whose virtues and vices are aesthetically pleasing or displeasing, or generally meliorating or exacerbating. These VV's have the common characteristic that they are forms of regard or lack of regard for the interests of others. They are, roughly, of two closely related classes: those that have to do with direct concern or lack or concern for the interests of other persons, and those that have to do with the unfair advantage one accords one's own interests over the interests of others.

While the meliorating virtues, and of course the instrumental ones, are *in* the interests of others, the unifying characteristic of the present lot of virtues is that they are traits concerned with the regard *the agent has* for the interests of others. What makes them virtues or vices is, essentially, the agent's acting or failing to act out of a certain sort of motive; but there is no motive that provides a distinguishing characteristic of meliorating VV's.

What many people must have in mind when they speak of the demands of morality, and when they deplore the absence of 'moral' qualities in others (and in themselves), is that people take unfair advantage in their own interest or in the interest of those with whom they identify. Most of what are regarded as moral qualities can be seen to involve some form of this failing or sin.

Various forms of non-deceptiveness – honesty, sincerity, truthfulness – fall under the general requirement that no unfair advantage be taken. It is always in the power of the person conveying information to edit or distort it to his own advantage, if advantage is to be gained by misleading the hearer; the person trusted can always violate the trust to his own advantage; the insincere person can profess what he does not believe or hold, thereby hoping to gain advantage to which he is not entitled. The unfairness in these cases consists in this: that the dishonest, insincere, or untruthful person does what any other *could* do in his own interest, but what others generally refrain from doing to their own occasional loss. It is unprincipled opportunism that is unfair, if others, who have as much to gain, are principle-governed. This is a quality peculiarly evil in a system that depends on mutual trust, and on self-government rather than on coercion. This is one way of thinking of a moral community. Moral communities are notoriously fragile structures, and the unprincipled opportunist takes advantage of their existence while at the same time undermining them.[3]

HOW SOME CURRENT DEFINITIONS OF VIRTUE FAIL

I have tried in the present essay to provide a generous enough survey of the types of virtue to include most of the qualities that are or may be regarded as virtues or vices. I have seen this as necessary if we are to avoid definitions or understandings of virtue that are in any way reductive. Some contemporary definitions, while appropriate or insightful in other ways, do seem to me to have this consequence.

G. H. von Wright, for example, in his otherwise acute discussion of virtue in *The Varieties of Goodness* ([10], pp. 147ff.) tells us that "the role of virtue, to put it briefly, is to counteract, eliminate, rule out the obscuring effects which emotion may have on our own practical judgment, i.e., judgment relating to the beneficent or harmful nature of a chosen course of action" ([10], p. 147).

This definition seems to me to have the classical defect of philosophical definitions: it is at once too narrow and too broad. It is too narrow in that it would not count as a virtue those qualities that have little if anything to do with the overcoming of obscuring emotion. The person who is the happy possessor of the instrumental virtues of self-confidence and resourcefulness need be overcoming no obscuring emotion in exhibiting them; the same may be said for the mediating virtues of open-mindedness and tactfulness, the aesthetic virtues of gracefulness and dignity, and the meliorating virtues of gentleness and cheerfulness. It is too broad in that, if moral virtue is what is in question, and to the extent that it is true of moral virtue that it consists in the ruling out of obscuring emotion, then it does not help us to identify the emotions in question or to say why overruling them is peculiarly important to morality. I am not sure whether to call the want to take unfair advantage an emotion; but this want is at least a morally important one. The emotions that need overruling, from the present point of view, are those that would prevent the agent from taking proper account of the interests of others, and would lead him to give undue weight to his own interests.

Lester Hunt contends that "a person has a trait of character insofar as he holds the corresponding belief and holds it as a principle: insofar, that is, as he believes it, and acts on it consistently (for both of these things admit of degrees)" ([3], p. 183). To have traits of character, he tells us, is to have reasons for what one does, of the sort that brutes cannot have. But it is not clear what is to be gained by a definition that rules out such qualities as cheerfulness, imaginativesness, civility, gracefulness, and resourcefulness as qualities of character. For these qualities and very many others it is difficult to see that any special belief need be held for the quality to be present. This is to say nothing of the problem of identifying the corresponding belief.

Hunt comes closer to the mark if his definition is interpreted as applying to moral character. Then the principle in question might, as I have indeed suggested, be that one should take the interests of others seriously, and not take unfair advantage. But the difficulty here is in giving an account of having a corresponding principle that does not beg the question whether traits of moral character are necessarily accompanied by principles. Why a person cannot just be concerned for the interests of others, and be fair, without holding a principle, is not clear.

Maurice Mandelbaum tells us that "the traditionally acknowledged virtues are . . . precisely those traits of character which provide fitting answers to the ever-recurring demands which all men face" ([7], p. 150). 'Character' refers to the "relatively persistent forms which a person's motivation takes" ([7], p. 141). So traits of character are, for Mandelbaum, essentially dispositional, as opposed to actional, meaning, in his usage, that the actual feelings and motives of a person are relevant to our judgment as to whether he possesses the attribute ([7], p. 144).

The problem here is again that there does not seem a good argument for ruling out as a part of a person's character what does not depend so heavily on the motivation of the agent. There seem to be plenty of mediating, aesthetic, meliorating, and instrumental VV's that are not so heavily dependent.

More seriously, even when we move to moral virtues and vices, the crucial issue of what is to count as a demand, and what sorts of demands are to count, must be faced. There are a great many recurring demands, the meeting of which seems to have little to do, on the face of it, with morality. I have suggested a way of setting off one sort of demand, that there be concern for the interests of others, or at least the not taking unfair advantage of others, as having particular importance for morality.

Alasdair MacIntyre offers as a 'partial and tentative' definition: "A virtue is an acquired human quality the possession and exercise of which tends to enable us to achieve those goods which are internal to practices and the lack of which effectively prevents us from achieving any such goods" ([5], p. 178). While MacIntyre amplifies and extends this definition in ways that I find intriguing, still it has an uncomfortably reductive sound. It is not clear to me in the end why, given the full spread of what may with justification be counted as virtues, we should be limited to those qualities that have the appropriate relation to the goods that are internal to practices. The distinction between goods external to and internal to a practice is perhaps best indicated by reference to MacIntyre's example of the child, the candy, and chess. A child is motivated by a good external to the practice of chess if he is offered money to play to win against an adult who offers the child money and agrees to play in such a way that the child can, with effort, win. This may eventually bring the child to the point that he is able to appreciate and be motivated by the goods that obtain in the playing of chess, "goods specific to chess, in the achievement of a certain highly particular kind of analytical skill, strategic imagination and competitive intensity" ([5], pp. 175–176). These are goods internal to the practice. While the notion of a practice is a very broad one for MacIntyre, still it is not clear why we should regard that notion as centrally definitive in the concept of a virtue. The issue, to put it in my terms, is whether there should be but one sort of quality, defined by the sort of reason it provides for preference of the bearer, or whether there may not be many sorts, sorts that are mutually irreducible one to another.

The issue between MacIntyre's and my way of understanding virtue is, I think, this: MacIntyre's conception takes a particular value as prior. We first determine what is valuable – the values internal to practices – and on that basis determine which qualities are virtues and which vices. MacIntyre leaves the matter suitably vague by acknowledging that *ex*ternal values also count: the wealth, worldly success, recognition, say, that can result from engaging in practices. The issue, which I would resolve in MacIntyre's favor, whether internal or external values are most worth cultivating, is irrelevant here. (A closer look at that subject would require at least that we distinguish those practices, like the dance or painting where external values are a kind of barbarian intrusion, from practices like the conduct of war or the management of industry in which the satisfactions gained from the practice are at least arguably secondary to those gained from the consequences of engaging successfully in it.) On my own, functional, view, we look for those qualities that serve as reasons for preference in the ordinary and not-so-ordinary exigencies of life. Given that much of life, whether in complex or tribal society, consists in engagement in practices, it might seem as if there would be very great overlaps between a functional view and one that makes the status of qualities as virtues contingent on their contribution to the internal values of practices. That may be so, but the basis of designation as a virtue is importantly different. On a functional view, it is the tensions, tendencies, pleasures, and pains of common life including the engagement in practices, that lead us to value or disvalue that quality as responding well or ill to what we go through together. On MacIntyre's view it is the particular satisfactions of engaging in practices that set off those qualities that are virtues from those that are not. This is, from the functional point of view, so far an arbitrary designation. It is not clear that the value we place on the internal satisfactions of engaging in practices *determines* what is to count as a virtue. As far as I can see, there is no inconsistency in the conjunctive assertion that one holds a given quality to be a virtue but that one does not regard that quality as a necessary or a sufficient condition of enjoying the satisfactions of engaging in a practice or practices.

James Wallace ([11], esp. Chs. I and IV) offers a naturalistic conception of virtues and vices. Those qualities are virtues that, roughly, are necessary or desirable for human flourishing, a kind of flourishing that is set off from other kinds primarily by its taking place in convention-governed communities. Vices are qualities that stand in the way of flourishing. Much of what Wallace has to say about flourishing derives from a biological and Aristotelian perspective. What it is for a human to flourish is to be understood in essentially the same way that a biologist determines what it is for a jellyfish or a coyote to flourish. We observe the beings in their typical circumstances of life and pick out those that are in some ways unhealthy and abnormal from those that are normal and healthy.

This approach *is* a functional one, a biologically functional one. It recognizes that there can be different qualities that contribute in different ways to human flourishing. At the same time, it is observer-oriented rather than agent-oriented; it is aimed more at explanation than at practical wisdom.

The agent, the person living in the convention-ordered community, is not primarily interested in the qualities in other people and in himself that are conducive to, or constitutive of, health and flourishing. He is primarily interested in whether he can trust a person, whether he can count on the person's remaining constant in his attitudes, whether he is likely to be cruel or unjust or cowardly. That he would not be cowardly is a good reason to prefer him as a delegate to a bargaining session with the administration; that he is just is a good reason for nominating him to a judgeship. Good character, from the agent's point-of-view, is not so much character that is seen to be functionally necessary as character that one wants in the choices one must make between persons, and also between possible selves. What one wants when one must choose persons may be, in the last analysis, persons whose qualities are necessary for flourishing; but that would remain to be shown.

The class of practically desirable virtues is, I suspect, wider than, and inclusive of, the class of virtues necessary for flourishing in a biologically-understandable health/normality sort of way. There are different conceptions, inevitably, of what it is to flourish, once one moves beyond the biological level. To confine the virtues to the qualities conducive to health or normality is to suppose that the concern of those people who use the language of virtue is confined, or ought to be confined, to the commendable ends of encouraging health and normality in the human community.

One could go on to an examination of Peter Geach's defense ([2], p. 177) of the traditional narrowing of the frame, or to Philippa Foot's suggestion [1] that virtues are correctives to human nature. But perhaps enough has been said to show that the burden should be on any such reductive definitions to justify the exclusion of wide categories of dispositions that serve as reasons for preference of persons who have them over persons who do not.

CONCLUSION

Philosophers who would limit the virtues to one or a very few, as Plato and Kant do, or who would characterize the virtues as dispositions that have this or that function or special character, as many contemporary writers do, are taking the position that only some of what appear to be virtues are such. Only some of the apparent grounds for preference and choice of persons are grounds. This is reasonable enough, but what is needed is an argument that the excluded grounds are only apparently such. That is what is lacking in contemporary, and was lacking in much ancient, discussion. It is simply not clear why some philosophers should exclude aesthetic or instrumental qualities, e.g., from virtue, whereas other philosophers would make them stage-center. It is this characteristic of discussions of virtue, I think, that has alienated most moral philosophers from the ethics of character, and has led those psychologists and philosophers interested in moral education to ignore what Kohlberg [4] has called the 'bag of virtues' approach in favor of emphasis on rules, principles, rights, duties, and obligations.

I have suggested that the subject of virtue need not be one that is either arbitrarily simplified or hopelessly amorphous. It is possible to approach the problem of classifying and evaluating the relative weight of virtues by examining the most general grounds on which it is possible to base preference for or avoidance of persons, thinking of persons not as quasi-legal abstractions or causal forces for good or bad, but as beings that have certain sorts of disposition. I have also suggested why the 'moral' virtues have seemed especially important to philosophers and educators, and why justice has a central place among the moral virtues.

APPENDIX I

*Some Personality Traits**

able
affable
affected
affectionate
agreeable
alacritous
amiable
ambitious
argumentative
arrogant
avaricious
belligerent
benevolent
bilious
boastful
brainy
brave
brooding
bullying
calculating
cantankerous
callous
careful
cautious
changeable
charitable
cheerful
claustrophobic
clever
civil
cold
complacent
complaisant
conceited
condescending
conscienceless
conscientious
contrary
co-operative
cosmopolitan
courageous
courteous
courtly
covetous
credulous
cruel
curious
dastardly
decent
dedicated
dependable
devout
dictatorial
dignified
disciplined
disgraceful
distinguished
distant
dominating
domineering
ebullient
energetic
enigmatic
enterprising
envious
equable
euphemistic
excitable
exhibitionistic
fair-minded
fearful
flattering

* List compiled by Robert Audi and Edmund Pincoffs.

foolish
foresighted
forgiving
frank
friendly
frugal
gallant
gay
generous
gentle
good-living
good-natured
good-tempered
grouchy
helpful
hasty
honest
hospitable
humble
humorous
idle
imperturbable
impulsive
independent
inflexible
ingenious
ingenuous
inscrutable
intelligent
interesting
irritable
jingoistic
jocular
jovial
just
kindly
knightly
lazy
learned
lighthearted
lively
lustful
magnanimous
manly
mannerly
meek
mercurial
methodical
mercenary
mixer
moderate
modest
morbid
nefarious
neurotic
odd
openminded
opinionated
orderly
patronizing
peevish
persistent
pessimistic
petulant
phlegmatic
picayune
polite
pompous
prescient
pretentious
presumptuous
proud
prudent
quarrelsome
queer
querulous
quick-witted
quizzical
reasonable
refined
reliable
religious
repressed
reserved
respectable
respectful
revengeful
secretive
self-confident
self-contained
self-controlled
self-disciplined
self-indulgent
selfish
self-pitying
self-reliant
self-respecting
self-satisfied
sensitive
serious
shy
silly
simple
simple-minded
sensible
sinful
slavish
sober
spoiled
stimulating
stubborn
stupid
submissive
suggestible
sympathetic
tactful
tender
tender-minded
tense
thoughtless
thoughtful
threatening
thrifty
timid
tolerant
torpid
truthful
uncomplaining
undemonstrative
understanding
uxorious
vigorous
vindictive
virile
vivacious
waspish
watchful
weak
weak-minded
watchful
well-intentioned
whimsical
wise
withdrawn
witty
worrying
xenophobic
yielding
youthful
zany
zealous

APPENDIX II

- *Personality Traits*
  - *Virtues*
    - *Instrumental*
      - *Agent-instrumental*: persistence, courage, alertness, carefulness, resourcefulness, prudence, energy, strength, cool-headedness, determination
      - *Group-instrumental*: cooperativeness, 'practical wisdom' and the virtues of leaders and followers
    - *Non-Instrumental*
      - *Aesthetic*
        - *noble*: dignity, virility, magnanimity, serenity, nobility
        - *charming*: gracefulness, wittiness, vivaciousness, imaginativeness, whimsicality, liveliness
      - *Meliorating*
        - *mediating*: tolerance, reasonableness, tactfulness
        - *temperamental*: gentleness, humorousness, amiability, cheerfulness, warmth, appreciativeness, openness, eventemperedness, non-complainingness, non-vindictiveness
        - *formal*: civility, politeness, decency, modesty, hospitableness, non-pretentiousness
      - *Moral*
        - *mandatory*: honesty, sincerity, truthfulness, loyalty, consistency, reliability, dependability, trustworthiness, non-recklessness, non-negligence, non-vengefulness, non-belligerence, non-fanaticism
        - *non-mandatory*: benevolence, altruism, selflessness, sensitivity, forgivingness, helpfulness, understandingness, super-honesty, super-conscientiousness, super-reliability

NOTES

[1] This essay includes parts of a previously published paper: See [8]. The author gratefully acknowledges permission given by *The Monist* to reprint the relevant portions of the paper.

[2] Cf. Appendix II.

[3] Cf. Herbert Morris's suggestive remarks on fairness [6]. It is for me still an open question whether there is any satisfactory way of setting off 'moral' from other virtues. In the paper referred to in Note 1, I have given reasons for suggesting that the disposition not to take unfair advantage is at least morally central.

BIBLIOGRAPHY

[1] Foot, P.: 1978, 'Virtues and Vices', in *Virtues and Vices and Other Essays in Moral Philosophy*, University of Calfornia Press, Berkeley and Los Angeles, pp. 1–18.

[2] Geach, P.: 1977, *The Virtues*, Cambridge University Press, Cambridge.

[3] Hunt, L. H.: 1978, 'Character and Thought', *American Philosophical Quarterly* 15 (July), 177–186.

[4] Kohlberg, L.: 1970, 'Education for Justice: A Modern Statement of the Platonic View', in N. F. and T. R. Sizer (eds.), *Moral Education: Five Lectures*, Harvard University Press, Cambridge, Massachusetts, pp. 56–83.

[5] MacIntyre, A.: 1981, *After Virtue: A Study in Moral Theory*, University of Notre Dame Press, Notre Dame, Indiana.

[6] Morris, H.: 1968, 'Persons and Punishment', *The Monist* 52, 4, 475–501.

[7] Mandelbaum, M.: 1969, *The Phenomenology of Moral Experience*, Johns Hopkins Press, Baltimore.

[8] Pincoffs, E.: 1980, 'Virtue, the Quality of Life, and Punishment', *The Monist* 63 (April), 38–50.

[9] Ryle, G.: 1949, *The Concept of Mind*, Hutchinson's University Library, London, New York.

[10] von Wright, G. H.: 1963, *The Varieties of Goodness*, Humanities Press, New York.

[11] Wallace, J. D.: 1978, *Virtues and Vices*, Cornell University Press, Ithaca and London.

University of Texas,
Austin, Texas,
U.S.A.

KAI NIELSEN

CRITIQUE OF PURE VIRTUE: ANIMADVERSIONS ON A VIRTUE-BASED ETHIC

I

Goal-based ethical theories, duty-based ethical theories and rights-based ethical theories have all been well represented and well canvassed during the modern era. But it has also become evident, particulary since the extensive examination of Rawls', Dworkin's and Nozick's views, that none of these accounts are without very fundamental difficulties – difficulties which are not just difficulties in detail but difficulties in the basic structure and the programmatic intent of such theories. Just as with the deadlock in ethical theory of some twenty years ago there were scattered voices telling us to go back to Kant, so in our present circumstances it is understandable that some should try to return to a virtue-based ethics.

Virtue-based ethical theories in a way go back to Aristotle. We have with them a turning away from an ethics of principles, including an attempt to find the supreme principle of morality such as we find in Kantian or utilitarian theories. Kantian and utilitarian theories take the central task of moral theory to be the formulating and justifying of fundamental moral principles or principles of human conduct which would guide both individual and collective choice. A virtue-based ethics, by contrast, seeks to delineate the ends of human life (the good life for man) and to characterize what it is to be a good person. On such an account we find out what it is to be a good person and what are the ends of life by by finding out what the distinctive human virtues are. This is the key, we are told, to discovering what human flourishing is. It is because of this that such an ethics of ends is called a virtue-based ethics. Where, in a goal-based theory or a duty-based theory, we have an ethics of principle, virtue is an ancillary concept. Virtue, on such an account, is characterized in terms of the disposition to act on principles of right conduct. Virtue-based theorists, following Aristotle, are distrustful of such gestures in the direction of precision. What we need instead is a theory of the virtues explaining the good for man and what it is to be a good person. We, in turn, will, in many circumstances at least, come to understand right action in terms of what a good person would do.

There has of late been a sprinkling of newly minted virtue-based theories:

Earl E. Shelp (ed.), Virtue and Medicine, 133–150.

James Wallace's careful and insightful *Virtues and Vices*, Phillippa Foot's lead essay in her collection of essays with the same title, and Peter Geach's *The Virtues*, But to my mind, the most significant and the most challenging of them all is Alasdair MacIntyre's *After Virtue*. It is a historicized Aristotelianism jettisoning Aristotle's metaphysical biology and his conception of the function of man. Employing a distinctive moral methodology, it uses, much more than traditional moral philosophy, historical analysis, a narrative method and the human sciences to first critique the dominant goal-based, duty-based and rights-based traditions in ethical theory and then to present his own positive alternative account – his historicized Aristotelianism.

It is with this positive account that I shall be concerned here. Since I am not inclined myself to take a virtue-based turn, though I am not disinclined to use some elements of it, I turn to a critical examination of MacIntyre's account as constituting what I take to be the most significant attempt, with which I am acquainted, to develop such a theory.

II

Before I turn to critique let me set out the bare bones of his account with the warning that this can hardly begin to convey the nuance and the subtlety of MacIntyre's view.

MacIntyre believes that not only moral philosophy but morality itself in our time is in disarray. Indeed, the disarray of morality and moral philosophy go hand in hand, for MacIntyre would have us believe, we cannot properly understand a moral philosophy without understanding its social embodiment in a culture. Morality, for the Greeks, for the Icelanders represented in the Sagas, and for the Medievals was, MacIntyre believes, whole, but in our culture it is no longer whole and our moral philosophers in their attempts to understand morality are, like philosophers trying to understand science after, because of some great catastrophe, a scientific culture has disappeared for several centuries. Such philosophers, living after its disappearance, would be trying to piece some understanding of it together from the fragmentary accounts still available to them of what it was like. Our moral philosophers, MacIntyre believes, are people with analogous disabilities; they have available to them no more than fragments of a conceptual scheme which has lost its context – a context which once made that conceptual scheme intelligible but which we now have lost.

To try to make it intelligible our philosophers invent moral fictions like natural rights or utility. In such fragmented conceptual schemes, we come, naturally enough, to use moral utterances to express our emotions and the very idea of moral knowledge becomes a Holmesless Watson. MacIntyre claims that with this employment of moral discourse, we show, and indeed further instantiate, how we have lost our grip on the distinction between treating people as ends and manipulating them. And these conceptualizations in turn have their social embodiment in the bureaucratic manager and the therapist, both elitist paternalists, dedicated, though in different ways, to manipulation to achieve certain ends which themselves are never, and never can be, rationality defended.

MacIntyre thinks that there is but a slight chance for us to escape this cultural condition, but to the extent that there is a way, it is, he believes, through recapturing something of the Aristotelian notion of the virtues. We have lost our firm sense, a sense that came naturally to the Greeks and the Medievals, of what the virtues are. MacIntyre develops the notion of a practice – a cooperative activity in pursuit of goods internal to that acitvity – to explicate the virtues and their role in the moral life. Our various social roles, when they actually are engaged in, are practices such as being a parent, a teacher, a partner or ombudsman.

We not only have practices which, with their internal goods, define virtues, but we need as well some conception of a human life as a whole which like a narrative would have some unity. The making sense of our life as a whole comes to seeing its *telos* as it is revealed when we come to see the narrative unity in our lives. And this means that we need, as well, to recapture an understanding of tradition in which we see that we are what we are in large part because of our history, though this does not mean that we cannot be critical about the traditions which mold us, though we must also recognize that the very direction our criticality can take us is in turn determined by these traditions.

The virtues are necessary in the sustaining of traditions, traditions that in turn make possible a life in which the good for man is realized. But what is this good for man? It is a life with the unity of a narrative quest, a life, which, as MacIntyre puts it himself, is "spent in seeking for the good life for man, and the virtues necessary for the seeking are those which will enable us to understand what more the good life for man is" ([3], p. 201). To make sense of our lives, to make sense of morality, we must, in a way that is almost impossible for people caught in the culture of liberal modernity, see our lives as a unity, see our individual lives as a whole. To do this we need to have a full-fledged narrative understanding of our lives; with such an understanding, it is possible, though for us extremely difficult, to come to an understanding of the good of a human life as a whole viewed as a narrative unity.

III

In the preceding section I gave you the core of MacIntyre's historicized Aristotelianism. I now want to turn to reflective commentary and critique. I am inclined to believe, where it is really crucial, where MacIntyre really needs to deliver, he doesn't deliver the goods, that his account is as empty or at least nearly as empty as the liberalism he despises. It may be, however, that I am asking too much, expecting something which is too determinate where that expectation is unreasonable.

The above remarks without any elucidation are a cluster of dark sayings. I will try to make them clear as I go along. In querying MacIntyre, as I am about to proceed to do, I want to make one thing perfectly clear at the outset. I think he asks the right questions or, to put it both more guardedly and more adequately, I think he, where for years we have neglected these questions, forces us to ask some very old and some very important questions that contemporary moral philosophy has been the poorer for not asking.

MacIntyre maintains that what we need to articulate and persuasively defend is some reasonable account of an "overriding conception of the *telos* of a whole human life" ([3], p. 188). We need to start with a recognition of how practices define the virtues, but to gain an adequate understanding of morality and the place of the virtues in morality we need to go beyond a careful attention to practices and even to traditions to an understanding of the ends of life. We need, if we can get it, some reasonably determinate conception of the good of a human life conceived as a unity. Without this being the case, MacIntyre contends, both "a certain subversive arbitrariness will invade the moral life" and it will also be the case "that we shall be unable to specify the context of certain virtues adequately" ([3], p. 188). Moreover, he further contends, we shall not have provided any viable alternative to the typically goal-based but sometimes duty-based or rights-based Enlightenment tradition and to liberalism, traditions he has argued are bankrupt.

It may well be, MacIntyre to the contrary notwithstanding, that not everything is lost if we cannot articulate some common conception of the good, for it may be that ethics in the form of a system of coordinative guidelines will still be of a not inconsiderable import in enabling us to forge forms of cooperation that will give some coherence to our lives together even though we do not have much in the way of any common conception of the good. But still a lot would be lost if we are incapable of specifying, and making a social reality, both a reasonably determinate and a rationally vindictable conception of the good of a human life conceived as a unity. To achieve this, MacIntyre argues, we must understand human action. And to do this – to render human action intelligible – we must provide an alternative, more holistic understanding than the reigning atomistic conception which tries to analyze actions in terms of some conception of 'basic action'. To make actions intelligible we need to see them as a part of an ordered narrative sequence at least in part understood by the agent. In understanding this narrative sequence, it is important (a) to recognize the agents will have some primary intentions and (b) for us, the spectators, to understand what those intentions are. It is principally these primary intentions which give both the narrative and the actions which are part of it a teleological cast. We need, to make sense of our actions and our lives, to see them as having a narrative unity, including some image of the future in terms of which our actions tend to be ordered. It is important for each of us to know the stories of which we find ourselves a part. Our personal identity is a social identity in which we find ourselves in some enacted narrative of which we are a part.

MacIntyre thinks, or at least seems to think, that if we come to accept his view of what intelligible actions are (with its rejection of atomism), come to accept his view of personal identity, his views on how our lives are enacted narratives and his views on the importance and role of tradition in morality and in life more generally, we will come to believe that he has explained to us in what the unity of human life consists and how it is that there is a distinctive human *telos*. I am inclined to accept *something* like his account of the above matters but I do not see how they are sufficient to give us a sense of what the unity of human life consists in or of what our human *telos* consists in, if indeed we even have that sort of thing. What I am suggesting is that we can agree with him about his characterization of human action and personal identity and still be very skeptical about whether that will do much to solve his problem about giving an objective characterization of what the good for man is or even help make plausible that there is such a thing. We might even agree with him about his very general conception of what the unity of a human life consists in and still doubt that he has given us any determinate theory or even a conception of what the good for man consists in or what our human *telos* is.

The unity of a person's life, according to MacIntyre, would consist in "the unity of a narrative embodied in a single life" ([3], p. 203). What is good for that person is how she could best live out that unity and bring it to a completion. To ask, 'What is the good for man?' "is to ask what all answers to the former question must have in common" ([3], p. 203). To see what the human good consists in (what the end of human life is) would come to giving the correct answers to this question. (We must not forget, in examining this question, that there will be a not inconsiderable number of people who will either deny that there is something called 'the human good' or be skeptical about its reality.)

However, even with MacIntyre's appreciation of the import of tradition, even with his holistic understanding of what an intelligible action is and his understanding of personal identity, how are we to specify in any reasonably determinate way what this human good is? How can we, or can we, even specify what the good of a human life is?

Suppose I try for myself. After all, I should know myself better than anyone else. I view my life as a narrative, I ask myself what have I been doing and with what intentions, how have I been relating to others and what is the point of these various activities and the various relations into which I enter, what were my primary underlying intentions in engaging in such activities, what kind of unity do they have? How am I to sum them up and bring them to a completion to give unity and point to my life?

Suppose I do put something like this quite personally and non-evasively to myself as I, or anyone else, would have to to make the question at all real, to make, that is, the question have any real thrust or point. But what am I to say? There are a number of primitive certainties with which I could start. There are a number of people around me who regularly in one way or another enter into relations with me. Do I respond to these people or relate to them in a decent way and with kindness, understanding, and with a genuine caring for them as persons or am I largely indifferent to them or do I manipulate them or treat them with callousness or arrogance? (What I just called 'primitive certainties' could just as well have been called, à la Rawls, 'very deeply embedded considered convictions'.) A lot of evaluative terms are coming into play here and *sometimes* their meanings are *somewhat* troublesome and certainly we would have a lot of trouble, in every case, with their definition. But remember that useful definitions are about the last thing we can give after we have fully mastered not only operating *with* the terms expressive of these concepts but after having mastered operating *upon* them as well. It is a Platonic fallacy to think we do not understand a concept until we can define it. So I use in the above remarks terms like 'decency', 'kindness', 'integrity', 'caring', 'indifference', 'manipulation', 'callousness', and 'arrogance'. In some contexts these concepts can be tricky but I think in the context in which I used them I could in most instances in most situations perfectly well know whether I had acted in any of these ways. There is, of course, room for self-deception but that is also corrigible. It is one of the primitive certainties (*our* primitive certainties, if you will) that callousness, arrogance, manipulation, and even indifference should be out and that kindness, decency, caring, and understanding are required of a human being.

If I really do these things, if, that is, I act in the way I described above, I have given a certain unity and purpose to my life. But only a certain unity, for I could do those things and still be a lost human being utterly astray in Eliot's Wasteland. I could be a drunk or even a person thoroughly hating myself and convinced, and perhaps rightly so, that my life was a loss and still so relate. Moreover, it is not true that everyone whose life has had the unity of an enacted narrative, not everyone who has lived such a unity and who has brought to a completion with integrity and purpose her life, has lived something that can be correctly called a good life. Some pretty unsavory characters here had such a unity to their lives. Think here of Hitler, Franco, or Stalin. They have lived lives that have had the unity of a narrative quest. They have violated some of these primitive certainties but then, in evaluating these lives, it is these primitive certainties that are carrying the day in our moral evalution and not the fact one's life has the unity of a narrative quest. One's life could be through and through evil and still have such a unity and it could, in certain respects, be a good life and lack that unity.

It could be countered, the 'in certain respects' gives the game away. Suppose I look at my life again and convince myself that I treat those around me with decency, kindness, and integrity. But I know full well that I could, that notwithstanding, still be 'a lost soul'. My life, for all of that, could still lack anything like the unity of a narrative quest. So I ask myself, as you might ask yourselves, how best am I to live my life to give it such a narrative unity and to bring it to an appropriate completion? But there are so many ways I can go here. I have, in a society like ours, with a history like ours, so many role-models. I have nothing like the certainty of the people portrayed in the Icelandic Sagas or even that of the turn of the century Quebec farmers around Lac St. Jean portrayed in *Marie Chapadaline*. I have been a university professor for the greater part of my adult life and that concrete particularity gives me a few additional primitive certainties. I know I must try to teach my classes with integrity. That is, I know, I must try accurately to understand and comprehensively master the subject matter I am trying to teach and then try to convey it in comprehensible and truthful ways. These virtues are goods internal to the practice of teaching. And I also know I must treat my students fairly and, it should go without saying, that the earlier mentioned primitive certainties about kindness, decency, and integrity must obtain in my relations here as well.

Is this enough, if I can really carry it out, to give my life the requisite narrative unity? Some would say so if the other primitive certainties continue to obtain in my family life and the like. Many university professors have so seen themselves, have so picked out such a unity of the narrative quest. Others, and I am one of them, have also seen themselves as intellectuals, as members of the intelligentsia, and have seen this as a central part of their vocation and as determining certain roles – determining certain ways to act and how to relate to others. But not all university professors so view themselves. Some see themselves merely as professionals, members of a certain profession with a certain expertise such as an engineer, an M.D., or a lawyer would view herself.

If the particularity of your life is being a university professor, and, let us say, a philosophy professor at that, how in that area of your life should you view yourself to fill out the narrative of your life? Which way should you fill it out for it to have the narrative unity of a good life? In this domain I have no doubts, subjectively speaking, how I should try to fill it out. But I know that there are others, at least as well educated as I, who see themselves simply as professionals. Which way do we have to go to best live out the unity of a narrative embodied in a single life if we are philosophy professors working in a university in the second half of the twentieth century in North America? I opt for trying to be an intellectual and not merely being a professional with a certain expertise. I think this is essential for an adequate self-definition for a person placed as I am placed, but what reasons could I give for this and how objective would they be?

Let us run with that a bit. I would say I was teaching and trying to understand philosophy and to develop some philosophical notions and I would further contend that the attaining of these things does not merely come to the having of a certain expertise – I am not just around to make distinctions – but very centrally involves the attempt to see steadily and as a whole how things of some human importance and social significance hang together, what sense we could make of our lives together and what it would be like to have more adequate societies and ways of relating to each other such that our lives together would be better lives. Beyond that, I would want to know, if I could, what steps we need to take to achieve such a truly human society. These hedgehoggy questions are not technical questions, though it may be that the answering of some technical questions are not irrelevant to the answering of those questions. These questions go beyond anyone's domain of competence and technical expertise. There is no expert we can turn to go grind out an answer here. It is not at all like asking what conceptions of necessity are essential for understanding modal logic or how material implication is related to our ordinary notions of if then or how is entailment to be understood. These are technical questions and technically trained professionals can come up with the proper answers to them. Yet it is these non-technical questions (the questions about life and society) I raised above that are at the nerve of my own impulse to do philosophy.

Even if, in facing such questions, the last word we could with clarity and honesty give is that such 'questions' do not admit of any kind of genuine answer and that we are only mystifying ourselves and others if we give to understand that, at the end of some long inquiry or some long quest, perhaps carried out over many generations, we would, or at least could, attain answers to them like someday we might find a cure for cancer. But even if we on reflection judge that to be the proper response, we still take it as the response of a certain kind of intellectual and we also recognize, if we know anything, that it is but one of many responses and not, by any means, the only response we could give and that, at any rate, what is the proper response here could not be determined by any profession or even be a matter of some professional expertise.

Since it is intimately a matter of my own self-definition to try to face such perplexities felt as questions, this is an intimate part of my search for a narrative unity in my life. But what if someone says resolutely to me, 'Nothing like this is built into the role of being a philosophy professor. Philosophy is not the name of a natural kind. Look about at your colleagues. They have, to put it mildly, not an inconsiderable variety of rather different conceptions of their role. Why should your conception of your way of living out and completing the narrative unity of your life be the right one? To think that it is is both foolishness and *hubris*.'

If I reply, 'Because it is my life with my enacted narrative so I should be the one to decide how best I might live out that unity and bring it to completion', I have embraced just that individualism and liberalism MacIntyre so detests and thinks, not without reason, is so intellectually and morally bankrupt [4]. Moreover, it also seems to be a false claim, for it does not seem very probable that we are *always* the best judges here of what would be the best for ourselves. It is not very likely that we always best understand what would be the best life for ourselves. We are not always even the best judges of what is in our own interests even on a particular occasion. No matter how anti-paternalist we are resolved to be we need to recognize that. Why should it be the case, or indeed is it the case, that we are always, or perhaps even usually, the best judges of how best to live out our lives so that we, severally, could give our lives a unity and, like a narrative, bring it best to completion? That is a much more complex question than even the rather complex question of judging what in some determinate but fairly complex situation is in our own interests. That each of us, no matter who we are, and how we are situated, could best judge what would give our lives as a whole unity and integrity is, to put it mildly, highly improbable. (What morally we should do in the light of this is another question.)

If, in turn, it is responded 'Oh no, it is not, for there is nothing to be known here or warrantedly believed or reflectively assessed, for such matters are really matters of just *deciding* how we are to live and what sort of persons we are to try to become', we have now fully embraced the non-cognitivism of the tail end of the enlightenment project, a non-cognitivism and decisionalism that MacIntyre was concerned to reject as the confused end product of a fragmented morality [4]. He does not want to say that the unity of my life is whatever I decide to make the unity of my life. He does not want to say that however I forge the unity of my life and bring this unity to completion, then, if that is done with integrity, that is the best life for me. He does not want to have anything to do with such rampant individualism, liberalism, and non-cognitivism.

Still, MacIntyre tells us that "the only criteria for success or failure in a human life as a whole are the criteria of success or failure in a narrated or to-be-narrated quest" ([3], p. 203). But what are the criteria of success here? I am not just a university professor and a philosopher but I am a husband, a father, a Canadian, a socialist, an atheist, an owner of a dog, a writer, and a lot of other things besides. What are the criteria of success of my human life? In answering this I would have to put these various activities into some unity and see them as being woven together in some narrative which would have some appropriate unity and ending. Perhaps I can put this together in a way that I find satisfactory or at least in a way that does not seem to me wildly wrong or alienating. But when I reflect on it in a non-evasive manner I can also see that I could have gone in other directions here, have taken other paths in a yellow wood. I might even have a sense of sorrow that I could not travel them both and be one traveller. But, as I reflect, I would also be aware of a myriad of paths that could be taken, of the many different ways of ordering and completing a narrative. Would I not have further to ask myself 'What reason do I think we have for believing that we have anything even close to an objective criterion for success or adequacy here?' MacIntyre talks of criteria for success or failure of a narrated or to be narrated quest. But he never gives us any sense of what these criteria are or could be. I have taken just one segment of my life, namely, my being a professor of philosophy, in thinking about how it fits into the narrative quest of my life. But even with this one tolerably determinate sequence there seems at least to be no tolerably objective criteria about how I should fulfill that role. Moreover, surely somewhat earlier in my life there were other things I could have been. Perhaps there are other things that I still could be. Besides being a university professor, I am also a Marxist committed to a socialist transformation of society. I care very much about what I am doing in doing philosophy but I not infrequently wonder if I should have done political economy instead (notwithstanding that it bores me) or whether I should have become an M.D. or an engineer and have gone off to some place like Angola and built bridges or spent my time doctoring in the backcountry. If I had it to do over, I am not so sure that I shouldn't have done these other things rather than what I am doing now. (Again, let me ask, as an aside, is it at all plausible to believe that each person, no matter who that person is, can best answer such a question for himself?)

Even with the particulars of my life reasonably well stamped in, I sometimes wonder whether, in my situation, I should abandon or cut down on doing the academic Marxist work I do and become a more directly political creature spending more time involved in actual concrete political struggles in my immediate environment or whether, when I was younger, I should have chucked up academic life altogether and tried to organize workers or to have become a soldier in some liberation army? Some of these, given who I am and what I can do, may be far-fetched but at least some of these are possible ways of narrating out my life. Which of these various possible activities would narrate out, or would have narrated out, my life best and give it (would have given it) the best unity, integrity, and completion?

It is possible to doubt that there are any objective answers here while very much wanting something with some objectivity, if it is to be had. But what would it be like to obtain anything making even a reasonable approximation to objectivity here? MacIntyre does not give us even a hint as to how such an answer is to be found. Here I have been talking about one person, namely, me. When I reflect on what is as obvious as obvious can be, namely, that my life is but one token of a type of thousands of types of ways a human life could be narrated out and given unity, it is possible to get very nervous indeed about 'true narratives' here or about the 'truth of narratives' or about the having of any even remotely adequate criteria for objectivity here.

IV

MacIntyre is not insensitive to such problems of contextuality (to call them problems of relativity begs some questions and exploits some ambiguities). He writes: "What it is to live the good life concretely varies from circumstance to circumstance even when it is one and the same set of virtues which are being embodied in a human life. What the good life is for a fifth century Athenian general will not be the same as what it was for a medieval nun or a seventeenth century farmer" ([3], p. 204). That is all well and good, but if we are to have some determinate conception of a *final telos* for human beings, something MacIntyre agrees with the medievals in thinking we need, we must also be able to ask and answer such questions as these: will a society and an assemblage of human lives which has the role of a fifth century Athenian general, a medieval nun, a seventeenth century English farmer be a better society than one without these roles or with altered roles or quite different roles? Being a nun or a general or a slave or a serf or a proletarian or a lumpen-proletarian or a capitalist – having that possible cultural space – goes with a certain kind of society with a certain set of practices and, as MacIntyre stresses himself, carries with it certain internal and external goods and rather different conceptions of the good life for humans. Would a world without nuns and/or without capitalists be, in conditions of productive abundance, a better world than a world with them? I think, and MacIntyre at one time thought, and perhaps still thinks, that, at least in circumstances of productive abundance, a world without them would be a better world. But, if our judgments are to be nonarbitrary here, we need criteria for such judgments (or so, at least, it would seem), but it is just this that MacIntyre does not provide us with or even make a gesture at how we might discover or construct. But surely answers here are necessary if we are to give an answer to what is the good life for man.

Am I quite right in saying he gives us no hints? Let us examine some very key paragraphs on page 204 of *After Virtue*. We have to be able, he remarks there, to in some reasonably determinate way answer the question, 'Quest for what?' if we are to make any sense out of the notion of the good life having the unity of a narrative quest. And this means MacIntyre avers, that the medieval Aristotelians were right in believing that we must have "some at least partly determinate conception of the final *telos*" ([3], pp. 203–4). MacIntyre believes that this conception of the good for human beings is to be drawn from the questions we ask and what we learn from our "attempt to transcend that limited conception of the virtues which is available in and through practices" ([3], p. 204). When we examine practices, we learn, MacIntyre argues, that they all require trustworthiness, courage, and justice and also, knowing we very much need practices, we rightly conclude that a good life for human beings must contain these characteristics as virtues. It also becomes apparent to us – that is apparent in the history of development of ethics and of moral philosophy – that we are "looking for a conception of *the* good which will enable us to order other goods" ([3], p. 204). But we are also looking for "a conception of *the* good which will enable us to extend our understanding of the purpose and content of the virtues . . . " ([3], p. 204). Thirdly, and lastly, we are looking for "a conception of *the* good which will enable us to understand the place of integrity and constancy in life" ([3], p. 204). Plainly and understandably, as his discussion of Jane Austen makes plain, MacIntyre wants, in addition to trustworthiness, courage and justice, to add integrity and constancy to the list of virtues which must be a fixed part of a good life. Some might say that constancy overstresses the value of a certain unity of the person. Why not give greater weight to the having of intrinsically valuable experiences at a given time and perhaps to the maximizing of such experiences and less to constancy? But even if we do give such weight to constancy and integrity (our considered judgments are likely to pull us along here), it can come to very different things in different contexts. The Inquisition sometimes showed considerable constancy and integrity and so did the Conquistadors even when they brutalized in almost unimaginable ways the Andean and Mexican populations. And similar things have been said about the Black Angel of Auschwitz and a similar case might even be attempted for Hitler or Stalin. Even the virtues of trustworthiness, courage, and, by their own lights, justice could be exemplified in the lives of Inquisitors and Conquistadors, even in those Conquistadors who slaughtered Indians all over the place, melted down their silver and gold religious objects, and drowned while crossing a body of water literally under the weight of the plundered silver and gold with which they were laden. They had a conception of the good and they had these central virtues. Admittedly, these are extreme cases and MacIntyre, no doubt, as much as any other morally reflective person living in our time, or perhaps any time, would reject these things as gross immoralities which could not be a part of the good life for human beings. But, putting him on the side of the angels does not gainsay the fact that the various virtues he has been able to show a rationale for are all capable of being exemplified in such behavior. They all can be seen as being a part of such narrative histories and as being a part of such narrative quests. Recall that for him justice is nothing more than the getting of what you deserve. He rejects Nozick's account, Rawls' account, and more radically egalitarian accounts of justice. Justice, as he characterizes it, could come to many different things in many different contexts. It is, on his understanding at least, a very inderterminate essentially contested concept. Moreover, these extreme cases aside, there are, over cultural space and historical time, and even in our own moral cultures, plenty of exemplifications of situations in which we could have these virtues in place and still have radically divergent and often deeply conflicting conceptions of the good for humans. Moreover, we can and do have very different orderings of the various goods and schedulings of the different virtues.

We indeed would reflectively want to be able to extend "our understanding of the purpose and content of the virtues" ([3], p. 204). We would indeed want to do this in order to have a conception of the good life and of what a critical morality could come to. MacIntyre does give us *something* of that here, though it has a certain daunting vagueness about it. But what he does not do is give us a sufficiently clear understanding of the purpose and content of the virtues so as to give us a reasonably determinate conception of the end for man (man's distinctively human good) even when we bring in the concept of a moral tradition.

MacIntyre might respond that his concept of a quest for the good is not something that should be thought to be adequately characterizable all at once. It is something like a *Bildung* which would emerge, as a kind of moral education, that occurs in the course of the quest in the face of all the "particular harms, dangers, temptations, and distractions" that we will encounter along the way. It is a kind of pilgrim's progress or a Wilhelm Meister's apprenticeship. It is in this way that we gain our moral education and through such an apprenticeship, as our self-knowledge grows, the goal of the quest finally is understood.

Should not the response be this: Such moral education has been going on for a long time and, except where we have had very sheltered and homogeneous societies, e.g., in our reconstruction of the Heroic Age and in certain, but no means all, primitive communities, we have not attained a consensus about the good for man, we have not obtained a consensus, let alone anything close to what we could characterize as a rational consensus, about what, if anything, our final *telos* is where that notion is given a reasonably determinate content. And the various *consensi* of limited communities have just been such limited and varied *consensi*, local affairs both temporally and spatially.

The virtues, it is surely at least plausible to maintain, are "those dispositions which will not only sustain practices and enable us to achieve goods internal to practices, but which will also sustain us in the relevant kind of quest for the good, by enabling us to overcome the harms dangers and distractions we encounter . . ." ([3], p. 204). Though this may be how to characterize virtue, still MacIntyre's virtue-based moral theory has not told us what the good is. The increasing self-knowledge we gain from our increasing understanding of the virtues is supposed to give us a better understanding of the good and it indeed does give us an understanding of some elements (the virtues I have been adverting to) of the good, but, as we have also seen, it is still far from taking us to a knowledge of the good that is also a knowledge of our final *telos* or of just our *telos sans phrase*. We still do not know what that is or what it would be like to attain a knowledge of such a *telos*. Indeed we cannot even be confident that such a conception makes sense. So when MacIntyre remarks that we "have arrived at a provisional conclusion about the good life for man: the good life for man is the life spent in seeking the good life for man, and the virtues necessary for the seeking are those which will enable us to understand what more and what else the good life for man is," he has not told us very much ([3], p. 204). It may even be a mistake to place such a weight on the *seeking* instead of the *having*. A good life for humans might very possibly be one in which there was not much to be done on the questing side for what was taken to be the good life was (sociologically speaking) fairly secure. Given that security, a person could turn her·creative powers to other things. But that plainly contentious point aside, without a better idea of what successful seeking would come to here than MacIntyre has been able to give us, we are looking at best for the holy grail and at worst for the color of heat. MacIntyre understandably wants something more determinate by way of the knowledge of the good than what the reigning liberalism and individualism has been able to give us. But here at least he has not been able to deliver on that.

V

In arguing as I have, I have not rejected MacIntyre's insightful understanding of the role of tradition in morality and in our social-political life and his claims about the need to start from the particularities set in part by our varied traditions. (The particularities in question will, to a not inconsiderable extent, vary with what tradition we are in and with where we stand in that tradition and with other contextual features distinctive of our cultural and historical situation.) Nothing I have said in the previous sections commits me to a search for a *purely* universal conception of the good life for human beings that would try massively to set aside distinctive historical identities in determining the good for human beings. But I have maintained that, along with these contextually variable elements, there must, for such an appeal to be viable, be a sufficiently universally determinate conception of the good for humans so as not to so mire us in a historicism such that we are deprived of any critical vantage point in accordance with which we can assess societies or whole moral traditions ([3], pp. 205–6).

Also, nothing I have said would commit me to siding with J. L. Austin against MacIntyre over the following central consideration:

> It has often been suggested – by J. L. Austin, for example – that *either* we can admit the existence of rival and contingently incompatible goods which make incompatible claims to our practical allegiance *or* we can believe in some determinate conception of *the* good life for man, but that these are mutually exclusive alternatives. No one can consistently hold both these views. What this contention is blind to is that there may be better or worse ways for individuals to live through the tragic confrontation of good with good. And that to know what the good life for man is may require knowing what are the better and what are the worse ways of living in and through such situations. Nothing *a priori* rules out this possibility; and this suggests that within a view such as Austin's there is concealed an unacknowledged empirical premise about the character of tragic situations ([3], p. 208).

I do think there are tragic confrontations between goods and there are also tragic situations in which our best moral choice is the lesser evil. But, as MacIntyre concedes, there are better and worse ways to respond in such situations. We are not left here with utter incommensurabilities. In saying this I do not mean to disagree with MacIntyre that there are tragic situations where we must just choose between evils. My complaint is that he has not given us a sufficiently determinate conception of the good for humans to give us much of a basis for any beliefs we might come to have about what those better and worse ways are when we have to choose between evils. We do not know, from what he tells us, how they are even remotely to be determined here. Sometimes the choice of the lesser evil also involves the choice of what in that situation is the greater good. We indeed should recognize in such situations that "both of the alternative courses of action which confront the individual have to be recognized as leading to some authentic and substantial good," but we do not have a sufficiently determinate conception of a core concept of the good for humans for us to use it to determine in such a circumstance which of several responses that we characteristically make is the more appropriate.

VI

In spite of what I have argued is a central failure of MacIntyre's Aristotelianism, I would not want to maintain that it is as centrally and as irretrievably flawed as is traditional Aristotelianism with its metaphysical biology and its conception of the function of man. Perhaps someone working out of that tradition, demythologized in something like the direction in which MacIntyre has demythologized it, perhaps supplementing it with a theory of needs, could articulate and rationally defend a more determinate conception of the human good that was neither ethnocentric nor as empty as MacIntyre's conception. I do not see any *a priori* objections against it, though it is also reasonable to entertain considerable skepticism about the likelihood that such a research program will pan out. However, I think anyone trying to work it out or anyone setting himself to do moral philosophy period would do well to accept the following core claims of MacIntyre:

> . . . if [as it does for MacIntyre] the conception of a good has to be expounded in terms of such actions as those of a practice, of the narrative unity of a human life and of a moral tradition, then goods, and with them the only grounds for the authority of laws and virtues, can only be discovered by entering into those relationships which constitute communities whose central bond is a shared vision of and an understanding of goods. To cut oneself off from shared activity in which one has initially to learn obediently, as an apprentice learns, to isolate oneself from the communities which find their point and purpose in such activities, will be to debar oneself from finding any good outside of oneself ([3], p. 240).

I think this is right. Anything else would hardly lead to or leave us with any moral understanding at all. Indeed it would not even allow us to have what MacIntyre calls a powerful Nietzschean moral solipsism. But while what MacIntyre characterized above is essential for moral understanding and moral culture, it will not give us, as I have argued, anything even remotely like an objective conception of the good for man. But, that notwithstanding, it will provide us with a good starting point.

BIBLIOGRAPHY

[1] Foot, P.: 1978, *Virtues and Vices*, University of California Press, Berkeley.
[2] Geach, P.: 1977, *The Virtues*, Cambridge University Press, London.
[3] MacIntyre, A.: 1981, *After Virtue*, University of Notre Dame Press, Notre Dame.
[4] Nielsen, K.: 1981, 'Linguistic Philosophy and "The Meaning of Life"', in E. D.

Klemke (ed.), *The Meaning of Life*, Oxford University Press, Oxford, pp. 177–204.
[5] Wallace, J.: 1978, *Virtues and Vices*, Cornell University Press, Ithaca, New York.

University of Calgary,
Calgary, Alberta, Canada

GILBERT MEILAENDER

THE VIRTUES: A THEOLOGICAL ANALYSIS

There is no single best way to think about or explicate the concept of virtue; there are many perspectives which may be useful for one who seeks to understand or (what is far harder) develop virtue. But one way, a way which both affirms the importance of the virtuous life and challenges our attachment to virtue, is the way of theological analysis. It is from that perspective, from the viewpoint of a Christian theologian, that I seek to consider the place of virtues in the moral life – to appreciate both the self-mastery which the virtues promise, but also the sense in which for the Christian it is always dangerous to think of our virtue as a possession.

For Christian theology there can be little doubt of the authority of the Bible. Even if we stress the authority of the sovereign God or the risen Christ, the language Christians speak about God and the beliefs they affirm about Christ have their root in the Scriptural word. Even if we emphasize the centrality of Christian experience, that experience is itself shaped by biblical narrative and teaching. Even if we seek to learn from and respond to the needs of our world, we view and interpret that world through the prism of the Bible. Even if we emphasize the indispensability and authority of the Church's tradition, it is still a tradition which seeks to unfold and develop what is conveyed in the biblical record.

To say that the Bible is an authority which the Christian theologian must heed is not to say that this authority can solve all theological difficulties or by itself provide an answer to every question. It would be almost as true, in fact, to say that many of the important theological problems are raised by the Bible. They become problems precisely because they are found in an authoritative text and so must be taken seriously by the Christian. Without such authorities we have few difficulties; we think whatever we please. But to submit one's reflection to the authority of the Bible is to confess a willingness to consider certain issues as problems just because the texts raise them. Something like this is true of Christian reflection about virtue.

Perhaps no book of the Bible han been more constantly in use among Christians than the Psalms. Important in the liturgical life of the ancient Hebrews, it has been no less significant for Christians. Used regularly in the Church's worship, a source of inspiration for much Christian prayer

Earl E. Shelp (ed.), Virtue and Medicine, 151–171.

and poetry, often committed to memory – the language of the Psalter has been of inestimable influence in Christian piety and theology. It is, therefore, of some importance to consider the image of human virtue which many of the psalms convey.

> Who shall ascend the hill of the LORD?
> And who shall stand in his holy place?
> He who has clean hands and a pure heart,
> Who does not lift up his soul to what is false,
> and does not swear deceitfully.
> He will receive a blessing from the LORD,
> and vindication from the God of his salvation (24: 3–5, RSV).

The virtuous human being is one who can with some confidence stand before God, whose character can withstand the penetrating judgment of the Almighty.

"Thou hast upheld me because of my integrity, and set me in thy presence forever" (41:12). Here again is the authentic voice of many a psalmist, the affirmation that it is virtue which fits one for entry into the divine presence. And this affirmation, in the mouth of one who seriously proposes to pursue such virtue, can suggest a confident self-mastery which even God must recognize.

> Vindicate me, O LORD,
> for I have walked in my integrity,
> and I have trusted in the LORD without wavering.
> Prove me, O LORD, and try me;
> test my heart and my mind (26: 1–2).

Even more striking is the sense the psalmist sometimes conveys that there is no hidden corner of the self in which vice lurks, the confidence that he can see himself, as it were, with the eye of God.

> Hear a just cause, O LORD; attend to my cry!
> Give ear to my prayer from lips free of deceit!
> From thee let my vindication come!
> Let thy eyes see the right.
> If thou triest my heart, if thou visitest me by night,
> if thou testest me, thou wilt find no wickedness in me;
> my mouth does not transgress.
>
> My steps have held fast to thy paths,
> my feet have not slipped

As for me, I shall behold thy face in righteousness;
When I awake, I shall be satisfied with beholding
thy form (17: 1–3, 5, 15).

Reading these and similar psalms we might wonder how the Psalter could have played such a prominent role in Christian prayer and worship. Such a confident affirmation of meritorious achievement might seem inappropriate on the lips of those whose central affirmation is "While we were yet helpless, at the right time Christ died for the ungodly" (Romans 5:6). From this perspective any virtue we possess can hardly amount to a righteousness which would fit us to stand before God. St. Paul writes that Christ died for the *ungodly*. If, therefore, we can stand within God's presence, it is because in Christ God has come to sinners and shown himself willing to count Christ's virtue as our own.

One might respond by suggesting that virtue in the New Testament is something quite different from virtue in the Old Testament. In the psalms we have cited, virtue involves self-mastery and such development of character as makes us fit to withstand God's judgment. In the passage from Romans, virtue means our acceptance by God for Christ's sake. However, to set Old and New Testaments against each other in this way would be mistaken; for there are other voices and accents in the Psalms than those we have noted thus far.

When our transgressions prevail over us,
thou dost forgive them.
Blessed is he whom thou dost choose and bring near,
to dwell in thy courts!
We shall be satisfied with the goodness of thy house,
thy holy temple (65: 3–4).

Here the psalmist does not stride confidently into the presence of God; he is conscious of being forgiven, chosen, brought near by God. He is satisfied not with his integrity but with the goodness of God's house. Again in Psalm 51, a penetential psalm often used by Christians, the psalmist recognizes that God desires "truth in the inward being" (v.6) and that such an inner virtue must be God's work. "Wash me thoroughly from my iniquity, and cleanse me from my sin" (51:2). In Psalm 17 we found a confidence of freedom from any hidden vice; in Psalm 19 we find a recognition that we can never see ourselves as God does – whole and entire. "But who can discern his errors? Clear thou me from hidden faults" (19:12). From this perspective there can be little confidence that we *possess* virtue; instead, we live in hope.

If thou, O LORD, shouldst mark iniquities,
LORD, who could stand?
But there is forgiveness with thee,
that thou mayest be feared.

O Israel, hope in the LORD!
For with the LORD there is steadfast love,
and with him is plenteous redemption (130: 3–4, 7).

The most causal examination suggests, then, that it would be far too simple to play off Old against New Testament. On the contrary, a Christian theologian examining the concept of virtue, who acknowledges the Bible's authority and wants his thinking to reflect that acknowledgement, now has a problem. He must come to terms with two seemingly different understandings of human righteousness. He must ask whether human virtue consists in the gradual development and mastery of character traits the goodness of which even God must admit, or whether our virtue is simply that a forgiving God dwells with us even in our sinful condition.

To be a person of good moral character is not merely to have certain virtues, though it is that; the idea of character suggests something less piecemeal, a fundamental determination of the self for which the agent is to be praised and for which he can take a certain amount of credit ([2], pp. 12ff). This suggests that nothing could be more important than that we should – like Socrates – tend the soul and examine our life. Thinking about virtue directs our attention inward upon the self and its capacities for self-mastery and self-realization. The stability that comes with the development of character – and characteristic behavior – may suggest that virtue becomes a possession of the person who has achieved such self-mastery. Yet, as we have already noted, some central Christian themes challenge this emphasis upon the examined life, suggest that it may ultimately be a self-defeating endeavor, and hold that not self-mastery but grace is the prerequisite for virtue.

The same ambivalence appears if we focus upon the *development* of character. One of the recurring themes in the literature of moral education is that, if character involves habitual behavior, it can be developed by a kind of moral exercise. Our being is shaped by our doing. By hitting the baseball often enough I may become proficient at it. Similarly, by facing danger regularly I may become courageous. There are, of course, alternative theories – Rousseau disliked habit about as much as Aristotle recommended it – but it is hard to deny that all our efforts at moral education draw on the importance of habit to some degree. Yet, there is a gap between the deed and the person, between habitual behavior and character. Facing danger regularly may teach me to discipline the inclinations which urge me to flee, but can it guarantee that I stand firm not because I fear the shame of being branded a coward but because I love the good I defend? This seems to call for a more fundamental transformation of character, and it is not clear that any amount of doing can create such being.

To see what is theologically at stake here we will examine how one theologian dealt with the concept of moral virtue, attempting to take seriously its two quite different meanings. For Martin Luther the issue we have posed was of central importance, and we might summarize the central theme of his theology as a claim that the examined life is not worth living. We will consider two of his important writings: his treatise "Against Latomus", and his exposition of Galatians 3:10–14 in his magisterial 1535 Galatians commentary. We see in these writings what a *theological* analysis of moral virtue requires, the central issue it raises. In both writings Luther subjects an ethic of virtue to serious theological challenge; the examined life is given its place, but only within a larger context in which divine grace rather than human virtue is preeminent.

TWO THEORIES OF VIRTUE

In 1519 Luther's writings were condemned by the theologians of Louvain, a condemnation to which Luther replied in the following year. In 1521 one of the Louvain theologians, known as Latomus, published a defense of the original condemnation – to which Luther in turn responded in his treatise "Against Latomus". A good bit of the argument turns on several propositions affirmed by Luther and denied by Latomus: That sin remains in the Christian even after baptism, and that every good work done by a Christian while still in this life is sin. Of these two the latter is obviously the stronger and, on the face of it, more paradoxical claim. We can begin with it.

Part of the debate turns on the question of how properly to interpret certain biblical references. We need not rehearse the intricacies of the debate, but it is worth noting an example in order to appreciate the case Luther is prepared to press. In Ecclesiastes 7:20 we read that "there is not a righteous man on earth who does good and sins not." Our first inclination might be to suggest that this means simply a kind of commonplace truth: No one manages never to fall into sin. Everyone falls short somewhere along the way. Luther insists, however – and we need not worry here about the adequacy of his exegesis – that the passage makes a considerably stronger claim.

It says, he holds, that a righteous person sins even when doing good. There is no one "who sins not when he does good" ([3], p. 183). Our good works are not, in his view, an entirely "spurious righteousness"; they are good "before men" and in the forum of our consciences ([3], pp. 173ff). But they cannot stand the divine judgment; hence "our good works are not good unless His forgiving mercy reigns over us" ([3], p. 172).[1]

This is Luther's famous anthropological maxim: that the Christian is *simul justus et peccator* – simultaneously saint and sinner. And here this maxim does not mean: partly saint/partly sinner. It means entirely saint and entirely sinner – and both at the same time. Our deeds, taken in themselves and subjected to divine scrutiny, must be seen as the deeds of a sinner. But seen as the deeds of one who is 'in Christ', they are the actions of a saint with whom God is well pleased. It is worth noting that one of the reasons Luther holds this view is because he thinks not just in terms of acts but in terms of character. He is constantly concerned not to isolate the deed from the person, to evaluate not individual slices of behavior but the total self. The fruit exhibits the nature of the tree, he says, using one of his favorite examples ([3], p. 209). And again, "a man doing good is a subject which has sin as its attribute" ([3], p. 187). One could scarcely ask for a greater emphasis upon *being*, not just *doing*. This might be said, in fact, to be the old thesis of the unity of the virtues, stated now in the categories of Reformation theology.

At the same time it is true that the way in which Luther stresses *being* is not conducive to some of the central themes of an ethic of the virtues. For what one is, one's being, is determined by the divine verdict. That verdict admits of no gradations – one is either a sinner (when seen and judged in oneself) or a saint (in Christ). And it makes good sense of a sort to say that one is purely passive before this divine verdict. Here there is no stress upon self-realization, much less self-mastery. Virtue is not in any sense one's possession; it is simply a verdict of Another upon whom one must rely. Any stable characteristics which a self may display, any continuities within one's character, are relatively unimportant. Life if not the gradual development of a virtuous self; it is a constant return to the promise of grace. The examined life, if honestly examined, will reveal only that the best of our works are sin – for the fruit exhibits the nature of the tree. And indeed, as Luther thought he had learned by personal experience, too much attention to the examined life can be dangerous; it directs our gaze inward rather than outward to the promise.

The *simul justus et peccator* maxim, taken as literally as Luther sometimes means it, leaves little place for the concerns of an ethic of character. The whole of life is taken out of the human agent's hands by the divine verdict, and there is no real space left for judgments about gradual development of character. The divine judgment is an either/or pronouncement made upon the person, not just upon particular deeds or traits. Yet, even Luther could not always talk this way. There were things he needed and wanted to say which required a different understanding of what it means to be *simul justus et peccator*. There were moments when for Luther this meant: partly saint/partly sinner – moments when he had to recognize that the unity of the virtues was a goal of life, a present reality only in hope.

We can see this when Luther argues against Latomus that sin remains in the Christian even after baptism. Latomus' claim is that after baptism there may be *imperfection* but not *sin* in the Christian. Luther holds that sin remains in the Christian but no longer *rules* the Christian. He wants only to insist that this remaining but not ruling sin should be called by its proper name. When we call it sin we make clear that it still warrants God's condemnation and that we must flee to the promised grace. Thus he can write, still very much in the either/or mode:

> You will therefore judge yourselves one way in accordance with the severity of God's judgment, and another in accordance with the kindness of His mercy. Do not separate these two perspectives in this life. According to one, all your works are polluted and unclean on account of that part of you which is God's adversary; according to the other, you are genuinely pure and righteous ([3], p. 213).

The interesting thing about these sentences is that they combine the two different ways of understanding the *simul justus et peccator* maxim. There is a 'part' of the Christian which is God's adversary and a 'part' which is not – the partly saint/partly sinner model. But since God judges whole persons, not just deeds or character traits, the Christian is also entirely saint (in Christ) and entirely sinner (in himself).

We may notice, however, that only the one way of talking – partly saint/partly sinner – can make place for any talk of development in the moral life. Luther, still thinking and talking both ways, says that baptism removes the *power* of all sins but not their *substance*. "The power of all, and much of the substance, are taken away. Day by day the substance is removed so that it may be utterly destroyed" ([3], pp. 208f). 'Day by day' – here there is room for gradual development of character, room for the Christian's assault, aided by grace, upon the fat relentless ego, room even for the self-mastery which makes progress in virtue possible.

In a very clear and well developed section Luther sets forth both ways of talking ([3], pp. 223ff.). We can chart his position in the following manner:

	God's two ways of dealing with sin:	
	LAW reveals	*GOSPEL reveals*
the person viewed in 'parts' (doing)	corrupt nature	gift of infused faith
the person viewed whole (being)	wrath of God	grace of God

God deals with sin, Luther writes, in two ways – through law and gospel. These two divine verdicts each reveal something about our human nature and something about God's judgment on that nature (and what is shown in the gospel answers a need revealed by the law).

In the law we learn the corruption of our nature. We come to recognize the sinful impulses which lurk within us. This does not mean that we see no good in ourselves; it means only that we recognize the sin that is there and know our guilt. In the gospel, by contrast, we learn of and receive the gift of faith. This faith, which Luther is willing to call an infused gift, begins the internal process of healing the corruption of our nature. "Faith is the gift and inward good which purges the sin to which it is opposed" ([3], p. 227). Faith is that 'part' of the self no longer opposed to God, and it is committed to purging that other 'part' which continues to be the adversary of God.

We see ourselves in part – both in the light of the law, which reveals our sinful nature, and the gospel, which infuses faith and begins the healing process. God, however, also sees us whole. And the contrary judgments of God – judgments not just of deeds or virtues but of persons – we also learn in law and gospel. In the law we learn that God punishes sinners; we learn of God's wrath. This is, Luther writes, more terrible than to learn the corruption of our nature; for this is not a partial verdict, not a verdict which finds something to praise and something to blame. Rather, the judgment of God's wrath is that there is "nothing profitable" even in what seems good in our nature ([3], p. 225). Indeed, Luther specifically notes that this judgment applies even to our virtues – there is "nothing profitable" in our prudence, courage, chastity, "and whatever natural, moral, and impressive goods there are" ([3], p. 225).[2] In the gospel we learn that God's grace is opposed to his wrath, that for Christ's sake he is wholly and entirely favorable toward sinners.

When we think in the first way – of our corrupt nature and the gift of faith which begins to heal that corruption – we can, without denying the importance of God's grace in this process, make room for progress and development in the moral life, for the gradual achievement of moral virtue through effort and discipline, and for the careful examination of our lives and 'tendance of the soul' which Socrates commended. When we think in the second way – of the contrary divine verdicts of wrath and grace upon our whole person – there is room only for a continual return to the word of promise which assures us that the wrath of God has been overcome by his favor. The first model encourages us to think of life as a grace-aided journey and of virtue as a gradual possession which we may come to acquire through the moral discipline to which faith is committed. The second model encourages us to think of life as a perpetual dialogue between the contrary verdicts of wrath and favor, to see the self as passively determined by these verdicts, and to understand faith as the continual return to the promise that grace has triumphed. On the first model we may grow in faith as that 'part' of us which is God's adversary is gradually purged. On the second model faith is always the same – a naked trust in the promise. Luther, himself, offers finally a marvelous summary of his position, a summary worth quoting at length.

Now we finally come to the point. A righteous and faithful man doubtless has both grace and the gift. Grace makes him wholly pleasing so that his person is wholly accepted, and there is no place for wrath in him any more, but the gift heals from sin and from all his corruption of body and soul. It is therefore most godless to say that one who is baptized is still in sin, or that all his sins are not fully forgiven. For what sin is there where God is favorable and wills not to know any sin, and where he wholly accepts and sanctifies the whole man? However, as you see, this must not be attributed to our purity, but solely to the grace of a favorable God. Everything is forgiven through grace, but as yet not everything is healed through the gift. The gift has been infused, the leaven has been added to the mixture. It works so as to purge away the sin for which a person has already been forgiven . . . . In the meantime, while this is happening, it is called sin, and is truly such in its nature; but now it is sin without wrath . . . . To be sure, for grace there is no sin, because the whole person pleases; yet for the gift there is sin which it purges away and overcomes. A person neither pleases, nor has grace, except on account of the gift which labors in this way to cleanse from sin. God saves real, not imaginary, sinners, and he teaches us to mortify real rather than imaginary sin ([3], p. 229).

From either vantage point God's grace is essential for virtue. From the perspective of the first model grace is necessary as an enabling power, infusing the gift of faith which struggles against sin. From the perspective of the second model grace is necessary as a pardoning word, which sees the sinner whole in Christ and therefore sees him as righteous.

What is at stake here – both for Luther and for our own understanding of an ethic of virtue? Luther himself recognizes that many might regard his dispute with Latomus as mere verbal disagreement ([3], p. 236). Latomus says imperfection remains after baptism; Luther says sin remains; both agree that what remains does not condemn the believer in Christ. Why then all the shouting? Because, I think, each sees danger in the position of the other. Latomus finds in Luther one who imperils the examined life, who – because he brands everything as sin – is unable to take seriously the small disciplined steps by which virtue struggles to root out vice from the Christian's life. Luther – while granting the importance of this daily, bit by bit, struggle – finds in Latomus one who fails to see that selves, not isolated deeds or character traits, are what count before God. Luther speaks of sin, not just imperfection, in the Christian in order to make clear that any hope a person may have before God must lie in the undeserved favor of God. All the particular deeds and the developed virtues we may claim do not add up to a righteous *person*. And, perhaps paradoxically, Luther suggests that it is his position rather than Latomus's which will foster true virtue. If we fail to see ourselves as God sees us, he writes, "we cheapen Christ's grace and minimize God's mercy, from which necessarily follows coldness in love, slackness in praise, and lukewarmness in gratitude" ([3], p. 240). That is, the virtues of love, praise, and gratitude are fostered precisely to the degree that we do not try to foster them but, instead, look outside ourselves to the promised mercy. The examined life is not worth living – not only because it can give the conscience no peace before God but because it does not really issue in virtue. When we stop trying to tend our own soul, when we hand that soul over to God and realize that we cannot bring our virtue to God but that he must (in Christ) bring virtue to us, then we experience a kind of liberation which makes possible true virtue, the virtue which comes not from the slow, disciplined transformation of character through development of virtuous habits but from the liberating acceptance of the sinner by a loving God.

This is Luther's fundamental claim. We may perhaps be struck by the fact that, however different the language, the claim is not unlike an understanding of virtue whose pedigree extends back at least to Plato. Plato's thesis that the virtues are one is essentially the observation that character is not a matter of bits and pieces, not a matter of isolated virtues and vices. That thesis is paradoxical, as is Luther's language, but both see a deep truth about virtue. Both see that our virtues are *ours* – and that if the self is not whole, the virtue is specious. Where Luther and Plato differ, the difference lies primarily in the fact that Luther goes farther. Plato's entire system of education sketched in the *Republic* is intended to discipline the soul and bring the person out of the cave and toward the vision of Good. Yet, even at the highest level of Platonic dialectic there remains a gap. The person is brought toward the Good as his character is gradually shaped and disciplined, but the shaping is in bits and pieces while the person is not. Hence, there always remains a question whether dialectic is powerful enough to help a person see the Good. Our attempts to develop virtue can be only partial, piecemeal. But, as Plato realizes, really to see and love the Good one must already be virtuous. It is not clear, therefore, how the gap which separates being from doing can be bridged unless – as Luther thought had happened – the initiative should come from the side of the Good itself. "God in his grace has provided us with a Man in whom we may trust" ([3], p. 235). And so, having been made virtuous in that man, we now love virtue. All our doings, all our bit by bit progress, could not bridge the final gap, could not create a being wholly dedicated to virtue. The tree cannot be made healthy by spraying the fruit. Thus Luther against Latomus: However much piecemeal progress toward virtue we sinners make, ours is still the virtue of sinners – hence, sin. First the person must be virtuous. *We* may develop worthy traits of character; only the divine initiative can create a virtuous self. Character depends finally not upon self-mastery but upon a moment in which one is perfectly passive before God.

THE NEED FOR BOTH THEORIES

Luther's theological standpoint is not just unintelligible paradox. It answers to a deep need in any ethic of virtue. It recognizes our inability to get from virtues to virtue, from traits of character to a transformed self. We should not forget, however, that Luther speaks in two ways. His theology in its most radical moments might suggest that moral education is futile, that the only worthwhile 'moral education' would be a constant return to the promise of divine grace, that there could be little point in the development of virtuous habits of behavior since these could never eliminate the need for divine initiative. Why not, then, simply start with and constantly return to that divine initiative? Why bother ourselves with virtues and the gradual shaping of character?

But Luther does bother. In the treatise "Against Latomus" he had bothered. He had written of gradual progress in faith, of gradually purging that 'part' of the self which is God's adversary. We can see again how important it is for Luther to speak in both ways if we consider a small portion of one of the masterpieces of his mature theology, the 1535 Galatians Commentary. In expounding Galatians 3:10 Luther notices something peculiar in St. Paul's argument. Paul writes that "all who rely on works of the law are under a curse". As support for this claim he cites the Old Testament passage "Cursed be every one who does not abide by all things written in the book of the law, and do them." Peculiar support, one is inclined to think. The Old Testament passage says that anyone who fails to do what the law requires is cursed. And on the basis of that passage St. Paul claims that all who rely on doing what the law requires are cursed. We might suppose that St. Paul can move from

(1) anyone who fails to do what the law requires is cursed, to

(2) all who rely on doing what the law requires are cursed, only by means of some concealed premise such as

(3) no one can do fully what the law requires.

Luther, however, makes a different interpretive move. The move suggested above would imply that *if* anyone could in fact do fully what the law requires, he would be in a position to rely on his doing of the law. The only reason one should not rely on it is because none of us can meet the law's requirements. For Luther, however, the problem goes deeper than this. To 'do' what the law requires while relying on this 'doing' for one's standing before God would not in fact be to do what the law requires. Any true doing of the law calls for faith in the mercy of God, and such faith cannot be present when we rely on our virtue. Thus Luther writes:

> There are two classes of doers of the law. The first are those who rely on works of the Law; against these Paul contends and battles in this entire epistle. The second are those who are men of faith . . . ([4]. p. 253).

And again:

> Therefore 'to do' is first to believe and so, through faith, to keep the Law. For we must receive the Holy Spirit; illumined and renewed by Him, we begin to keep the Law, to love God and our neighbor ([4], p. 255).

And once more, picking up a metaphor prominent in "Against Latomus":

> But because there are two sorts of doers of the Law, as I have said, true ones and hypocrites, the true ones must be separated from the hypocrites. The true ones are those who through faith are a good tree before they bear fruit and doers before they do works ([4], p. 257).

Thus, when we read in the Old Testament "Cursed be everyone who does not abide by all things written in the book of the law, and do them", the *doers* who, it is implied, will not fall under the curse are true doers, what Luther calls 'theological doers'. Their doing presupposes faith.

True virtue, therefore, is not possible as human achievement; it cannot be thought of in terms of self-mastery. Indeed, it requires once again that moment of naked faith in which the self is perfectly passive before God. Our vices, Luther writes, have their source in our refusal

> to be justified by a divine blessing and formed by God the Creator. It [our hypocritical self] refuses to be merely passive matter but wants actively to accomplish the things that it should patiently permit God to accomplish in it and should accept from Him. And so it makes itself the creator and the justifier through its own works . . . ([4], p. 259).

Even the language of 'theological doing' is slightly suspect and easily misunderstood. The faith which is presupposed by all theological doing of the law is a gift, the creation of God. It is naked trust in the promised favor of God, trust abstracted from all consideration of our deeds and virtues.

> Refer doing to the Law, believing to the promise. As widely as the Law and the promise are distinct, so far apart are faith and works – even if you understand 'doing works' in a theological sense . . . . [For] the Law, whether it is done morally or theologically or not at all, contributes nothing whatever to justification. The Law pertains to doing. But faith is not of this sort; it is something completely different . . . ([4], p. 272).

The faith of which Luther speaks here quite clearly admits of no development over time. It is not a virtue gradually developed and strengthened bit by bit. It is a mathematical point: the self passive before God – and the self a self, whole and entire, only because it is so passive before God ([1], p. 81). The wholeness of this self is the product not of its piecemeal doings but of the gracious verdict of God, who sees the self whole.

When we compare this discussion with "Against Latomus", written almost fifteen years earlier, we must say that, if anything, Luther has sharpened and intensified his emphasis upon the divine initiative. And this in turn suggests still greater urgency for our question: Why concern ourselves with virtues and the gradual shaping of character? Our being – which is wholly the work of God – precedes any doing. What place, then, for character formation?

Luther seems relatively unperturbed by the problem. "Christians do not become righteous by doing righteous works; but once they have been justified by faith in Christ, they do righteous works" ([4], p. 256). That is his view. And yet, the very next sentences suggest that he is perfectly aware of the alternative understanding of virtue and willing to grant it a limited place. "In civil life the situation is different; here one becomes a doer on the basis of deeds, just as one becomes a lutenist by often playing the lute, as Aristotle says. But in theology one does not become a doer on the basis of works of the Law; first there must be the doer, and then the deeds follow" ([4], p. 256). This theme recurs throughout his exposition of Galatians 3:10–14. Philosophical doing and theological doing must be clearly distinguished. The former seeks to shape character through virtuous action. The latter holds that virtuous action is possible only if character has first been transformed.

Philosophical doing is fine – in its place. But its place is emphatically not to determine the status of the self before God, the persons we really are. It will do for everyday attempts to develop our own character or that of others; there doing may precede and shape being. But by itself it can fashion us only in bits and pieces; it can at best make us selves who are partly saint and partly sinner. Before God, who sees us whole, that is not enough; for God desires fellowship neither with deeds nor virtues but with persons.

We find, once again, that Luther seems simply to have two different ways of talking about our virtue and character. We could, without much difficulty, find similar distinctions in many of his other writings as well as the writings of other Christian theologians.[3] But there would be little point in continuing to pile up evidence. The point now is to try to understand why Luther seems compelled to talk in these two ways about virtue and to ask whether the two can be held together in one theological system.

When we consider human character, its virtues and vices, we seem driven to think both in terms of particular traits of character – delimited virtues and vices – and of character in a more general sense as a fundamental determination of the self ([2], p. 16). And, if nothing else, Luther's willingness to speak paradoxically makes clear that these are quite different ways of thinking about the self. On the one hand, he displays what we may term a *substantive* understanding of virtue. Our virtues are traits of character which can be developed over time through habitual behavior. They provide the self with a certain continuity even in the midst of its growth and development. They suggest, when relatively well developed, a kind of self-mastery, a moral agent upon whom we may rely, who can be depended upon to display the virtues in his behavior. From this perspective a certain confidence in moral education – the attempt to shape being by doing – is warranted. Substantive virtues can be developed and sustained by those who take seriously the examined life.

On the other hand, Luther displays what we may term a *relational* understanding of virtue. From this perspective the virtuous man or woman is simply one accepted by God, one upon whom the divine favor rests. We are not just bundles of deeds or virtues; we are men and women made for fellowship with each other and, ultimately, with God. We need to be loved and accepted, not just to have our deeds or virtues commended. Hence, 'character' may suggest not just particular traits but the person – the person who is not made to be an isolated monad but to exist in relation. And there is no way to get from particular traits of character, from any ensemble of virtues, to a person who is accepted. All our doings may shape our being, but we can never see the self we have made whole and entire. We may shape our being, but we can never create a self. We may make ourselves lovable, but we cannot thereby establish a relation with or claim upon God. If therefore we are to have character in the fullest sense, if the fundamental determination of our self is to be right, we must be willing to be passive before God. Renouncing our claims to self-mastery, recognizing that the self we are, wholly and entirely, is known only to God, we simply listen to the divine Word which announces and thereby establishes a relation of love between creature and Creator. From this perspective our efforts at moral education are relativized – for what is needed is not development of virtues but a continual return to the divine initiative. Our efforts at moral education may add to our virtues, but they cannot bring about the relation that establishes our selfhood. And hence, one might say that the examined life is not worth living; for at its very best – which can be good indeed – it offers only an ensemble of virtues, not the relation we seek and for which we are created. Indeed, the examined life may even prove dangerous. Focusing our attention on our own self-development, we may, in the name of moral effort, lose that sense of liberation which true virtue requires. Focusing our attention on that sense of self-mastery which results from the serious attempt to shape being by doing, the examined life may lead us to think of virtue as our secure possession upon which we can rely.

Luther's view seems to be, then, that the examined life is necessary but not, finally, worth living – that we should make what progress we can in virtue but always without anticipating that it could make of us what we want to be. Can these two attitudes toward virtue form a coherent theological whole? If we think of virtues as substantive traits, we are thinking of the person as partly saint and partly sinner – as one who seeks to progress and develop in a way that will diminish the part that is sinner and augment that part of the self that is saint. If we think of virtue as a relation of acceptance with God, we are thinking of the person as simultaneously and entirely saint and sinner – in one's self a sinner, in Christ a saint – and as one who needs constantly to hear the divine word of favor. On the first model life is a journey, a gradual progress toward virtue. On the second life is a continual return to that perfectly passive moment, that mathematical point in which the self is whole and wholly virtuous before God.

No one could deny that these two views stand in some tension. On the one hand, our substantive virtues may be few, yet we may be accepted and righteous before God. On the other, our substantive virtues may be many, yet if we rely on them we may lack the faith which *is* virtue before God. There need be, it would seem, little correlation between our virtue understood substantively and our virtue understood relationally. Indeed, Stanley Hauerwas has described just this as the 'Protestant concern': to show that the actual shape of our life has no significance for our virtue [relationally understood] ([2], p. 4). How can one manage to speak of virtue in both ways?

In part, I think, the answer lies in the framework which the Christian story – beginning in creation and moving steadily toward a providentially determined end – supplies. The divine Word announces what we, caught up in the successive moments and piecemeal progress of our temporal existence, cannot see: that we are whole and wholly virtuous. But that same divine Word, spoken by the God who rules our temporal history, announces his own commitment to display one day the truth of his verdict. That is, God is committed to transforming people who are partly saint and partly sinner into people who are saints *simpliciter* – who are substantively what they already are in relation to him. The narrative of the Christian story which provides the contours for Christian living envisions a day when these several evaluations of our character meet, are reconciled, and no longer stand in tension. Until that day, however, we live within the constraints of a temporal narrative – adding virtues piecemeal, shaping being by doing, unable to see ourselves whole. And hence, until that day the virtue of hope must be the *leitmotif* of Christian existence, the fundamental shape of Christian character – hope that God can make of our piecemeal virtues more than we can see or know and thus bring the plot of this story to a happy ending. The tensions involved in any attempt to think both ways about virtue – the tension between the self-mastery of moral virtue and a self perfectly passive before God; the tension between a virtue which we can claim as our possession, upon which we and others can rely, and a virtue which must be continually reestablished by divine grace; the tension between a self which can see itself only in part and a self whole before God – these are the tensions important for any theological analysis of virtue. We see them clearly in Luther's theology because they are so central to his concern, but they are crucial any time we try to think theologically about virtue. The tensions are nicely captured by St. Paul in one short verse: "Not that I have already obtained this or am already perfect; but I press on to make it my own, because Christ Jesus has made me his own" (Philippians 3:12; RSV). Both ways of talking about virtue are necessary and will be necessary for as long as we remain pilgrims, caught up in a story which begins and ends in God but which moves toward its final curtain in the bit-by-bit medium of time and history.

THEORY AND PRACTICE

Even if the discussion above provides a satisfactory analysis of the place of virtue in the Christian life, we need not deny that it may be difficult to translate theory into practice and find a way to do justice to both senses of virtue in our lives. Since we have used Luther's writings as the prism through which to examine a theological understanding of virtue, we may use the movement which bears his name to illustrate this difficulty. In a fascinating book, *Luther's House of Learning*, Gerald Strauss has suggested that Lutheran attempts to indoctrinate the young were largely a failure.[4] The Lutherans were heirs of an educational tradition which held that, in the words of Plutarch, "moral virtue is habit long continued" ([5], p. 63). They were heirs, that is, of a tradition which attempted to shape being by doing. Yet, no theme in Luther's theology had greater urgency than the belief that only God could finally transform the heart, transform character.

The difficulty faced by Protestant educators was, therefore, to "forge a motivational link" between the external discipline of habit and moral education and the inner spirit ([5], p. 237). Committed to the view that virtues could be shaped by habit *and* to the view that true virtue was the work solely of divine grace – committed, that is, to the two understandings of virtue we have isolated in Luther's writings – these Protestant pedagogues labored to shape an educational practice which made place for both. The word *einbilden* became a key term in Protestant educational theory.

> How could religious and moral precepts be imprinted so lastingly on men's hearts, minds, and characters as to redirect their impulses? Protestant theology and pedagogical practice clashed on the answer to this question .... All agreed that only God could turn the individual's heart .... Nontheless, Protestant educators proposed to bring about just such a fundamental change in men's nature. Luther himself pointed the way toward this transformation. The word of God, he suggested, can be impressed (*eingebildet*) upon the hearts of men to allow the divine spirit to do its work there ([5], p. 152).

A moment's consideration will suggest, of course, that this is more a statement than a resolution of the difficulty; for it leaves unexplained the link between strenuous educational efforts to inculcate piety and the claim that these efforts are entirely unable to bring about the fundamental transformation which is needed.

Strauss concludes that if the Reformation's purpose in Germany was "to make people – all people – think, feel, and act as Christians, to imbue them with a Christian mind-set, motivational drive, and way of life, it failed" ([5], p. 307). And he explains this failure by suggesting that Lutheran educational efforts were hamstrung from the outset by ambivalence. "Torn between their trust in the molding power of education and their admission that the alteration of men's nature was a task beyond human strength, they strove for success in their endeavors while conceding the likelihood of defeat" ([5], p. 300). The goal – true virtue – was so lofty that it could do little more than call into question the worth of the everyday methods of habituation to which they were also committed. Rather than finding some way of holding both senses of virtue in fruitful tension, they were paralyzed in their educational efforts by their commitment to true virtue. Seeking the virtue which only God could work but mindful of its rarity, they committed themselves – but only halfheartedly – to the external disciplines of moral education.

> Lutherans seem to have lacked the temper, or the stomach, for such measures. They had opportunities for creating social pressure, but they failed to turn them into a molding process. They insisted that everyone in society should be subjected to internal and external coercion, but they shunned police-state methods (for which they lacked the means, in any case) and treated with resigned tolerance the widespread deviation from their exalted norms ([5], p. 301).

Of course, the tension and ambivalence which Strauss detects are not a Lutheran problem alone, though it may be that – because Luther's theology focused so emphatically on character and the meaning of human righteousness – within Christian history the problem comes to its clearest expression in Lutheranism. The tension must always be present in Christian theory and practice, and theological analysis can sometimes do little more than unveil it. Christian character requires a kind of soulcraft, an ability to transform and reshape the person at his innermost core. And this, Christians believe, must be the prerogative of God alone; for he alone sees the self whole and entire. At the same time, Christians are committed to cultivation of the virtues, if only in piecemeal fashion. These virtues do at least foster human life together and fashion human behavior – if not character in the full sense – in a way which more closely approximates God's will for human life. We cultivate the virtues, knowing the dangers which always plague such efforts and aware that there is a certain futility in our attempts, since the tree determines the fruit.

The tension between these several views of virtue cannot, I think, be removed from the Christian perspective. Its theoretical resolution lies in the narrative Christians tell and retell – a story, not yet finished, in which God is graciously at work transforming sinners into saints. But that story, because it is not yet finished, must be lived. The theoretical resolution explains but does not remove the tensions of the practical life. Within human history

Christians are committed to the attempt to shape character by inculcating virtues. But they are committed also to the belief that this attempt is safe – protected to some extent against the moral paralysis which comes from focusing upon the self, from our illusions of self-mastery and our tendency to claim virtue as our possession, from the temptation to step across the gap which divides inculcation of virtues from shaping of the soul – only when we do more than just attempt to transform being by doing. Our attempts at moral education are safe only when we also gather to worship, when we continually return to hear the Word which announces that the end of the story is present now in hope – the Word which makes present the grace of One who alone sees us whole and who has both authority and power to transform character and shape the soul. To tend the soul is a high calling. But it is done safely only by those who know the truth that the examined life is not worth living.

THE PHYSICIAN'S VIRTUE

It is not hard to think of virtues which may be needed by both physicians and patients. Physicians will need prudence and humility; patients will surely need courage and hope (to mention one of the cardinal and one of the theological virtues). And indeed, the lists could just as easily be reversed – patients will need humility, and physicians should not be lacking in courage. We are not likely to think of many virtues which would be of no benefit to physicians or patients. Whether, when we are patients, we will come with the virtues we need depends, for the most part, upon the seriousness with which our culture takes the inculcation of virtue – a question too broad to consider here. But it may be useful to think briefly of the virtue of a physician from within our theological perspective.

The understanding a physician needs is not always simply technical competence, and the judgments a physician makes – though they may certainly be right or wrong – will not always involve the straightforward application of a rule to a clear case. What the physician needs, therefore, is not only technical competence but also virtue – the fidelity, prudence, humility, and truthfulness which open our eyes to the truth of life and the needs of others. Consider, for example, the judgment that a patient has begun to die, a judgment which will affect the sort of care deemed appropriate. David H. Smith has offered a 'parable' which suggests the way in which virtuous judgment goes beyond any matter of technique. The good doctor is, we may say in parable, like the 'good host'.

A couple invite friends to dinner. Food and drink are pleasant; the conversation bubbles. The good host is hospitable and courteous to his guest, no matter what his shifts in mood. But there comes a time when the party 'winds down' – a time to acknowledge that the evening is over. At that point the good host does not press his guest to stay, but lets him go. Indeed he may have to signal that it is now acceptable to go. The one thing that being a good host will not allow is forcing a sleepy cantankerous guest to leave.[5]

Perhaps some people are to the manner born and quite naturally make good hosts; maybe the same is true of physicians. More probably, both had good role models from whom to learn. But at any rate, if this kind of 'good' judgment cannot exactly be taught, virtues like faithfulness and truthfulness can be inculcated – and from them such judgment may flow. Whether the requisite virtues can be inculcated within a program of medical education is, perhaps, doubtful. For the development of such virtue, what a physician learned on his mother's knee may be more important. But if it cannot be developed easily within medical education, virtue can at least be prized. One does not become a good host overnight; the bit-by-bit development of such ability calls for time and opportunity to live the examined life, to reflect on questions not purely technical in nature, to be challenged to discipline and control that part of the self not yet habituated to virtue.

Is that enough? It is, to be sure, a lot to ask. Perhaps, though, the theologian is not wrong to expect even something more, not wrong at least to hope that the physician's own experience will also suggest the need to move beyond the examined life. The struggle to be a good host must surely pale by comparison with the effort to be a good physician. The bit-by-bit progress in truthfulness, fidelity, and humility may not be sufficient when the physician struggles to see the truth about a patient's condition and to communicate the truth. When the physician stammers at such moments, when one's vision is clouded and one's self-mastery suddenly doubtful, when one's activity accomplishes little, then we may need a theological analysis of the virtues which can turn us away from the examined life and remind us that the wholeness of the self is not finally the product of its piecemeal doings but of the gracious verdict of God who, alone, sees the self whole. And from such a turn may flow true virtue, a renewed fidelity and humility; from such a perfectly passive moment before God may flow action which is effective and powerful – this may be paradoxical, but it will not surprise the physician who has been given opportunity to ponder Luther's case against Latomus!

NOTES

* Parts of this paper also appear as Chapter 5 of the author's *The Theory and Practice of Virtue*, published by the University of Notre Dame Press (1984).

[1] Nor is this merely the 'introspective conscience' which it is fashionable to charge Luther with having foisted upon Western Christendom. In the same context Luther quotes St. Paul's "I am not aware of anything against myself but I am not thereby acquitted" in support of his own position ([3], p. 190).

[2] Luther also writes that God himself does not deny that these virtues are good. That is, taken by themselves and considered in isolation from the person as a whole, some of them merit praise. And, says Luther, God does indeed reward such virtues with temporal benefits ([3], p. 225). This is worth noting because in an interesting way Luther here inverts the old problem of whether a good man can be harmed. One way to deal with the problem was to assert that, whatever a good man might suffer in this life, he was bound to be rewarded for his virtues in the next. Luther suggests the opposite: Before God judgments are made upon selves, not upon their virtues taken in isolation. But in this life, before the world, God is content that such isolated virtues should have their reward.

[3] For example, for a consideration of an analogous problem in the theological ethic of Karl Barth, see [6].

[4] In terms of an older but still important and influential generation of Luther scholarship, we could say that Strauss tends to support the claims of Ernst Troeltsch rather than Karl Holl. Lutheranism did not manage to shape a society. Before that could be done, Protestants would have to become more willing than early Lutherans were to commit themselves to the examined life.

[5] In an unpublished paper titled, "What Should We Do?" I cite it here with David Smith's permission.

BIBLIOGRAPHY

[1] Elert, W.: 1962, *The Structure of Lutheranism*, Concordia Publishing House, St. Louis.

[2] Hauerwas, S.: 1975, *Character and the Christian Life*, Trinity University Press, San Antonio.

[3] Luther, Martin: 1958, 'Against Latomus', *Luther's Works*, Vol. 32, Fortress Press, Philadelphia, pp. 133–260.

[4] Luther, M.: 1963, 'Lectures on Galatians, 1535: Chapters 1–4', *Luther's Works*, Vol. 26, Concordia Publishing House, St. Louis.

[5] Strauss, G.: 1978, *Luther's House of Learning: Indoctrination of the Young in the German Reformation*, The Johns Hopkins University Press, Baltimore and London.

[6] Werpehowski, W.: 1981, 'Command and History in the Ethics of Karl Barth', *The Journal of Religious Ethics* 9 (Fall), 298–320.

Oberlin College,
Oberlin, Ohio,
U.S.A.

SECTION III

VIRTUE AND MEDICINE

MARX W. WARTOFSKY

VIRTUES AND VICES: THE SOCIAL AND HISTORICAL CONSTRUCTION OF MEDICAL NORMS

> "What will help deliver us now that shame has turned into honor and vice into virtue?"
>
> (Martin Luther, *An die Pfarherrn wider den Wucher zu Predigen*, Wittenberg 1540)

INTRODUCTION

What happens to the medical virtues when there are conflicting styles and norms of medical practice, or when the disparity between moral traditions and moral practices in medicine becomes great? Is there an essential core of virtues in medicine, which remains invariant through socio-cultural and historical transformation? Or are there 'medical revolutions' in which different sets of virtues and vices replace each other discontinuously? Do old virtues become vices and vices virtues? Does historical and social change in the practice of medicine transmute base metals into gold and gold into dross?

There is no doubt that there has been a large and unsettling transformation in the theory and practice of medicine in the last half-century, and perhaps even more radically in the last twenty years. How have the aims and values of medicine fared through this change? Indeed, was there anything like a coherent core of medical virtues against which transformations may be measured, or which suffered stress under the radical changes in medical practice? In order to answer these questions, or at least to broach them, I would like to consider three issues in this paper: (1) How may one characterize the virtue of medicine, and the medical virtues? (2) Have the norms of medicine changed? And if so, what are the causes and what is the nature of the change? (3) On what grounds is a critique of the norms of medicine possible? Given what the virtues and vices of medicine presently are, or have become, how would one argue about what they *ought* to be?

My general approach to these questions may be stated at the outset: I believe that the norms of medical practice are neither carved in stone, nor are they whatever a given social-historical form of medical practice takes them

Earl E. Shelp (ed.), Virtue and Medicine, 175–199.

to be. Thus, I will eschew both an ahistorical essentialism, on the one hand, and an historical relativism, on the other. This leaves a very narrow space to work in, but it is a space I should like to enlarge. It is the space of the historical development of norms, measured by the growth of medical knowledge, as well as by the growth of social responsibility *for* medicine and the social responsibility *of* medicine. I am not arguing that history confers legitimation on the present cumulatively, nor that there is some orthogenic evolution of medical virtues. I will argue, however, that the recent developments of medical knowledge and technique, the rapidly increasing socialization of medical practice and the deep involvement of medical practice with the social and political welfare of humankind, with the consequences of war and the possibilities of nuclear holocaust, all create the opportunity and the demand for a wider and deeper construal of the virtue of medicine and of the medical virtues than has heretofore existed. Concomitantly, medical vices are likewise possible now on a scale previously unimaginable, in terms of what harms of commission and omission can be effected by medicine. The general context for my discussion here is therefore the social and historical construction and transformation of medical norms. But since this is a consideration of 'virtues' and 'vices', it may be well to begin by getting straight on this quaint usage.

SOME PRELIMINARIES ON VIRTUE-TALK

Talk about virtues is old-fashioned. It belongs with the nostalgic memories of a younger, purer time, when the use of expressions like 'true-blue' and 'straight-shooter' would not have been embarrassing or campy. This younger, purer time was one in which everyone ostensibly knew what was right and wrong, and moral skeptics, like atheists, were run out of town on a rail (whatever *that* is.) Virtue was what a Good Girl preserved, against the day of her marriage; what true believers aimed at during their short sojourn in this world, in order to earn their reward in the next; what honest, courageous, truthful, merciful and just men exemplified – (mainly men, since these were 'male' virtues); and what thrift, modesty and self-sacrifice were instances of. Virtue was its own reward: it couldn't be bought, and though it yielded the benefits of deserved fortune and happiness to the lucky or to God's elect, it was not valued for its consequences, but for its own sake.

Medical virtue, in this rose-tinted memoir, was horse-and-buggy virtue. Its paragon was the caring, wise, and patient physician-friend who rode through the midnight blizzard for no fee, but in pursuit of a scared duty and trust – young Dr. Kildare grown gray, as painted by Norman Rockwell.

Now this has all the characteristics of a cultural myth. Yet, cultural myths both express and help to shape the self-conception of an age, even if only in terms of an idealized (or indeed, an ideological) construction. Such myths demarcate the arena within which virtues and vices are acknowledged, and at the same time exclude from consideration (or hide from view) the large area of non-relevant practices. The older nostalgic view of medicine may never have been true of the actual practices of medicine in the past. Yet, it may accurately reflect how both the medical profession and the public *conceived* of medicine, and of the medical virtues and vices, and thus it may tell us how the profession presented itself, what practitioners believed about themselves and what was believed about them by the public at large. This image may have very little to do with what the actual practices of medicine were (just as, for example, the manifest political morality of imperialist nations has little to do with their rapacious practices, or the Sunday Sermon to which the profiteer gives a hearty Amen has little to do with his weekday behavior). Yet this very misrepresentation and self-delusion may have been part of the practice of medicine, part of what constituted the practice as what it was. A historical reconstruction of medical practice which reveals such disparities between myth and reality, between self-conception and actual practice, should lead to a critical judgment of the practice, and to a moral as well as sociological criticism either of the professed virtues of medicine or of the failure to embody these virtues in practice. But beyond this, it should lead to some explanation of the failure as well.

Where revisionist histories reveal serious differences between self-conception and actual practices, they point to the incoherence of the moral professions of an age and suggest, indeed, that the mask of morality may serve to hide the face of incompetence, inadequacy and vice. Thus, for example, the striking revisionist history of morals that Alasdair MacIntyre presents in his recent book, *After Virtue*, is revealing in just that way, whatever one makes of his own conclusions. For in that work, the professions of virtue characteristic of modernism are laid bare as incoherent fragments of an earlier moral system, fragments which no longer have purchase, nor, in fact, any reference to the practices of which they are ostensibly the norms. Similarly, if one takes a dry-eyed view of the professed and imagined virtues of horse-and-buggy medicine, it may be seen that, after all, bedside manner and the virtues of character and conscience were small compensation for the vices of technical incompetence, scientific ignorance, and clinical hocus-pocus. On the other hand, the traditional virtues of medicine (to which we will shortly turn, and which go beyond the limits of the nostalgic myth) seem unexceptionable. They seem to transcend the local limits of any given historical form of medical practice and medical knowledge, and to express both a humane and scientific ideal which retains its relevance beyond the vagaries of cultural myth or professional ideology. The question is whether we can sort out such historically invariant and perduring virtues, and make sense of them outside the contexts of their specific interpretations in historically and socially situated actual practices; or whether, instead, we are deceived by the fact that diverse and even incommensurable practices, taken as virtuous, have been denominated by the same *names*, and that virtue-talk simply hides these fundamental discrepancies from view.

It is disturbing when ideal and reality, norm and practice, diverge, or even conflict. But more disturbing still is the deliberate dichotomy in which ethical norm and medical practice are not even brought to bear on each other. This is the situation in which the moral and the technical-scientific components of medicine are taken as constituting two distinct, non-overlapping domains, and where the practice of medicine comes to be conceived of as, on the one hand, pure technique or clinical know-how (with its growing scientific and technological component); and on the other, 'medical ethics'. Where medical skill and moral judgment are understood as two discrepant constituents of medicine, medical ethics becomes a specialty like cardiovascular surgery or rheumatology. Yet, this is where the development of modern medicine has tended, for the demands of cognitive and technical adequacy become all-consuming, as specialization grows; and so too do the new demands upon moral accountability and sophistication, as social, legal, and religious issues and jurisdictions intrude upon the previously autonomous provinces of medical judgment.

In this setting, talk about virtue, though old-fashioned, has become fashionable again. And this fashionableness is odd, for its principal source is not a *new* construction of the virtues, nor even a reconstruction or reintroduction of the old virtues. Rather, it represents a stark and almost sudden awareness that the moral foundations of specific current practices – among them, politics, war, business, the natural sciences, and medicine – are crumbling or cracked or otherwise in such disrepair that utter confusion reigns. Moral authority is not certain, either as to its seat, or as to the consistency of its utterances, or as to the limits of its jurisdiction. Suddenly, we have been made aware by a certain articulate group of social, cultural and philosophical critics, that it is *After Virtue* time; and though it has been a half century or more (some say three centuries!) since the virtues have been coherently understood and embedded in our practices, we have just come to notice their absence. This radical claim is not intended by the critics to mean that no one has acted morally through this time, or that there are no virtuous folks left, but rather that our professions of virtue have little to do with what we do, and that what we think we are doing is at variance with our actual conduct. Worse yet, the severest of these critics (MacIntyre) argues that there is no available way to resolve moral conflicts coherently in the present, and that this leaves us with incommensurable moral claims and conflicting 'virtues', choice among which is condemned to irrational emotivism or reduced to technical or bureaucratic procedure.

Let me begin by granting the incoherence of our current virtue-talk, and by agreeing with the social and philosophical critics to this extent: We have as yet not constructed a moral framework adequate to the new forms of practice – specifically in medicine, but surely also in politics, and in international matters of war and peace, and in the uses of science and technology. And the older forms of practice to which our virtue-talk may have been relevant have been dramatically and radically changed in the last half-century at least. What has happened is that our practices have outrun our older norms, so that what remains viable and what is no longer so have not been systematically sorted out. What has also happened is that, under the pressure of events – of social, technological and scientific change – the practice of medicine, like other social and institutional practices, has made *ad hoc* adaptations to these changes. There has been a pragmatic and piecemeal attempt to fix things up as the occasion demanded. Thus, for example, medical ethics has jumped around from crisis to crisis – abortion, brain-death, euthanasia, life-death decisions in allocation of scarce new technological resources or of available medical personnel, informed consent, organ-sales, use of human experimental subjects, use and testing of new drugs, etc., etc. – as the range and efficacy of medical intervention increased dramatically with new knowledge and new technology.

It is not clear that this *ad hoc* approach, for all its flexibility and responsiveness to concrete problems as they emerge, is adequate. Nor is it clear that what the new situation demands is simply an updated moral theory which would provide some systematic and encompassing framework for this range of issues, and set things straight once again. Perhaps what is needed first is an intensive and critical study of the emerging forms of practice in medicine, and of the corollary modes of practical judgment which are developing in these contexts. Any moral theorizing or moral criticism which starts from the top down, or from ready-made formulas, or from the frameworks of traditional ethical theory will miss the mark. For it will tend to frame the perception of current medical practices within one or another *a priori* understanding of the virtues derived from a different social world which no longer exists, and will therefore tend to blur just those features which are new in the current situation, and which are therefore crucial to any adequate formulation of new theory.

On the other hand, if new theory or a new moral framework is to be derived inductively, from a review and generalization of emerging practices, then such theory will fail to have the normative force that moral principles require. It will be no more than a description or a distillation of what in fact the current styles of moral judgment are in medicine; and if no more than this, then acritical in its content. But if the reconstruction or creation of the medical virtues is to proceed neither on *a priori* nor on inductivist-empirical grounds, how is it even to begin?

There are, it seems to me, two desiderata here: one is a critical review of the traditional medical virtues; the other is a critical assessment of the current modes of moral judgment in medicine, of the goods to which such judgments appeal, and the medical virtues which are commensurate with these goods. But here, the deep methodological problem presents itself, which will be the focus of the final section of this paper: How is such a critique of either traditional norms or of current practices possible, unless the criticism itself is grounded in some systematic conception of the good of medicine and of the medical virtues? And if it is, then that conception has itself to be defended and in turn remain open to criticism. It seems that we would then be caught in an infinite 'critical' regress, or in a 'critical' circle: every new ground for critique becomes itself the subject of critique, in that spiralling dialectic which leaves nothing whole and no foundations secure.

What is the alternative? It seems to me there is none. We are condemned to begin *in medias res*, with the analysis and critique of current norms, and with the responsibility for constructing new ones where the old ones are found to be inadequate. There is, however, nothing desperate about all this. For norms have their genesis in the theoretical and critical reflection on practices. They emerge as distillations (or inspirations) of the particular and concrete circumstances of an historically evolved form of medical practice, and as institutional judgments about what that practice requires for the realization of its professed goods. There is nowhere else to go to seek their justification. The Holy Grail is in the backyard. To begin *in medias res* is, however, not to begin from scratch: the creation and transformation of the norms of a practice begins with the critical assessment of the inherited norms, and from the recognition of limits without which transcendence of the present is impossible.

If this historicity of norms and virtues is in fact the very condition of their genesis and existence, this suggests their malleability and adaptation, or even their rejection, as the historical forms of a practice change, and as new needs emerge. Contrariwise, it suggests also the perdurance of some norms in the face of criticism, if it happens that the goods of a practice remain historically invariant through the social or cognitive or technological transformations of that practice. Together with an openness to the newer forms and problems of medical practice, there is nevertheless a need to recognize whatever deep continuities there may be with the traditional medical virtues in these contexts. Eschewing an ahistorical essentialism does not therefore commit one to any cult of the current, or to a view of virtues as fashions, in this year and out the next. What it does require is some reconstruction of the historical continuities in medical norms which ought to be preserved, as well as a critique of those norms and practices which ought to be abandoned. The question is: how to choose? Is a critique of norms in medicine possible, from within the perspective of medical practice itself? To this question we will return. But first, how shall we talk about the traditional medical virtues?

THE VIRTUE OF MEDICINE AND THE MEDICAL VIRTUES: FROM THE ABSTRACT TO THE CONCRETE

In what follows, I will proceed with the first of the three questions that I broached at the outset ('How may one characterize the medical virtues?') by a series of approximations, or, as the Hegelians might say, by a progression of ordered 'moments', moving dialectically from the abstract to the concrete. With each successive characterization of the medical virtues, I hope to approach more closely to the current situation, and thus to the second question ('Have the norms of medicine changed?'); and then from considerations concerning the historicity of norms, to the third question ('Is a critique of norms possible?')

Let me begin with some stipulations concerning usage. The term *Virtue* has no strictly canonical usage, though there are systems of ethical thought – e.g., the Aristotelian, the Thomist – in which the term is well-defined. I am not going to hew to any such systematic usage. Rather, in the ensuing 'approximations', I hope to give, in effect, an implicit definition. Yet, I want to eschew, or at least to qualify, one standard distinction: that between 'virtues' as moral qualities of agents or persons, as against 'goods' as moral properties of actions or of the consequences of actions. I have no hard and fast alternative distinction to propose here; but I will want to talk about the 'virtue' *of* medicine in terms analogous to those concerning the 'virtues' of agents or persons, namely, as the excellence or perfection of that entity (whether an agent or an institution or a practice) in realizing its 'end' or its characteristic 'good', i.e., that for the sake of which it exists or functions. This is, of course, a straight-out teleological characterization of virtue; and I want to preserve it for ascription to institutions or practices (e.g., the State, or Medicine), not in order to reify such abstractions as 'persons' or 'agents', but rather to take these abstract entities as no more than what is constituted by the agents or persons who define them, or who give them their existence by their social relations and their moral qualities. Thus, I will take medicine to be a social practice whose virtue is the teleologically defined excellence or perfection for the sake of which it is carried on by its practitioners. That this approach therefore requires a *theory* of the nature of the entity (e.g., of the person, or of the institution or practice) in order to make sense of virtue-talk is just to the point; for this then makes central the question 'What is medical practice (e.g., in its new forms) for the sake of?', and makes it also a normatively critical question. The virtue of medicine will then define what are the medical virtues, by which I will mean those excellences of character, competence, and conscience which are required to realize the virtue of medicine by its practitioners. In this sense, the medical virtues are entirely derivative from the virtue of medicine, and subordinate to it. Now this may seem perverse, in that it doesn't derive the distinctively medical virtues from more general human virtues, but rather assesses the human virtues only from the point of view of how they subserve the ends of medicine. Yet, the perversity vanishes, I think, when it is seen that the practice of medicine is itself defined by an overarching human good, namely, that one which it is the specific aim of medical practice to realize. Oscar Wilde remarks somewhere: "That a man is a poisoner says nothing against his prose." But of course it does, if literature is seen as a moral practice, *and* if moral character can be shown to affect the morality of one's literary practice. But not otherwise. That medicine is an ineluctably moral social practice is part of what I want to argue here; and further, that it is so in all of its reaches – cognitive-technical as well as what is usually regarded as explicitly ethical. In this sense, the answer to the question 'What is medical practice for the sake of?' defines both the virtue of medicine and derivatively, the medical virtues. Thus, though one may not speak of the virtue *of* an action, but rather of its ends or aims, one may speak of virtuous actions as those which serve to realize the good of a practice (like medicine) through the agency of persons whose virtue consists in their "settled habits or dispositions to do what

[they] ought to do . . . " ([1], p. 235)[1] where the 'oughts' follow from the virtue of medicine.

First Approximation: Abstract Essentialism and Its Limits

To talk about medical virtue is to talk about the true end of medicine. That has an essentialist ring to it. It presupposes that medicine has a 'true end', that there is some specifiable good for the sake of which the practice is pursued; or that there is a set of goods, which together constitute the aims or excellences of medicine, and which in effect define it as a normative, teleological practice. The practice of medicine may be conceived here as an institution – a socially organized activity which has both explicit, socially recognized norms or rules which identify its proper form and content, as well as implicit models of procedure or rules of art which are exhibited in its practices and which it is expected will be learned and understood by its practitioners. The physician, in this rather classical model, is what he or she is *as* physician by virtue of acting in accordance with, and on the basis of knowledge of and respect for the norms and rules of that practice, i.e., by acting so as to realize the good which defines that practice, after having been certified in some accepted institutional way by the guild or profession or by the society as competently trained to do so.

Criticism of the First Approximation

What is immediately wrong with (or at best inadequate about) such a view is its abstractness and emptiness. It has, as yet, no specific *medical* content; and with the substitution of the name of any other practice or profession – e.g., law, physics, tailoring, truck-driving – the very same account of the 'virtue' of that practice may be given, with little or no alteration, and with at least formal plausibility. What gives any account of the virtue of medicine substance is precisely the concrete content of the practice, i.e., that which makes it medicine, rather than, say, biochemistry or social work. Thus, we need to specify what determines the virtue of a certain practice by specifying its distinctive good, or goods, in the service of which the practice is pursued.

Second Approximation: The Specific Virtue of Medicine and Medical Virtues

Traditionally, the good of medicine has been taken to be the health of the human being who is the object of the practice, or the well-being of society insofar as this is associated with the health of its members. More specifically, the service to health, as a positive state or as a human, and social good may be defined in terms of the diagnosis, treatment, and cure of disease, injury, or (bodily or mental) defect (i.e., of 'unhealthy' or morbid states), the alleviation of suffering due to ill health, the care or maintenance of those who are ill or diseased, the prevention of disease, injury, and untimely death, and the promotion of sound health by hygiene, diet, or regimen. This list could be expanded or amended, certainly. Thus, one could add to it the 'social' parameters of medicine, to include medical regulation of the environment for the prevention of disease, injury, or genetic defect or the diminution of other risks to life and health; or for prevention of the communication of disease, etc.

To this, we may append the account of the standard medical virtues as those traits of character, competence, and conscience which practitioners would need to realize the virtue of medicine optimally. Thus, we may include here such virtues of character as veracity, honesty, integrity or trustworthiness, confidentiality, courage, autonomy; such virtues of competence as mastery of the knowledge in one's field of medicine and of the requisite skills or techniques (e.g., diagnostic, technical) as well as the maintenance and development of these competences with the progress of medicine; sound judgment and an ability to reason well; such virtues of conscience as dedication to one's work, availability when needed, as well as such more general moral qualities as fairness or justice in treatment or in the allocation of medical resources, respect for the autonomy and dignity or personal worth of patients and colleagues, etc., insofar as these virtues are necessary to realize the virtue of medicine, or the good of the practice. Older lists include such traits as loyalty, piety, modesty, and almost all include benevolence and caring.

Criticism

But this, once again, remains too formal, vague and abstract to yield an adequate account of the virtue of medicine and the medical virtues, not because it fails to name them, but because it is no more than a naming of them. We still remain at some distance from the real content of the medical virtues, though we may have most generally defined their scope or the domain of their practice. It is a moot point whether, having named the virtues, we have any understanding of what has been named, or exactly what understanding we have of it. For it is precisely the interpretation of such terms as health and disease, or diagnosis, or cure, or care that is at issue in determining what the virtue of medicine is. Furthermore, unless it is explicitly made clear why such virtues as veracity, trustworthiness, autonomy, etc., are essential to the realization of the goods which define medicine, it is easy to read them simply as a pious importation of general human virtues which it would be nice for doctors and nurses to have as well. Of course, the list is not quite as formal or as empty of interpretation as I suggest here. We bring a background of social and cultural interpretation and understanding to bear on these terms, in this context. It seems clear what we mean by "disease" and "health" and why one is bad and the other good. (Or is it?) It seems obvious that doctors ought to be benevolent, and that they should be autonomous. (Or is it?) The problems begin to emerge when we recognize that some of these virtues, in concrete circumstances and in particular historical readings, are in conflict; and that what is a sometime virtue may become an othertime vice. (The argument here has been made by Alasdair MacIntyre in his instructive essay "How Virtues Become Vices: Values, Medicine and Social Context" [2] and I will simply point to it here, but not pursue it further.) The payoff of this naming of virtues comes only with its contextualization, only with the specific interpretation of these virtues in particular modes of the practice of medicine, in concrete social and historical settings.

Third Approximation: How Are the Medical Virtues Understood in Contemporary Practice?

One further step is needed then in this progression from the abstract to the concrete. It would be to examine what understanding of the virtue of medicine, and of the medical virtues exists *in fact* within the institutional forms of contemporary practice or in the minds of practitioners (or what understanding existed within historical forms of the practice in the past. For this, see some of the other essays in this volume). Beyond this, since the practice is defined not only by its practitioners, but also by the society within which the practice is carried on, we also need to examine what a given society or culture takes to be the appropriate domain of medicine, what it demands of the profession, and what its wider theory and practice of the virtues is, within which the medical virtues take their place and are interpreted. In this way, we would be able to reconstruct what is concretely understood to be the good of the practice, what its virtues or excellences are in the eyes of those who constitute the profession of medicine, and of those who constitute the domains of the practice whether as patients or prospective patients; or as legislators or judges or administrators who make decisions about the practice or regulate it; or as taxpayers or fee-payers or those who otherwise support the practice and influence its direction or its attention. Such a description of the socially constituted norms of the practice and of its ends, as these are understood by the profession, by the society at large, and by its critics, would begin to approximate that concrete account of the medical virtues we have been pursuing. It would tell us what medicine is taken to be in the present, what the self-conception of the physician is, what medicine is understood as by those whose needs it meets, or is expected to meet – in short, what the professional and social idea of medicine is, as this idea shapes the actions and expectations of a community. It would be a first serious step towards the reconstruction of the medical virtues *in vivo*, so to speak, as they describe – and prescribe – the current norms of the practice, and therefore, it would be the precondition for any comparative investigation of changes in these norms, and for an appreciation of their historicity. In this sense, such a reconstruction in the concrete would also be the precondition for critique.

Thus, for example, if we study the Hippocratic Oath in its social contexts, we may distinguish, in retrospect, what is distinctively local and historically dated in its precepts – (e.g., the medical emphasis on dietetic regimen, the social emphasis on sexual injunctions regarding both male and female patients, and both slave and free patients) – and what seems to span the ages in its relevance – (e.g., the definition of the benefit of the sick as the aim of the practice, the virtues of confidentiality, of intraprofessional loyalty and care, the injunction against abortion). By situating the document it its circumstances, we may read back from it, and from a background knowledge of ancient Greek society, what the values were which defined the practice of medicine, and what its social integument was. Similarly, if we read the AMA Principles of Medical Ethics, and of the changes in them, we are struck with the preoccupation with the 'business' of medicine – (e.g., Section 5, on the physician's right to 'choose whom he will serve', Section 7, on fee-splitting) – and the stated concern for the welfare or well-being of both the individual and the community (Section 9 and 10). Clearly, such modern documents as the Nuremberg Code, the Declaration of Helsinki and the DHEW Statements on biomedical experimentation using human subjects make sense only against the specific background of the Nazi concentration camp 'experiments' in World War II, the proliferation of drug-testing by the pharmaceutical industry in the contexts of Federal regulation in the U.S., the development of statistical methods and therefore, the use of prisoners or other cloistered populations as experimental subjects. All of these examples evidence the historical genesis of norms and of changes in norms, i.e., the ways in which a given historical or social context defines the practice of medicine in the face of changes in medical knowledge, or technology, or of the uses to which the profession is put, or of the institutional or economic forms which the profession adopts, or those within which it exists.

We may proliferate such examples of the historicity of norms and of virtues, to discover both their variability and the persistent continuities through change. But without some deep and complex historical interpretation which situates the moral abstractions in the specific contexts which give them their living significance, what are we to make of such relatively empty virtue-terms as *benevolence*, *conscientiousness*, or even *veracity*, except as general moral qualities whose relevance to medical practice remains unanalyzed, or depends on an uncritical tacit understanding from within a given culture? What indeed are we to make of the varying socio-historical construals of the goods of 'health' or 'well-being', or (to quote from Section 3 of the AMA Code) "a method of healing founded on a scientific basis"? (It is enough to mention that what constitutes 'a scientific basis' has been the central issue in the debate in the philosophy and history of science of the last half-century, and that there is no commonly agreed upon resolution of this issue at present.) However, when such vague heuristic norms and precepts or such abstract virtues are reconstructed in the contexts of their use, we begin to understand what medicine and the medical virtues are taken to be in those contexts, and how these norms and virtues express themselves in the actual practice of medicine.

Criticism

Even such a concrete interpretation and understanding of the virtue of medicine and of the medical virtues, and of their historicity, leaves unsolved problems and unanswered questions. Two such questions are crucial here. First: Is *what medicine is understood to be*, whether by its practitioners or by the society at large, what it actually *is*, in practice? Second: Is the practice of medicine, whether as what it is taken to be, or as what it actually is (if these are discrepant) what it *ought* to be? That is, is medicine as a practice pursuing its 'true end' or is it failing to do so? A critique of the medical virtues and of the virtue of medicine cannot remain satisfied either with a concrete interpretation of medical norms in a given historical context, or with an account of what its practitioners or the society at large *believe* these norms ought to be. Getting straight about all this may be the precondition for critique, but this does not yet constitute the normative critique itself. Still, this takes us as far as we can go in considering the first question, as to what the virtues of medicine are in their traditional or current acceptations. It also begins to answer the second question ('Have the norms of medicine changed?') But to answer this question in the concrete, again, is not simply to attest to the historicity of norms in general, as we have so far done, nor to give merely anecdotal examples of such changes or adaptations to changed circumstances and needs. This project would require a fuller and more systematic account of the changes in norms that characterize the current medical scene.

No such detailed account can be given here, of course. Nor is it to be expected that such an account would yield a coherent picture. As I will suggest (and have discussed elsewhere [3]), contemporary medicine is in a major crisis, effected in great part by the conflicting norms, expectations, and jurisdictions which characterize the present practice of medicine, and by conflicting models of the profession itself – of its roles, authority structures, social responsibilities. The image is therefore confused, full of conflicting and sometimes incommensurable elements, and difficult to reconstruct. What may be ventured, however, is a broad characterization of some of the main features of the crisis of contemporary medicine (in the U.S. at least) which will suggest the direction of ongoing changes. Concomitantly, we may suggest how these changes are affecting the prevailing norms and virtues, and creating stresses and conflicts among them.

The Crisis in Medicine and the Transformation of the Virtues

We may characterize the current transformation of medical practice as a revolution, for it is a deep and radical change in the fundamental structures – social, legal, economic, scientific-technological and cognitive – of that practice. The change has been so rapid and far-reaching in its effects, that it is fair to speak of a major crisis in medicine, as the new developments strain the older forms and norms of the practice to the breaking point.

Let me characterize these changes in five broad categories:

(1) Changes in the social and professional structures and roles.

(2) Changes in the political economy of medicine.

(3) Changes in the scientific and technological components of medicine and of their relation to clinical practice.

(4) Changes in the medical and social character of the population to which the practice is directed, and in the needs of that population.

(5) Changes in the legal-juridical and political status of the practice.

What can be offered here is only the briefest discussion of these changes, in order to suggest their depth and the stresses they create for traditional medical forms and norms. It needs to be said at the outset that the changes are not necessarily the introduction of utterly new features, but often the development of older features of the practice from marginal or occasional phenomena to dominant ones. At the same time, older and more traditional forms do not disappear, but become subordinated, or marginal, or are reserved for special services.

(1) There has been a major shift in the social and professional structures of medical practice from the predominantly private mode of physician-patient relations – one-on-one interaction, usually in the setting of the private offices of individual physicians, supplemented by the hospital and the clinic – to forms of what we may call mass-medicine: treatment in the contexts of groups of practitioners, whether in group offices, clinics or community health centers, health plan facilities or hospitals. The clinic or hospital which provided the public access to medicine for the poorest population in the past now becomes a central locus for a major part of medical practice. What develops is the multi-purpose medical center, a sort of medical supermarket or complex, centralized facility – the institutional form of what we will characterize below as the 'industrialization' of medicine, or of medical production. This general development may be described as the *socialization* of medical practice. I am not referring here to the forms of 'socialized medicine' in those countries where the economics of medicine has become public or government-supported, but rather to the organizational forms of the practice itself, as involving larger or smaller collectives of practitioners around common medical tasks. In a sense, this is not new: the surgical team is a traditional form of such 'socialization' and there is an implicit socialized mode of practice involved in the practice of consulting, and in the general cooperation among the medical specialties. What is new is the degree and scale to which such forms of cooperative practice have developed, and the correlate institutional forms which accomodate it. Because this model has a certain similarity to the structures of medical education, (where the medical faculty may be said to constitute the collective staff of the institution) many of the hierarchies and traditions of the medical school have been transplanted into the forms of the practice itself. So too, since this development exists side by side with the fundamentally privatized traditional modes of medicine, the social structures of this newer form of practice retain the hierarchies of power and authority of the older forms. In effect, there develops a strain between the

traditional guild forms of medical practice and the industrial forms of current medicine.

(2) By the political economy of medicine I will mean the forms of organization of the production, exchange, and consumption of medical services, and the social forms of control of the means of this production (ownership, management, modes of decision-making, etc.) The traditional forms of the medical economy were centered on the craft or guild character of the practice, in which each practitioner was certified or accepted into the guild by its existing members under one or another dispensation of certifying authority. This license to practice was therefore also a manner of control over the economy of medicine, since the guild rules specified the liberties, constraints, and privileges under which the practice could be carried out. The appearance of professional autonomy in such a situation was always belied by the varying extents to which the guild itself was permitted to exist in the society – e.g., under patents and privileges of royalty, or of parliament, under regulation by the law, and within the network of other institutions with which medical practice and the economics of medicine is involved: the banks, the courts, insurance companies, educational institutions, hospitals, etc. The traditional modes of practice nevertheless remained privatized under these limits, and in modern times, the practitioner was in effect a private entrepreneur, working for himself, in a market where fees paid by the medical 'consumers' constituted the income of the practitioner in large part. The typical capitalization of medical practice was itself private – family funds, bank-loans, etc. – and only occasionally also public or commercial, where physicians worked for institutions or health-plans or in government employ. The doctor's office was the locus of medical service, except for forays into the field, home visits, or hospital and military service.

Needless to say, all this has changed in major ways. Though the private practice is still a typical form of medical economy, it is itself much changed, as medical groups form to meet the expenses and take advantage of the economies of scale and joint technological and laboratory facilties. Moreover, the medical center and the hospital office relocate the practitioners in great concentrations. Income is in large part no longer by direct fee but through third party payments, as health insurance, Medicaid, Medicare, and other forms of subsidized and publicly financed payment for services develop. Capitalization of the new 'industrial' forms of the practice becomes more and more a matter of major investment not only by medical practitioners, but by professional private investment groups, by government grants for medical facilities, and by institutional health and welfare funds. Though the hospital as a publicly funded or supported institution or as a private corporation played some role in the capitalization of medical practice in the past, this changes both qualitatively and quantitatively in the present. As such institutional forms as the hospital, the health plan or the health insurance corporation, the medical research center, the testing laboratory become more central to clinical practice, their influence on the economy of medicine grows and their role in the capitalization of that practice becomes dominant. Thus, though the individual practitioner still has to have the capital resources for aspects of the practice, a larger and larger role in the capitalization of the production of medical services and facilities, as well as of medical education, falls to such social, institutional, or corporate resources, both public and private.

The political economy of medicine becomes 'Medbusiness': fixed costs for plant, technological equipment, etc., are of a scale no longer amenable to the older privatized forms of the medical economy. Research becomes part of the cost of the production of medical services, as the medical sciences and engineering play a radically increased role in the practice. Labor costs of medical personnel and of hospital workers and laboratory technicians become typically industrialized as hospital workers' unions and other labor and professional organizations enter into negotiation with cost-conscious medical management. Administrative functions – and costs – increase exponentially, and the paperwork revolution deluges the practice with requirements for record-keeping, reporting and accounting. In effect, ownership and management pass from the private entrepreneur to large medical corporations and partnerships.

The practice of medicine passes from its older guild-form, and from the concomitant economic mode of private small business to what may be characterized as an industrialized form. The strain here is that centralized modes of 'mass production', in what may by analogy be called medical 'factories', the rationalization of production and distribution of medical services, the socialization of the practice of medicine which mark such 'industrialization' come into conflict with the essentially individual nature of the medical situation: whereas epidemiological or public health aspects of medicine lend themselves easily to such 'mass-methods', as essentially stochastic contexts amenable to statistical methods and to standardization, curative medicine does not. The individual patient needs individual treatment, whatever the degree to which diagnostic and therapeutic functions may be rationalized on a mass-production basis, or assisted by computerization. In some of its central functions, therefore, clinical medicine is like barbering and some other professions: there are limits to the deindividuation or industralization of the procedures. The industrial model of the political economy of medicine therefore puts a great strain on some of the traditional, and perhaps essential, modes of the practice.

What happens to the relations between medical service producers and consumers, and to the traditional forms of economic exchange between them, reflects this depersonalization of the medical market, and its increasing domination by finance-capital, i.e., by banking and investment institutions. All of this takes place within a surprisingly unchanged guild-structure which adapts itself to the new economy by protectionist policies, major political involvement in regulatory legislation, resistance to socialized medicine, and investment guidance to its members.

(3) The marriage of medical practice with the advanced medical sciences and with medical engineering is a characteristic phenomenon of the current scene. This includes not only the remarkable developments in biochemistry, genetics, and pharmacology, and the proliferation of biomedical technology, but the effects of the computer revolution in diagnostics, record-keeping and information, and in the automation of many laboratory test-procedures. The practitioner is linked with a network of information and control facilities which become an integral part of the clinical practice, as well as of the administrative and data-managing aspects of the profession. The costs of these scientific and technological components are factored into the costs of medical services, though economies of scale and the rationalization and automation tend to lower the unit-costs of many procedures. The new technologies make possible the processing of greater numbers of medical consumers by fewer practitioners, on the one hand; but on the other, they increase the number of support-specialties required by the profession, and further depersonalize the doctor-patient interaction.

(4) In one sense, the medical 'population' – the consumers of medical services, or those who stand in need of health care – doesn't change: everyone is a prospective patient. The character of the population changes, however, in the following respects: (a) With the advent of health insurance and of health plans linked to employment, and with the introduction of Medicare, Medicaid, etc., large segments of the population who, for reasons of expense, rarely visited a doctor now do so more commonly and more frequently. (b) The increased public awareness of health and also the changes in what are regarded as reasons for medical attention have led many more people to seek medical services than was the case earlier. What counts as 'illness' or as 'treatable' undergoes cultural change with the changes in both medical

knowledge and in the public image of health and disease. (c) Changes in occupational patterns and in occupational hazards, in diet, in lifestyles, and in the environment give rise to new patterns of disease and injury. So too do the benign effects of medical practice itself: increased life expectancy generates a new and large geriatric population, and the concomitant proportional increase within the total population of the diseases and infirmities of age; increased life expectancy of sufferers from such diseases as diabetes, where there is a pattern of genetic propensity to the disease, and the increased ability to bear children by those who were previously unable to do so for medical reasons or because of early death, tends to increase the disease-population proportionately with time. Malignantly, some of the very conditions of modern medical practice may lead to the development of iatrogenic illnesses or infections, i.e., those that result from medical practice itself or from the conditions of medical care, e.g., hospital epidemics, or prophylactic breakdowns in mass facilities or procedures, or unanticipated side-effects of new procedures or drugs. In all of these ways, the profile of disease and health in the population changes, and thus, medical needs and demands on medical practice change as well. The very progress of medicine itself creates new needs, in a feedback loop.

(5) The traditional autonomy of the profession has become seriously affected by the intrusion of legal and juridical judgments in areas previously thought to be the sacrosanct domain of medical decision-making, e.g., in matters of abortion, euthanasia, brain death. Federal legislation and Federal grant-policy for medical research begin to influence the focus and direction of medical practice, and to make it sensitive to the political climate in a new way. The regulation of health insurance, the subsidization of medical facilities by government, and the spiral of malpractice suits and of malpractice insurance introduces legislation and the law into medicine in ways far removed from the precincts of older-style forensic medicine. The legal and juridical status of the profession and of its institutional structures – the hospital, the medical school, the Boards of Certification, etc. – changes, as the patients'-rights movement grows, and as the taxpaying public and the courts demand new modes of accountability. The professional guild organizations as well as new organizations of social concern (e.g., Physicians for Social Responsibility) become more fully involved in the political scene, of lobbying, petitioning, supporting political candidates and special legislation.

This rude and rapid sketch is not intended to be complete, but simply to suggest the changes in medical practice which call into question the older structures and ways of doing things. Do these changes, however, lead to changes in the norms of the practice, or in its conception of the good at which medicine aims, or of the medical virtues? Or do these norms simply persist unchanged as essential features of any forms of medical practice?

The answer to this question is difficult to give. It is not clear what, in fact, the answer is. What is clear is that the content and meaning of the traditional virtues requires reinterpretation within these radically changed circumstances. For example, how is benevolence, as a distinctively *medical* virtue, to be interpreted in those forms of the practice where the individual patient is literally not seen, as a person, but only through the mediation of the records, laboratory reports, or monitoring data in a computer network? Surely, technical competence and responsibility in making medical judgments in these circumstances conduces to the benefit of the patient. But 'the ruptured spleen in Room 409' or 'the infarct in Emergency' is the object of medical benevolence in a way which strains the traditional conception of that virtue, or at least gives it a new and exotic reading. Can benevolence be practiced as a virtue toward anonymous persons? There is an old interpretation of the virtue of charity (by Maimonides) which emphasizes just this point: that the highest form of charity is that excercised by someone who is neither known to the receiver, nor knows who the receiver of his or her charity is. Shall we adopt that interpretation for anonymous benevolence? Can benevolence be exercised abstractly as a general virtue of character simply by dedicating oneself to doing one's job well, or does this confuse benevolence with competence? If the motive of competence stems from the satisfaction of the exercise of one's abilities, with no regard for the effects of this competence on the well-being of others, can it be counted as 'unintended benevolence' since the objective effects conduce to the well-being of patients? Or is 'unintended benevolence' a contradiction in terms? If the outcome of a medical procedure is harm or injury to a patient, and can be shown to be due to a fault in a complex technological device or in a computer program, how is culpability to be assigned, and how is the virtue of responsibility and the vice of irresponsibility to be defined in such contexts? Clearly, in this and countless other instances, the interpretation of the virtues becomes a moral – and a legal – task of major dimensions. And this is precisely what faces medicine at present, and what accounts for the rapid and unprecedented concern with medical ethics and with the teaching and practice of the 'medical humanities'. The answer to the question of whether the norms of medicine are changing cannot be given, therefore, in any sweeping or *a priori* way. It can be traced, however, in the details of the present intensive discussion, in the legislation, hearings, judicial decisions and court cases, and in the institutional forms of

Ethics Boards, patients'-rights groups, and those organizations of practitioners which address some of the new questions of social responsibility that face the profession. I cannot attempt to answer the question here in this concrete way. What I have tried to do here is to pose it in a certain way, in terms of the historical and social contexts of the medical virtues.

Is a Critique of Norms Possible From Within Medical Practice?

The final question is perhaps the most problematic. I suggested earlier what some of the difficulties here are in establishing the normative basis for a critique. The question 'What ought the virtue of medicine to be?' or 'What is the true end of medicine?' and the correlate question 'What are the medical virtues?' do not admit of answers in terms of what the virtues are taken to be, *in fact*, by either the profession of medicine, or by the society at large. For this would be to report or describe what the current beliefs are, and not yet to take a critical approach to these very beliefs. The question is therefore not simply 'What do *I* think they ought to be' but rather the theoretical question *par excellence*: 'What is medicine?' But this would seem to suggest some essentialist approach which presupposes that medicine as such has a 'nature' apart from and independently of its social and historical contexts. If, by contrast, we take the 'nature' or 'essence' of medicine itself to be an historically emerging and developing one – i.e., one which changes through time, or one whose 'essential' features, though they persist, express themselves differently in changing contexts, then, though we may avoid an abstract essentialism, we would seem to be caught in one or another form of historical relativism. And then, what grounds can there be in the present for criticizing the present? Against what autonomous standard is the assessment of virtues to be measured, or the determination of what they *ought* to be, arrived at, if the critic is also historically and socially situated?

Some insight would seem to be afforded by the example of the law: Whatever justice is, the law tends to operate on the basis of precedent and principle, and to adjust, interpret, and amend these by some systematic process of argument and adaptation to current needs and beliefs. But it may be argued that the law represents either the interests of those who are socially, economically, or politically dominant in a given society, or by the adjudication of conflicting interests or beliefs for the sake of social stability. And if this is so, the normative force of law and of judicial decision derives from prior commitment either to the interests of a ruling class, or to the overarching good of social stability. In either case, there would seem to be no ground for the normative force of the law, other than the opinions, judgments, and agreements that come to prevail either as the result of social power or within some commonly accepted mode of procedure. And if the procedure itself is put in question, what then? It may be argued, instead, that that law embodies some essential and overriding principle – e.g., equity, or fairness, or some other view of justice – against which all subordinate judgments are to be measured. But then, too, the principle would either have to command consensus – in which case normative critique could only derive from the ways in which such consensus is arrived at – or it would have to appeal to some extra-societal and transcendental ground – in which case, *that* would become a point at issue. The law, then, provides at best an interesting analogy to the case of medical norms, but it would seem to be in the same boat.

The alternative appeal to an *immanently* 'essential' normative ground is that of *historicism*: the ground of critique lies in an understanding of the development of medicine itself, as a practice, and in its relation to its larger social setting. ("*Die wahre Kritik liegt in der Entwicklung selbst*," to quote Feuerbach's Hegelian aphorism – in this case, about the critique of philosophy.) The trouble here is that such an 'immanent' critique depends either on some notion of where medicine is heading, historically – and that can degenerate simply into forecasting – or on some notion as to where it *ought* to head – and this reduces once again to *a priorism* or to the earlier question, 'What *ought* the virtue of medicine to be?' or 'What *is* medicine?'

We come full circle to this question, then, once again. The problems of an infinite 'critical' regress, or of a 'critical' circle, which we introduced earlier, and which we said were not 'desperate' problems, leave us nevertheless with no recourse, I think, than to face this question fully, and to recognize it as a *theoretical* question, that is, a question whose answer would have to be a full and systematic theory of what medicine is as a practice. I will end with some brief suggestions as to how I would proceed to answer it.

First, any conception of what medicine is has to start, I think, with the fact that it is a social, human practice; and as such, it is one of the fundamental ways in which human beings order their existence and provide the means of their survival. The basic needs of human existence to which medicine answers have to do with the phenomena of disease or illness, pain and suffering, and birth and death. These are clearly transhistorical and transcultural phenomena of human life; and to the extent that medicine may be characterized as an historically continuous practice, from prehistory to the present, medical practice extends beyond the particular modes of the practice at any given time, and predates its professionalization in ancient cultures. This is an exceedingly broad context; yet, if medicine has a transhistorical or 'essential' virtue, it must be derived from the transhistorical needs which it meets and the values which it serves.

Second, as a social practice, medicine has historically become professionalized in various ways, as a result of the division of labor in society generally, and out of the needs of special education and training that the growth of medical knowledge made necessary. This professionalization, emerging out of the historical contexts of tribal folk-medicine with the development of special castes or guilds or secret societies of healers, places a value upon the self-regulation or autonomy of the profession, with respect to certification of competence and of control over who shall practice. The values of the profession as such therefore have to be added to the values that medicine represents for the society at large, i.e., the values of health and of cure or treatment.

Third, though it may be said that the practice of medicine is a response to a social need which transcends all historical and cultural differences, and that the development of medicine as a professional practice is a universal phenomenon of all cultures and historical periods, whatever the variations in the modes of the profession may be, I have argued earlier that what medicine is in any concrete instance is to be understood only in terms of the particular form which the social needs have historically taken in that instance, and the specific modes of professionalization or of institutionalization that developed in that instance. What medicine *ought* to be, clearly, is that mode of professional practice which best meets the prevailing social needs, or the newly emerging ones. The normative basis for an evaluation and critique of the practice then depends entirely on the determination of what the social needs are, or are likely to become in the near future, and on whether the professional forms of the practice are appropriate to meet these needs. But who is to determine this?

It seems to me that in a democracy, the people who ought to determine what are the needs that have to be met medically are those whose needs they are – i.e., the people at large, through their representative or participative forms. But this requires that the perception of these needs be medically well-informed. In a sense, then, the medical profession has a responsibility to inform the public of what its state of health is and ought to be, or better, to educate the public medically so that they will be able to make such judgments for themselves on an informed basis. On the other hand, the profession's conception of what the needs of medicine are may be strongly limited or influenced by its own inherited traditions, medical beliefs, and social ideology: the medical profession is not (yet) a good cross-sample of the population. In a sense, then, the public has a responsibility to inform the medical profession of what its needs are, so that the profession can respond to this perception. What I am suggesting, therefore, is that the answer to the question 'What ought the true aim of medicine to be' is an answer no one can give *except* as the result of a dialectic between those who practice medicine and those on whom it is practiced (including here not only actual, but prospective, consumers of medical services – i.e., the population at large).

The difficulties with such a proposal are several: (1) In the current political economy of medicine, those whose needs and whose perception of their needs are most clearly heard and responded to are those with the means to express these needs most effectively. This is not quite a tautology. Two things talk loudest here: money and the vote. Therefore, the expression of medical needs is tied to economic and political clout. This does not mean that medicine has ignored the needs of the poor and the politically weak entirely – for some practitioners, such service becomes a moral imperative in its own right – but it does mean that the priorities of medical response to public needs is skewed in this way. It is only the vote that has been effective in orienting the practice to the needs of the poor, for example; and thus depends on the efficacy of democratic procedures. (2) To the extent that the public is informed about its own medical needs, it ought also to be informed about what is necessary to meet these needs – i.e., what the needs of medical practice are, if it is to effectively meet the social needs. But then, the public is responsible, ultimately, for providing the medical profession with those instrumentalities required for adequate practice. And this seems to make the profession and *its* needs hostage to the public understanding of medicine.

Despite these difficulties, it seems to me that the determination of the virtue of medicine *in the concrete*, and of the derived medical virtues, has as its normative ground the informed judgment of the public. This says, in effect, that the virtue of medicine depends on the virtue of freedom – i.e., on the social and human virtue of free and democratic determination by individuals of what their needs are and how they should be met. But this freedom cannot be practiced effectively unless it is a well-informed freedom. This makes the responsibility for the education of the public in matters of health and illness one of the cardinal virtues of the profession. The 'practice of medicine' in this larger sense goes beyond the profession itself, to include the medical public as part of its 'practicing' constituency. In this rather peculiar sense, medicine is not simply practice *by* doctors *on* patients, but becomes instead a mode of social co-operation in pursuit of a common good. The virtue of medicine is thus a 'social' virtue in this deepest of all senses.

NOTE

[1] See there also the interesting discussion on Ideals, Virtue and Integrity in Chapter 8.

BIBLIOGRAPHY

[1] Beauchamp, T. L. and Childress, J. F.: 1983, *Principles of Biomedical Ethics*, 2nd edition, Oxford University Press, New York.
[2] MacIntyre, A.: 1975, 'How Virtues Become Vices: Values, Medicine and Social Context', in H. T. Engelhardt and S. F. Spicker (eds.), *Evaluation and Explanation in the Biomedical Sciences*, D. Reidel, Boston and Dordrecht, pp. 97–111.
[3] Wartofsky, M. W.: 1982, 'Medical Knowledge as a Social Product: Rights, Risks and Responsibilities', in W. B. Bondeson, *et al.* (eds.), *New Knowledge in the Biomedical Sciences: Some Moral Implications of its Acquisition, Possession and Use*, D. Reidel, Boston and Dordrecht, pp. 57–72.

Baruch College and the Graduate Center,
The City University of New York,
New York, N.Y., U.S.A.

EDMUND L. ERDE

THE VIRTUES OF MEDICINE: MEANING AND IMPORT

WHAT IS A VIRTUE? – PRELIMINARIES AND PUZZLES

What might 'the virtues of medicine' mean? If we start to answer this by asking next what does 'virtue' mean and if we try to answer this by reference to ordinary occurrences of the word, we find that it occurs almost exclusively in the context of the idiom 'by virtue of . . . '. This expresses causal or contextual explanation as in 'By virtue of his great weight, he was able to break down the door' or 'By virtue of much world travel, she was able to pass the geography test.' In this sense, a discussion of 'the virtues of medicine' might call for a discussion of what medicine can produce.

However, the context of this chapter is a book of moral theory. Thus, the far less common, perhaps archaic moral sense of 'virtue' is relevant to our exploration. Accordingly, I must consider the historic roots of virtue theory to explore the virtues of medicine. This is required because the rarity or absence of the moral uses of 'virtue' from ordinary discourse means that a reader cannot expect that I am using it in a sense he or she can take for granted.

Moral theorists, starting with the ancient Greeks, cast the virtues as constructive character traits or orientations of individuals. Aristotle's account is seminal. He begins by tentatively distinguishing the intellectual virtues – those concerned with understanding the nature of things – from the moral virtues – those concerned with influencing action (*Ethica Nicomachea* 1103a3–10 trans. Ross). In section three, below, I shall argue that these should be reunited under the account Aristotle gives of the moral virtues: that they are the constituent states of character or orientations of individuals[1] which dispose them to choose to act in balanced, appropriate ways – ways which can be defended as 'a mean' between extremes, ways chosen in accord with rational principles rather than disorganized feelings or passions (*Ethica Nicomachea*, 1106b36–1107a2). Such principles entail the ability (or excellence) to plan living one's life in and by means of a community in which a fulfilling human life can be lived [33]. Accordingly, the virtues structure how persons present themselves as choosers who take themselves seriously and take their community seriously.

Earl E. Shelp (ed.), Virtue and Medicine, 201–221.

Recently, Alasdaire MacIntyre has followed the Aristotelian model and outlined a relationship between the virtues and aspects of communities which he calls 'practices'. By 'a practice' he means a coherent (but complex) socially established cooperative endeavor which generates intrinsic satisfaction for those accomplished at living up to its demands of excellence; this accomplished living expands both the vision of the endeavor and the means of working at it ([41], p. 175). MacIntyre defines the virtues as those structures of an individual human's character which contribute to the good functioning and the progress of a practice by having the individual disposed to think and act in ways constructive for or commensurate with that practice; persons so disposed gain fulfillment from the practicing itself.

Let us agree that medicine is such a practice (though its nature will require explicit discussion below). By 'medicine' we shall mean 'the culture of doctors and/or the sciences and orientation they need in order to provide excellent health care (i.e., anatomy, physiology, cell biology, etc.).'[2] If we retain the current explanatory non-moral sense of the term 'virtue' *and* harken back to the moral sense, we find as the answer to the opening question three meanings of 'the virtues of medicine':

(A) the virtue(s) – the orientation(s) of characters – which the practice or the institution of organized medicine demands of its practitioners,

(B) the virtue(s) – the advantages – which organized medicine offers to a society or culture,

(C) the virtue – the orientation of character – which constitutes being a physician.

Questions stress the sense of each of these and make them puzzling. Regarding (A), how can an institution demand? Regarding (B), what has supplying advantages to do with virtue in the moral sense? Regarding (C), how can medicine be an individual's orientation to the world?

Although they are perhaps stressing questions, I believe each interpretation of the phrase 'the virtues of medicine' is valuable. Working through each will constitute the approach of this essay. A useful, unified understanding may emerge. At least the product will contribute to the enterprise of this volume. For I view this volume as a set of answers to the following instruction: "Use the classic moral concept of *virtue* as a filter or prism or lens, and apply it to various key ingredients in health care." Thus, there are chapters on the virtuous doctor, the virtuous nurse, the virtuous patient. The current chapter explores *medicine* at several levels of abstraction – as a subculture (as in sense A, above), as a practice in MacIntyre's sense (as in sense B, above) and as an aspect of minds (as in sense C, above). And, although I am sanguine about meaningful findings, I am eager that the vices of medicine not be swept from view by rhetoric which is too positive. For that reason, vices will be an integrated part of the discussion.

MEANINGS OF 'THE VIRTUE(S) OF MEDICINE' – PRELIMINARY INTERPRETATIONS

Meaning (*A*): *The Virtues Which Medicine Demands*

One meaning of 'the virtues of medicine' straightforwardly fits MacIntyre's analysis: the virtues of practitioners demanded by the practice. The ends or goals of medicine include returning or preserving health, and providing patient education [20]. The virtues demanded for its practice incline persons to seek biomedical knowledge: diagnostic, prognostic, and therapeutic know-how. To pursue such goods as health-care provides is to pursue what I call 'the fundamental logic' of a practice. However, many regretable habits of action might also foster the growth of the institution of medicine. In other words, medicine's growth might necessarily depend upon some vicious arrangements which are native or 'domestic' to that practice [19, 20].[3] The trouble is that we have difficulty sorting necessary and contingent non-virtuous norms which arise because of psychological and social history from virtuous norms. It is also the case that what subcultures prescribe cannot be considered ethical just because they prescribe it ([58], Chapter 4).

Nevertheless, this interpretation (A) of 'the virtues of medicine' – 'those virtues of individuals which medicine as a practice demands of its practitioners'[4] – bears a metaphysical difficulty mentioned above since the phrasing casts a peculiar, multicultural, odd historical enterprise as a unity which can demand. For (A) says that the practice is making the demand. Is such language merely metaphoric, is such thinking merely anthropomorphic?

We can, following French [27], recognize group action: crowds disperse – while individuals in them just move away; corporations dissolve – while individuals go elsewhere; surgical teams perform transplants – while individuals just cut and suture [19]. We recognize that groups (for example, unions) can even demand. But can the strange creature that is referred to by the term 'medicine' be said to do anything or be responsible for anything?[5] It could even be controversial as to who in the full past – including times before the granting of M.D. or D.O. degrees – is to be counted as a member of the group (cf. [38], Chapter 1) delineated by the term 'medicine'.

To find some actions or characteristics that apply to medicine as a practice, consider how Duffy [15] has shown that medicine had a bad name in the 19th century. Surely not all doctors had a bad name, but the way doctors squabbled among themselves, the lack of scientific base for their decisions and their uses of regrettable therapies gave not just them, but medicine a bad reputation.[6] To overcome this, the AMA emerged in 1846–7 urging among other things decorum of its members – and loyalty to them on the part of their patients!

Thus, a set of norms could be said to be urged or demanded by an organization (the AMA). But what of an entire practice? Members of the contemporary American practice could include individuals, organizations such as medical schools, local, state, and national societies, state licensing boards, and medical journals. These entities help give medicine an existence of its own, an ethos – so that attributing to it that it demands virtues of its practitioners (e.g., to be more decorous or courageous [25]) makes sense. In this framework, we might find that there are special virtues which medicine demands and no other practice demands. Perhaps we shall find that many of the various virtues of ordinary community life have a special structure when they are 'of' or 'in' medicine. I shall explore some of these possibilities later in this essay. Currently, however, I am trying to establish meanings or interpretations of the phrase 'the virtues of medicine'.

Meaning (B): Medicine as Virtue(s) of a Culture

Another possible interpretation of the phrase 'the virtues of medicine' is more feasible if we rephrase it as 'the virtue of medicine'. Through this phrasing, we could consider medicine to be a virtue of a culture as, say, honesty is of a person. There might, for example, be cultures which left therapeutics and its sciences out of its practices. A slight rewriting of the perverse novel *Erewhon* would exemplify this [10]. In *Erewhon*, people are arrested and punished if they are sick, whereas they are treated medically if they have committed a crime. In our proposed revision, the sick would just be ignored or exploited (as they are among the Ik [56]). Under this interpretation of 'the virtue of medicine', medicine is a means to social goods such as public health.[7] It could help in the understanding of the impact of an enviroment on the make-up of a culture's individual members. Some old studies attempt this. For example, one by J. M. Foltz contrasts the quality of life on Minorca as reported during the 'positive' administration of the British in the period 1744–1749 with the observed consequences of living under 'negative' Spanish administration 90 years later [24]. More contemporary moral analysis concerns the impact of carcinogens [29].

Medicine might also bolster or provide the means to act upon personal values. For example, medicine provides a culture with a rich context in which individuals can realize their sense of dedication to serve others, to answer the call or join the fight against human suffering [21, 30]. This bridges into the final meaning of 'the virtues of medicine' – the idea that a person in the field of medicine is a participant in a particular virtue – the virtue called 'medicine'.

Meaning (*C*): *Medicine as a Virtue of Individuals*

Pellegrino and Thomasma sometimes cast medicine in a seemingly problematical anthropomorphic way. For example, they declare that "it [medicine] must make right choices . . . " ([45], p. 148). At other times, they cast medicine as an aspect of a better known virtue – i.e., clinical prudence ([45], p. 136). And, at yet other times, they cast medicine as a synergistic union of science, of art (perfect making or craftsmanship) and of virtue ([45], p. 149). If medicine is either the virtue 'clinical prudence' or a strand of it along with art and science, then medicine can be understood to be not just a practice or an institution, but itself a trait within or an orientation of the character of persons who practice it – a trait or orientation of how, in particular cases, to choose to act based on responsible methods of knowing coupled with a commitment to certain ends such as health ([45], pp. 120, 122). Any chosen action should be tailored to the particular needs of a particular patient. The means for doing this – scientific knowledge and sensitivity to idiosyncracy aimed at health – constitute the virtue of medicine in this sense. This joining of general knowledge and particular judgment returns us to Aristotle's distinction between the intellectual and moral virtues.

MORAL AND INTELLECTUAL VIRTUES – DEPRECIATING THE DIFFERENCES

In Aristotle's view, the intellectual virtues are not concerned with action but with ways of pursuing truth. They arise not from habit but from teaching. They are structures or orientations of the person's soul or mind which allow grasp of universal metaphysical truths and/or details of natural science and the means for managing human life.

In her classic 'Modern Moral Philosophy', G. E. M. Anscombe argues that today we can barely understand or apply Aristotle's distinction because we feel inclined to distinguish the *morally* blameworthy from *intellectual* failures, and this over-draws and misses Greek understandings [1]. If she is right, we can now understand Aristotle's own retreat from the distinction (*Ethica Nicomachea* 1144b30–1145a30) and we can savor the insight in his persistent tendency to identify similarities between medicine – its theory *and* practice – and virtue (cf. [36])!

Consider how intellectual excellence connects with senses A, B, and C of 'the virtue(s) of medicine'. Under sense A, medicine demands of its practitioners that they be knowledgeable. Under sense B, a medical world-view can disabuse a culture of false beliefs about human life and public health. Under sense C – medicine as a perspective, a trait, or an orientation within the character of persons – interesting results are to be found. For I want to follow Anscombe and propose that taking Aristotle's distinction between the intellectual and the moral virtues radically (he did not) would be a mistake. Both kinds of virtue should be accorded epistemic status; both are ways of knowing *and* doing. I take this to be a central tenet of this chapter.

Philosophy since Aristotle has recognized instrumentalists' and pragmatists' insights. Articulate persons have come to argue that even our most abstract and theoretical concepts are devised with instrumental agenda [37]. I want to borrow Wittgensteinian arguments from Hanson [31] to diminish the distinction between intellectual and moral virtues. Then I shall test whether medicine fulfills the requirements of being a virtue – being a constructive structure or orientation of an individual's person or self (sense C). Finally, I shall try to show how this explains (but does not fully resolve) many articles of contention in medical ethics.

Hanson [31] argues that empirical concepts are like shared instructions from the culture for how individuals are to perceive the world. Seeing, he argues, is not a passive, bodily event followed by an active mental event of interpretation. Rather, seeing is an activity based upon culturally provided (taught) contextual understandings or background rules. Ideas and theory are built into our visual terms. Our visual concepts are ways of our being present in the world – not just taking the world in.

Let us consider the moral virtues as instructions or guidelines about how to be in the social world [34]. For, even if 'virtue' in its moral sense is absent from current discourse, many of the terms we use to characterize people reflect the virtue orientation of ordinary moral thinking. 'Be honest', 'be kind', or 'be firm' may not give details about how to act in a given context, but they give direction; they are not platitudes. To know what compassion is is to have some grasp on how to proceed.[8] This seems implied by Aristotle's account of the virtues: once a person has acquired the right habits *together with* the rational rules, he is constituted for social life – action and thought. But those rational rules embody (an) intellectual virtue; nonetheless, the rules have moral and practical elements (given the insights of instrumentalists). Thus, we may agree with Aristotle's own retreat from distinguishing the intellectual and moral virtues in a radical fashion.

Under this analysis, virtues – whether moral or intellectual – are a sort of readiness, a sort of being in the social world with a set of means to think about it and pursue elements in it. Medicine is a virtue in that it is a slant on the social-natural world, a perspective. It has or sustains an attitude – perhaps best symbolized by the wounded-healer – a person outraged by his or her own potential (or real) bodily suffering and that of others; such a person is eager to join as though in battle against the evil forces of disease, injury, and death [21, 30]. Medicine also provides some of the sensory concepts – for perceiving the normal and abnormal – the biologically benign and malicious [18].

This, of course, is the core of medicine's relation to science. Like physics, with which Hanson was dealing, medicine provides empirical concepts. But unlike physics (which is not a virtue in sense C), medicine deals both with individual persons and populations as its subject matter. And so medicine is a virtue (senses B and C), for it provides concepts through which to be observant for the good – especially bodily good – of people by way of the categories of *injury and disease*. These categories provide practitioners of medicine with some measure of practical wisdom – an understanding of what is best for the body.[9]

Medical science, then, can be thought of as a branch of practical wisdom. It contains information about which bodily *abnormalities*, if any, are likely to be compromising the present functioning of a particular person or which are likely to compromise its future functioning and how. These 'abnormalities' must be reportive first and foremost of what is undesirable, whether common or not. Medicine, then, as practical wisdom about the body, is the *science* of what may be called *bodily welfare*. It provides its practitioners with the empirical concepts which enable them to see the bodily as faulty or not. Contemporary medicine provides the means for observing or discerning the fine grain or details of what is dangerous, and for working to remedy the dangers.[10] Of course, the formulation need not be as dualistic as this – one can think of health of persons in a more holistic way and see the dangers as environmental, social, psychological, and biological.

HOW THE VIRTUES CAN BE VICIOUS

Unfortunately, this focus on the fine grain of welfare is often taken more as scientific than virtuous. This is not without its moral dangers. For, as Hanson shows, learning to see involves learning to ignore as meaningless other elements of the world. In medical education, for example, there is an expectation that the students will be remade or surrender.

Chester Burns, for example, claims that "The hallmark of being a professional person is that the individual is willing to undergo a transformation of his personal and social values . . . " ([9], p. 184). More generally Alasdaire MacIntyre argues that "It belongs to the concept of a practice . . . that its goods can only be achieved by subordinating ourselves to the best standard so far achieved, and that entails subordinating ourselves within the practice in our relationship to other practitioners" ([41], p. 178).

From anatomy [39] to surgery one of the virtues seemingly demanded is a skilled detachment. For example, Charles Bosk, in *Forgive and Remember*, reports how people in surgical training have to confess their errors so that they remember them as a way to avoid recommitting them; only on this condition can trainees be *forgiven* for their errors *by their teachers* [6]. Bosk describes four kinds of errors that have been devised by the supervising physicians to categorize mishaps: (1) technical errors are physical slips; (2) judgment errors are mistakes about what is happening inside a patient; (3) normative errors are acts of disobedience to general requirements of any service such as failing either to cooperate with nurses or promptly to inform supervisors about mishaps; (4) quasi-normative errors are either acts of disobedience to orders or acts contrary to habits of individual supervisors.

Bosk states that (1) and (2) are cast by supervisors as forgivable, since anyone (themselves included) can commit them in perfect innocence while (3) and (4) are easy enough to avoid. Returning to the contemporary distinction identified by Anscombe [1] between intellectual and moral failings, we may say that (1) and (2) are errors of skill or intelligence while (3) and (4) show weakness of character – true problems of moral virtue, *not* fixable by didactics or instruction. Indeed, the reactions of supervising physicians seem to be that such problems with disobedience are failures of virtue and barely fixable at all.

By contrast to the surgeons, Steven Hoffmann [32], in describing the moral stresses of being a medical student, downplays the importance of redemption in having learned from mistakes. He does not accept the idea that inadvertent harm to a present patient is justified by future skill benefiting someone else. He denies that exoneration follows from good remembering alone. Further, he seems to reject the idea that it is his teacher's pardon he most needs. Hoffmann feels that he must "suffer privately for the [patient]" ([32], p. 18). He rejects both adopting the ethos of medicine and following the models in roles ahead of the one he occupies (such as that of resident). Indeed, Hoffmann resists the transformation in Burns's remark and the surrender in MacIntyre's. His position coincides with MacIntyre's that neophytes need "to recognise [sic] what is due whom; ... to be prepared to take ... self endangering risks; and to listen carefully to what we are told about our own inadequacies ... " ([41], p. 178). The three imperatives listed in this quotation are, respectively, the commands of justice, courage, and honesty. Of these, Hoffmann's position emphasizes courage, for he argues that students must reserve their "freedom to question and explore and withhold ... commitment to attitude and style" ([32], p. 19).

Hoffmann rejects the idea of accepting decisions just because they are said to work. His ultimate aim in asking the questions and raising the challenges is that all students may "care sensitively for others" ([32], p. 19). He seems eager, then, to retain that which important aspects of medical education (such as anatomy) seem intended to dissolve or at least diminish [14]. In short, Hoffmann wants to retain intact some pre-professionalized moral virtues and resist some demands of the profession as voiced through educational structures.

Hoffman is morally conflicted over a clash between his commitment to ideals that might have brought him to a medical career and the psychological theory of those who would teach him how to relate to his patients. Perhaps we can see wisdom and morality on each side of the conflict. But medicine has also made demands of its practitioners that appear *less* moral. It has enjoined them to cater to certain social forces or trends which lack virtue. Duffy [15] shows how medicine is a follower of social trends such as racism and sexism. Further, having been largely uncommitted, American medicine, in seeking the favor of the establishment, came to offer arguments against abortion as a way of fostering the position of Protestants who were worried that Catholics might come to outnumber them.

Other examples of medicine's more questionable demands of its members include the demand for decorum among its practitioners as a way to foster achieving a monopolistic status [5, 51]. This may have nurtured some scientific and technical advances while causing serious cultural alterations – some charge damage – through coopting human independence and the sense of humanity's place in nature [35]. Consider, for example, a note in *The Lancet*

[59]. It discusses the conflict between lay-oriented movements and profession-oriented movements. Lay groups (which include some professionals) are eager to reduce parents' reliance on professionals' orders, mandates and superiority. By contrast, the President of the American College of Obstetrics and Gynecologists recognizes 'the siege of liberationists' and the need to take the offensive against those who would allow popular values to supplant hard-won, excellent technical care. Other controversies mentioned in the note include whether home delivery is child abuse, whether parents should determine the withholding of therapy for defective newborns and whether laws should be passed for controlling the risks pregnant women can take which could damage their babies.

The concern raised in this short note is whether health and therapies should be under the control of physicians. Perhaps the greatest contact between medicine and culture is where physicians have gained such control. They are monopolistic gatekeepers to such social goods as whether someone should be on worker's compensation, whether someone is able to serve in the military, whether someone should have access to medications and other therapies, whether someone should be exempt from ordinary obligations of productivity and instead obliged to be in the sick role and seek care. Surely this gatekeeping is often to the public good as with professional control of antibiotics.[11] But the heart of the moral dilemmas growing out of a control-oriented ethic of professional judgment must be shown unequivocally. Its chambers are (1) that such control affects and may undermine autonomy and responsibility for self care, (2) that it keeps many goods out of reach inappropriately, (3) that its knowledge base is shaky, and (4) its choice of parameters for measuring outcomes is often questionable.

Thus far in this subsection of the essay I have discussed how both the virtues medicine demands of its practitioners (e.g., obedience) and the virtues cast as cultural advantages of medicine (e.g., gatekeeping) might turn vicious. These virtues actually seem to merge in a way that compromises medicine as a virtue of individuals.

Recall Aristotle's account of the virtues. The choices he identifies as appropriate subjects for virtuous choosing include the possibility of selecting a mean relative to the self in a society. Aristotle holds that practical wisdom bears not just on universals such as 'the body', but on particulars such as Mr. Hughie Polloy. It has often been alleged that science has so structured medicine as to make it depersonalized or dehumanized – a victim of too much of a good thing. Indeed, both what medicine demands of its practitioners in the way of detached concern and what medicine offers to *this* culture in the way of support for the culture's concern which equality renders physicians depersonalized; they must either endorse their patients or tolerate them, never judge or punish them.

Medicine, then, seems to suffer in a general way from a condition which MacIntyre generally describes in "How Virtues Become Vices" [40]. He argues that moral inversion occurs when a practice gets detached from the original contexts in which the practice and its premises functioned.

At another level of depersonalization, medicine often seems interested in the probability of this or that affecting members of a population and neglects both the actual physical occurrences in individuals [50] and the non-biological implications of medical actions on an individual's life. That is, scientific medicine is dualistic: it is the science of bodily welfare apart from the individual self who is embodied. There are advantages to this kind of dualism [17], but a definition of 'health' should include both norms of others and norms of self [18]. When the self is included in medical thought, medicine ceases to be like physics – a science for viewing and understanding natural phenomena. It is then a virtue – a means for viewing other persons and acting in community with them. As medicine's current science gives its practitioners some structures for perceiving and some means for acting, it diminishes their ability to see non-biological phenomena.

This has been compellingly shown. Reiser [48] shows how technology, in enabling physicians to perceive into the body, has resulted in physicians ignoring much about the personal testimony of the patient. Baron [4] (referring to Reiser and others) shows how Sir William Osler's ability for sympathy and empathy in treating fevers with cool baths declined to a significant degree because of scientific categories. In the first edition of his classic *The Principles and Practice of Medicine*, Osler "sympathized with those who designate it [the bathing] as entirely barberous" noting it was done " in spite of piteous entreaties" and finding it "scarcely feasible" to use in private practice. By the fourth edition, a mere nine years later, Osler, explicitly impressed by the good results, cast the same symptoms of suffering as "not serious".

The point here is that Osler began to take outcome data as determinative of what to look for and therefore of what to see. Data are understood apart from the self who is embodied. They are collected in categories of values stipulated from a point of view of what is best in populations; the individual is thus neglected. Such is part of the reordering of medicine's connections with its premises – its diminished emphasis on individuals and heightened fascination with general knowledge.

The grasp of populations that the science (as opposed to the art or the virtue) of medicine provides can be put in moral terms through values that apply to individual members' *welfare*. These values seem platitudinously obvious. Nevertheless, I shall tease out a set of indicators of welfare by drawing upon Bernard Gert's *The Moral Rules* [28].

Gert identifies the following as evils: (a) death, (b) (undesired) pain, (c) disability, (d) loss of freedom or opportunity, and (e) loss of pleasure ([28], p. 45). He argues that humans have a very general set of beliefs required by reason. These beliefs give rise to rules we would all advocate, rules to the effect that (a)–(e) are evils since each of us wishes to avoid these for ourselves and those whom we care about ([28], p. 24). The moral rules, according to Gert, are the minimal interpersonal rules restricting one from inflicting any of (a)–(e) on fellow persons.

These rules should be distinguished from moral ideals which enjoin us to do good rather than merely resist doing evil. 'Preserve life', 'Relieve pain', 'Lessen disabilities', are such ideals. Gert declares that "In former times these moral ideals were the primary motives for those who went into the practice of medicine. Thus doctors used to be regarded as among morally best men" ([28], p. 130).

Medicine, then, is a means to welfare. Welfare includes (among other things) having life, being pain free, being fully functional, being free and enjoying oneself. Disease, especially serious disease, diminishes one's welfare. Medicine is the knowledge of the body for diagnostic, prognostic, and therapeutic purposes. And although this is not virtue (as defined by Pellegrino and Thomasma) but science, scientific competence is an intellectual virtue of practitioners as demanded by the practice (sense A) and it is the virtue which the profession offers to a culture (sense B). As such, certain sensitivities are lost. I shall further explore some of this loss in the next section.

SOME CONFLICTS BETWEEN ETHICISTS AND MEDICINE EXPLAINED

In addition to the abstract ideals of welfare, there are specific preferences, values, endeavors, and affiliations each person has. These may be termed 'interests'. The term is both misleading and felicitous. It misleads because the phrase '*best* interests' is used as synonomous with what I have called 'welfare'. On the other hand, the term 'interests' is felicitous because it captures the psychological, experiential subjectivity of personal values. I take interests, then, to be idiosyncratic – devised or adopted by a person subjectively. These constitute his personal values. Thus, it may be to one's welfare that he or she not smoke – given the cell biology of the lungs and the cardio-vascular system – but it may be important to a person in living his life that he have recourse to whatever good the smoking does. Thus we can see that sometimes one trades away aspects of one's welfare in pursuing one's interests.

Beyond welfare and interests we need the concepts *autonomy* and *self-determination* if we are to be providing a rich account of how issues in medical ethics arise. For my purposes here, these are best thought of in terms of *rights*. A system or what Flathman [23] calls 'a practice of rights' serves to protect rights-holders in two distinct ways: it protects from those who would help themselves to the goods of the rights-holder; i.e., from exploiters. It also protects from do-gooders; i.e., from the reformer in each of us. Do-gooders – paternalists – have the notion that they know the welfare of someone and they are licensed by that greater knowledge to disregard that person's interests. In short, they disallow as a bad deal the trades an individual may risk or make. A trade is strictly a value judgment about what is best from the point of view of personal interests, not from that of abstract bodily welfare. We need rights, then, to allow ourselves security in pursuing our interests against bad guys and do-gooders. As Feinberg [22] argues, rights are the special protective device recognized by society for an individual to employ *on his own behalf*.

But there may be times when a rights-holder's actions seem so immediately and extensively self-endangering, that those in the know have warrant to act paternalistically [13]. This is surely something the profession urges its practitioners to accept (sense A). Further, it is surely a moral benefit which medicine offers to society (sense B). Finally, it is a virtue of individuals to be constituted to act paternalistically (sense C). The trouble is, that even though it is an exaggeration to argue that medicine blinds its practitioners to perceiving the interests of others., it does seem to lessen their vision. As happened to Osler, bodily outcome rather than individual patients' interests is the dominant lens which medicine currently places over the eyes of its practitioners. Despite the great advantage to patients that their doctors perceive welfare, there is, also, the regrettable consequence that the doctor loses touch with the patient's interest and decides for abstract levels of welfare or even shallow guesses at a patient's interest based purely on stereotype [52]. At least this is the way to understand the critique of medical paternalism so prominent in recent medical ethics [8] and the way to understand the struggle to supplement an ethic of curing with an ethic of caring.

It is also the way to understand the bioethics backlash [11, 57]; doctors feel assaulted by bioethicists even when they are trying to act as utterly virtuous doctors. Bioethics has been criticizing the core unified 'virtue' (not the virtuosity) of medicine – the 'virtue' medicine demands of its practitioners, offers to the culture, and develops in individuals as their way of being in the world – namely the 'virtue' I here identify as *control aimed at health* (where control includes knowledge, know-how, detached concern, retaining the monopoly, and the power to act from paternalism).

The critiques have sometimes been of medicine's shallow insensitivity to non-health consequences for individuals – that is, to patients' interests; or the critique has focused on doctors' insensitivity to autonomy and rights. But neither rights theory nor consequentialism seems adequate to the needs of moral theory in medicine.

Rights theory lacks adequate theoretical formulation, and, it has been argued, allows for grim but easily avoided outcomes [2]. The theoretical foundation of rights theory is uncomfortable in typically resting on social contracts that never took place, in taking insufficient account of the kinds of surprising cases modern medicine makes us confront, in presuming that an ancient or fictional contract can bind a present generation and in hedging about how much agreement is required to legitimate a policy.[12] In short, the autonomy represented by a 'legalistic' or Kantian sense of personhood is a conceptually unrealistic banner. At the same time, much purported paternalistic consequentialism is self-serving, arrogant, and presumptuously omniscient. Thus, a place for wisdom, flexibility, generosity, and common sense in general moral theory is needed [34]. But, we find instead conflicts between ethicists and, say, physicians, which are sparked by the radical use of one approach (e.g., welfare oriented paternalism) by one group (the physicians) clashing with a radical use of another approach (e.g., rights) by the other group (ethicists).

Early on in contemporary medical ethics, the flaws in both consequentialism and rights occasioned appeals to the concept *virtue*. Roger D. Masters takes up the issue, 'Is Contract an Adequate Basis for Medical Ethics?' [42]. After an historical sketch of contract theory and an indication of its weakness, Masters reviews ancient Greek ideals. He cites Aristotle's point that moral standards exist prior to contractual arrangements, that social and political good are secured by virtues which are the prior natural morality. Masters appeals to Plato's idea that medicine be used only when a useful life is likely to be lived and to Aristotle's sensitivity about infanticide being proper for deformed babies, but not for population control. But Masters seems to believe, with Aristotle, in the (near?) infallibility of a virtue as a choice-making mental structure and he seems to believe, against Aristotle, that moral principles can be precisely articulated. (I disagree with them on the first point and I side with Aristotle on the second.)

As a social phenomenon, the appeal to virtue theory, by Masters and this whole volume, seems a statement of despair over the inadequacies of current ethical theory's approaches to consequences alone or rights and duties alone [16, 54]. As a philosophical phenomenon, the book in which this essay is but a chapter is the attempt to begin to articulate what is still too latent – the kind of persons we all should be.

THE VIRTUE(S) OF MEDICINE – A SYNTHETIC CONCLUSION

Although I have analyzed 'the virtues of medicine' into three distinct meanings, I believe there is a spirit common to each. Virtues are a socially produced means for a person to be a good and reliable member of his/her society. In the case of the professions, the professionalization process turns neophytes more into members of the subculture of a practice and reduces their kinship with the large society. The subculture demands special virtues (sense A), is franchised because it yields certain advantages (sense B), and contains individuals who perceive themselves and the world primarily through the categories of *welfare*, *diagnosis*, and *therapy* (sense C). Virtues arising out of this cluster include (no doubt among many others) seeking scientific truth, being of detached concern, fitting preconceived social norms so as to secure monopoly, and being inclined toward paternalism. As already discussed, each of these may have an extreme which takes it out of Aristotle's 'mean' and turns the virtue into a vice.

Further, each virtue – as a maker of the self of an individual – places a limitation on other purportedly rewarding aesthetic or hedonistic pursuits: (1) Detached concern limits over-reacting to bad outcomes and the likelihood of diminishing one's engagement in the practice. (2) Concern for monopoly explains, I think, why many practitioners who are against some aspect of medicine (e.g., abortion) refer a patient to a willing colleague rather than flatly refuse any help in meeting the patient's need: the restraint is on professionals to keep the monopoly actively fulfilling patient need even more than complying with some dictates of conscience. (3) The concept of *welfare*, operating as a character trait through which a person views others, happily limits the quest for scientific knowledge – witness Rothman's [49] finding that physicians involved in the controversial syphilis and hepatitis studies were much more concerned for the subjects' welfare than were pure scientists. Such is the virtue of medicine in the character of individuals.

This must be why many oppose physicians' participating in active euthanasia and/or capital punishment.

Interestingly, though, what medicine embeds in its practitioners actually distracts them from certain kinds of benevolence. Both at the level of medical training and reimbursement for services people are rewarded for action – especially highly technical, demanding action. In consequence, current professionalization obscures approaches to helping patients which are not of the technical sort.

Consider the 'Sounding Board' in the *New England Journal of Medicine* by DeWitt Stetten, a physician researcher at the NIH [53]. Dr. Stetten was diagnosed as having macular degeneration – an inevitably blinding disease. He sought attention from many ophthalmologists. Apparently these physicians considered their tasks as involving only bodily welfare – specifically the rectification of morbid anatomy and pathologic conditions. This seems so, for they lost all detached concern for him once they determined they could not rectify the problems in his tissues. They said nothing, apparently knew nothing, of community resources for the blind, nothing of devices that exist that could help along the way. As a physician and as a patient, Stetten was upset by physicians' lack of awareness of the resources and patients' needs to know about them. Surely part of the problem is that there is little economic reward for physicians to store information and services which they cannot market. But above this more venal level, even at the level of virtue, the lack of technical biological activity relating to these social supports makes it implausible for physicians to be interested in them.

The significance of rewards being for action occasions discussing medicine as a personal virtue under yet another distinction: that between moral and personal virtues [28]. Moral virtues orient one to others – to helping others or against injuring them and entering relationships in bad faith. For example, courage allows one to risk oneself on behalf of others. Personal virtues, by contrast, help one avoid evils befalling oneself. For example, fortitude allows one to hold up in the face of risks and dangers to the self.

Practitioners of medicine sometimes have to help patients because the patients' personal virtues (e.g., temperance) have failed. Practitioners of medicine also aim to educate patients into developing personal virtues. Moreover, practitioners of medicine acquire a set of virtues which start as cognitive moral virtues (diagnosis, prognosis, and therapeutics) and yield substantial extrinsic reward (money, status) in being practiced. Practitioners also come to *define themselves* via their skills and activities and get intrinsic reward in the actual doing of their work [30]. They crave the gadgetry, the responsibility, and the problem solving [7]. So their virtue-of-medicine, their control aimed at health, is its own reward. But these are the intrinsically rewarding activities which make Stetten's hope for changing the profession unlikely to be realized. Here, again, then, is another way in which a virtue can become vicious.

Ironically, the personal virtue *medicine*, which develops in individuals who are angry about suffering, may not be rewarding enough as suggested by the morbidity of being a doctor (defined in terms of suicide rate, substance abuse, and divorce rate). Nevertheless, medicine is a trait of persons, a way of being in the world. It arises from the neophyte's being immersed in a culture, being socialized as to what medicine requires of its practitioners (sense A) and how medicine serves or fits its culture (sense B). It culminates as a trait of existence in an individual human, i.e., as *a* virtue (sense C). There is trouble, however, if it is the only virtue of the individual. If there is such a virtue it is surely important that it be accompanied by other virtues. Our theoretical needs in this area are still a long way from being satisfied. Especially trying will be the endeavor of sorting true virtues (if any there be) from social and historical trends. The historical essays in this volume, and the historical and transcultural discussions of Veatch's recent book [58] show a tension between change and constancy at all levels we have considered: what the institution demands, what the culture demands and what the physician becomes.[13]

NOTES

[1] In [43] John McDowell argues that virtues are not like mechanisms or deductive structures which could crank out appropriate solutions, but are more like achieved orientations. Accordingly, he argues, the way humans have virtues is very much like the way humans understand practical mathematics. The practical understanding of mathematics is an orientation, not a structured or mechanical axiom system like those used to systematize mathematical knowledge. Because McDowell's argument is so interesting and relevant, I choose to write in terms of the disjunction 'states or orientations', etc. This becomes important later in my essay.

[2] This may be *narrower* than the common meaning of 'medicine' which is almost synonomous with 'health care' and thus would include nurses, allied health personnel, etc.

[3] For example, in America the domestic demands on members includes their protection of the monopolistic power of medicine [5, 51]. In contrast with the demands of the domestic logic, the virtues are those features demanded by medicine's *ideals* – such as the demand that practitioners should practice as a living but not for the money, also have tenacity, intelligence, and stamina.

[4] A superior practitioner may well have virtues beyond those demanded by the institution of medicine.
[5] The metaphysical issue is whether medicine is, in French's [26] terms, a conglomerate (a unified whole) or an aggregate (just a collection heaped together).
[6] Perhaps physicians tried to overcome this by exaggerating what they did through calling it 'heroic medicine'.
[7] Medicine as a vehicle of social goods is somewhat distinguishable from how medicine fits the patterns of the culture it is in.For examples of the latter, cf. [47] where Reich shows how social ideals regarding psychiatry fit the ideals of the time and country in which it is practiced – Russians use psychiatry to correct political dissidence while American attitudes about self-determination are fostered by its own approaches to psychiatry. Caplan [12] shows how researchers developed political support for renal dialysis by treating only white adult males because America focuses on productivity and 'cost effectiveness'. For a staggering analysis of medicine's non-Hippocratic tradition of trying to manage political outcomes for political reasons see Trever-Roper's essay [55] and compare Veatch ([58], pp. 58–62).
[8] This is not to say that, e.g., compassion will not lead some morally astray. Some might break important moral rules out of compassion for the villain who begs for mercy. In this case compassion would not be a virtue but a vice.
[9] The analysis seems insensitive to the uncertainty that pervades medicine. That uncertainty is ignored by practitioners, e.g., when they seek to harness the placebo effect, but is over-played when they seek to evade giving bad news. However, the uncertainty is present even at the simple level of perceiving ([3], Chapter 7), no less than at the level of theory [50].
[10] Aristotle makes this exact point about the medicine of his day: (*Ethica Nicomachea*, 1138b26–1138b31 and 1141b8–1141b16, trans. Ross).
[11] According to Duffy ([15], pp. 11–13), pressure to make venereal disease reportable was gradually spearheaded by physicians. Thus, they strove to shape the gate they might be keeping; and they did so, in this case, out of concern for individuals' welfare and public health.
[12] Veatch [58] argues that contract is the best foundation of ethics. He accepts Rawls's [46] fuctional, ahistoric 'veil of ignorance' over Nozick's [44] more historical and more plausible account of original contracts because, Veatch believes, Rawls gives a 'better' morality. Veatch hopes a broad-based contract will 'neutralize' biases ([58], pp. 120–1) and allows a discovery/invention of unbiased ethical terms of contract. My position is that such scenarios as Rawls's and Nozick's *illuminate* the conceptual contexts of democracy, liberalism, and libertarianism, but they do not *justify* these contexts. Each side begs some vital questions or destroys some crucial moral intuitions.
[13] I would like to thank Donald Light for constructive criticisms of an early draft of this chapter.

BIBLIOGRAPHY

[1] Anscombe, G. E. M.: 1958, 'Modern Moral Philosophy', *Philosophy* 55, 1–19.
[2] Appelbaum, P. S. and T. G. Gutheil: 1979, 'Rotting with Their Rights On: Constitutional Theory and Clinical Reality in Drug Refusal by Psychiatric Patients', *Bulletin of the American Academy of Psychiatry and the Law* 7, 306–315.

[3] Atkinson, P.: 1981, *The Clinical Experience*, Renouf U.S.A., Inc., Brookfield, VT.

[4] Baron, R. J.: 1981, 'Bridging Clinical Distance: An Empathetic Rediscovery of the Known', *The Journal of Medicine and Philosophy* **6**, 8–13.

[5] Berlant, J. L.: 1978, 'Medical Ethics and Professional Monopoly', *Annals of the American Academy of Political and Social Science* **437**, 49–61.

[6] Bosk, C. L.: 1979, *Forgive and Remember*, University of Chicago Press, Chicago.

[7] Bradshaw, J. S.: 1978, *Doctors on Trial*, Paddington Press Ltd., New York.

[8] Buchanan, A.: 1978, 'Medical Paternalism', *Philosophy and Public Affairs* **7**, 370–390.

[9] Burns, C. R.: 1974, 'Comparative Ethics of the Medical Profession Outside the United States', *Texas Reports on Biology and Medicine* **32**, 181–190.

[10] Butler, S.: 1872, *Erewhon*, Trubner and Co., London.

[11] Callahan, D.: 1975, 'The Ethics Backlash', *Hastings Center Report* **5**, 18.

[12] Caplan, A. L.: 1981, 'Kidneys, Ethics, and Politics: Policy Lessons of the ESRD Experience', *Journal of Health, Politics, Policy and Law* **6**, 488–503.

[13] Culver, C. M. and B. Gert: 1982, *Philosophy in Medicine*, Oxford University Press, Oxford.

[14] Diseker, R. A. and R. Michielutte: 1981, 'An Analysis of Sympathy in Medical Students Before and Following Clinical Experience', *Journal of Medical Education* **56**, 1004–1010.

[15] Duffy, J.: 1982, 'The Physician as a Moral Force in American History', in W. B. Bondeson, *et al.* (eds.), *New Knowledge in the Biomedical Sciences*, D. Reidel Publishing Company, Dordrecht, Holland, pp. 3–21.

[16] Engelhardt, H. T., Jr., with E. L. Erde: 1980, 'Philosophy of Medicine', in P. Durbin (ed.), *A Guide to the Culture of Science, Technology and Medicine*, Macmillan, Free Press, New York, pp. 364–461.

[17] Erde, E. L.: 1977, 'Mind, Body and Malady', *Journal of Medicine and Philosophy* **2**, 177–190.

[18] Erde, E. L.: 1979, 'Philosophical Considerations Regarding Defining "Health," "Disease," etc. and Their Bearing on Medical Practice', *Ethics in Science and Medicine* **6**, 31–48.

[19] Erde, E. L.: 1982, 'Logical Confusions and Moral Dilemmas in Health Care Teams and Team-Talk', in G. Agich (ed.), *Responsibility in Health Care*, D. Reidel Publishing Company, Dordrecht, Holland, pp. 193–213.

[20] Erde, E. L.: 1983, 'On Peeling, Slicing and Dicing an Onion: The Complexity of Taxonomies of Values and Medicine', *Theoretical Medicine* **4**, 7–26.

[21] Erde, E. L.: 1984, 'Pathos and Pathology: Physicians in *The Last Angry Man*, *M*A*S*H*, and *The Magician*', *Studies in Science and Culture*.

[22] Feinberg, J.: 1970, 'The Nature and Value of Rights', *The Journal of Value Inquiry* **4**, 243–257.

[23] Flathman, R.: 1976, *The Practice of Rights*, Cambridge University Press, Cambridge.

[24] Foltz, J. M.: 1843, *The Endemic Influence of Evil Government*, J. & H. G. Langley, New York.

[25] Fox, R. L. and J. P. Swazey: 1978, *The Courage to Fail: A Social View of Organ Transplants and Dialysis*, 2nd edition, The University of Chicago Press, Chicago.

[26] French, P.: 1982, 'Collective Responsibility and the Practice of Medicine', *Journal of Medicine and Philosophy* 7, 65–86.
[27] French, P.: 1982, 'Crowds and Corporations', *American Philosophical Quarterly* 19, 271–277.
[28] Gert, B.: 1973, *The Moral Rules*, Harper and Row, New York.
[29] Gewirth, A.: 1982, *Human Rights*, The University of Chicago Press, Chicago.
[30] Gregg, A.: 1955, 'Our Anabasis', *The Pharos* 18, 14–25.
[31] Hanson, N. R.: 1965, *Patterns of Discovery*, Cambridge University Press, Cambridge.
[32] Hoffmann, S.: 1981, 'Introduction to Ethical Medicine: A Student Examines Life on the Wards', *The Harvard Medical Alumni Bulletin* 55, 14–19.
[33] Homiak, M. L.: 1981, 'Virtue and Self-Love in Aristotle's Ethics', *Canadian Journal of Philosophy* 11, 633–651.
[34] Hudson, S. D.: 1981, 'Taking Virtues Seriously', *Australasian Journal of Philosophy* 59, 189–202.
[35] Illich, I.: 1976, *Medical Nemesis*, Pantheon Press, New York.
[36] Jeager, W.: 1957, 'Aristotle's Use of Medicine as Model of Method in His Ethics', *The Journal of Hellenic Studies* 77, 54–61.
[37] Jonas, H.: 1960, 'The Practical Uses of Theory', *Social Research* 26, 127–150.
[38] Kett, J.: 1968, *The Formation of the American Medical Profession*, Yale University Press, New Haven.
[39] Lief, H. I. and R. C. Fox: 1963, 'Training for "Detached Concern" in Medical Students', in H. I. Lief, *et al.* (eds.), *The Psychological Basis of Medical Practice*, Harper and Row, New York, pp. 12–35.
[40] MacIntyre, A.: 1975, 'How Virtues Become Vices', in H. T. Engelhardt, Jr. and S. F. Spicker (eds.), *Evaluation and Explanation in the Biomedical Sciences*, D. Reidel Publishing Company, Dordrecht, Holland, pp. 97–111.
[41] MacIntyre, A.: 1981, *After Virtue*, University of Notre Dame Press, Notre Dame, IN.
[42] Masters, R. D.: 1975, 'Is Consent an Adequate Basis for Medical Ethics?', *Hastings Center Report* 5, 24–28.
[43] McDowell, J.: 1979, 'Virtue and Reason', *The Monist* 62, 335–350.
[44] Nozick, R.: 1974, *Anarchy, State and Utopia*, Basic Books, New York.
[45] Pellegrino, E. D. and D. C. Thomasma: 1981, *A Philosophical Basis of Medical Practice*, Oxford University Press, Oxford.
[46] Rawls, J.: 1971, *A Theory of Justice*, Harvard University Press, Cambridge, MA.
[47] Reich, W.: 1981, 'Psychiatry's Second Coming: The Way We Analyze Now', *Encounter* 57, 66–72.
[48] Reiser, S. J.: 1978, *Medicine and the Reign of Technology*, Cambridge University Press, Cambridge.
[49] Rothman, D.: 1982, 'Were Tuskegee and Willowbrook "Studies in Nature"?', *Hastings Center Report* 12, 5–7.
[50] Spaeth, G. L. and G. W. Barber: 1980, 'Homocystinuria and the Passing of the One Gene – One Enzyme Concept of Disease', *Journal of Medicine and Philosophy* 5, 8–21.
[51] Starr, P.: 1982, *The Social Transformation of American Medicine*, Basic Books, New York.

[52] Stein, H. F.: 1979, 'The Salience of Ethno-Psychology for Medical Education and Practice', *Social Science and Medicine* **13B**, 199–210.
[53] Stetten, D.: 1981, 'Coping with Blindness', *New England Journal of Medicine* **305**, 458–460.
[54] Thomasma, D. C.: 1983, 'Medical Paternalism and Patient Autonomy: A Model of Physician Conscience for the Physician-Patient Relationship', *Annals of Internal Medicine* 98, 243–248.
[55] Trevor-Roper, H.: 1981/2, 'Medicine in Politics', *The American Scholar* **51**, 23–42.
[56] Turnbull, C. M.: 1972, *The Mountain People*, Simon and Schuster, New York.
[57] Veatch, R. M.: 1980, 'Medical Ethics', *Journal of the American Medical Association* **243**, 2191–2193.
[58] Veatch, R. M.: 1981, *A Theory of Medical Ethics*, Basic Books, New York.
[59] 1980, 'Whose Baby Is It Anyway?', *Lancet* **1** (Part II), 1284–1285.

Department of Family Practice,
University of Medicine and Dentistry of New Jersey,
School of Osteopathic Medicine,
Camden, New Jersey, U.S.A.

ALLEN R. DYER

VIRTUE AND MEDICINE: A PHYSICIAN'S ANALYSIS

The rediscovery of virtue as a concept worthy of consideration in medical ethics is welcome indeed. The field of medical ethics and what has come to be known as bioethics have been severely handicapped without the concept of virtue. The concept of virtue restores a lost dimension to medical ethics. In the past decade or two medical ethics has become a very impersonal enterprise, all but ignoring the moral psychology of moral agents who must make decisions (both physicians and patients). An analysis of virtues in the immediacy of the clinical setting offers a texture to moral life which neither an analysis of moral rules nor of utilities can offer.

In the remarks which follow, I will attempt to define the problem to which I see the reconsideration of virtue as a solution. I will suggest that the concept of virtue is fundamental to understanding what a profession such as medicine really is. Furthermore, the prevailing accounts of professions, which see professions as defined primarily in terms of technical skills and organized to serve economic self-interest, have failed to appreciate this. The prevailing accounts of ethics, lacking an adequately developed moral psychology and failing to attend to considerations of virtue, have failed to set the matter straight.

THE LIMITS OF BIOETHICS

Consider the following anecdote as an example (rather extreme) of the kind of misunderstandings that arise in ethics without recourse to the concept of virtue. A prominent ethicist, seeking experience in medicine, went onto an orthopedics ward and asked the residents if they had any ethical problems. Thinking he meant terminal patients on respirators or some of the usual headline issues often spoken of as ethical dilemmas, the residents replied that they currently had no ethical problems. The ethicist then examined the charts of two patients with similar fractures and found that one patient had received Demerol for pain and the other had received only Darvon. He then concluded that these were not medical decisions but ethical decisions *instead*, which the physicians had failed to recognize as such.

The physicians acted on their values unreflectively; the philosopher

Earl E. Shelp (ed.), Virtue and Medicine, 223–235.

abstracted ethical reflection from the context in which an action was to be taken. In attempting to demonstrate the centrality of ethics in clinical judgment, he regretably separated thought from action. A valid point could be made that choice of analgesic involves human values, attitudes toward pain and suffering, and compassion for the patient. To say that these concerns are not medical reinforces the prejudice that medicine is merely a technical enterprise and that science is value neutral. It further implies that a physician may defer ethical reflection to a specialist in ethics. This view literally depersonalizes both the practice of medicine and ethics as well.

The point which needs to be stressed is that 'clinical judgment' involves ethical choices even in mundane examples. This is an important realization in medical education because there is a real tendency for medical students and physicians, confronted by the hostile critics amongst ethics specialists, to attempt the strictly impossible: to limit their concern to technical matters, which can be carefully defended.

When an ethicist such as Robert Veatch argues that physicians have little training in ethics, but are much involved in making ethical decisions, we are easily led to his conclusion that this is a 'generalization of expertise', the translation of expertise from a technical area to a moral area. We are too ready to accept that the physician should refrain from making moral judgments because we accept the specialization of forms of knowing and accept that there could be such a thing as a value-free technical judgment independent of moral considerations. Physicians and non-physicians both participate in this *folie à deux* because it is part of our common cultural heritage. We ask for a precise clarification of the issues, when we know that such is not possible. Yet neither physician nor patient would dare say very much in the public arena about the personal ingredients that might bear on a particular decision when such personal disclosures might be subject to critical and judgmental scrutiny.

Thus matters of personal ethics very quickly get translated into matters of public policy. We are relatively more comfortable analyzing issues than we are in looking at the ingredients of personal decision, indeed looking at ourselves and our sometimes conflicted motives for action. Our culture inhibits moral reflection at a personal level. It encourages us to alienate our moral authority to an expert or system of rules. In a culture both skeptical and (morally) perfectionistic, ideals are confused with imperatives. Legitimate moral ideals are not recognized as goals toward which one might strive, but rather are held as imperatives to which one must adhere.

An explicit attention to ethics might be seen as a corrective humanizing force in medical education. In fact, however, medical ethics and 'bioethics' often have their own paradoxically dehumanizing effect on medical practice. By removing ethical reflection from the clinical arena – from the setting of actual practice where those inalienable unspecifiable elements of the humane physician's attention to his patient continue to appear – to the arena of public policy and attempting to establish ethics as a rival rather than a coordinate discipline, one in which the physician is said to have no 'expertise', even thinking about ethics is estranged from actual decision-making.

The irony of this development lies in its circularity. The charge which can legitimately be leveled against the technocrat is that of shortsightedness, in which decisions are based on private expertise, unconscious of value conflicts, rather than on socially reflective judgments. The ethicist may rightly repudiate this arrogation of power, but cannot at the same time lay claim to the same prerogatives as the true expert in ethical reasoning – and certainly not moral authority! It is often true that physicians are poorly trained in ethics and may assume that the warrants of their judgments lie in their special training or 'expertise'. As unfortunate as this misunderstanding may be philosophically, it is largely mitigated by the personal concern most physicians demonstrate for their patients; however, the same arrogance in a philosopher may have devastating consequences. Physicians and philosophers may thus seem naive to each other: physicians for deciding with a paucity of ethical reflection; philosophers for arguing at a level of abstraction or criticism so far removed from the complexity of real-life situations as to appear ultimately hollow.

THE PLACE OF ETHICS IN THE DEFINITION OF A PROFESSION

Ethics without considerations of virtue is left to try to deduce the moral rules or principles which might be applicable in particular situations or in trying to determine some utility calculus of the greatest good. Such an account of the tasks of ethics leaves a profession defined merely by its technical expertise. Once the rules are established, any technician can carry them out. There is another understanding of profession which is more personal and is defined by the ethics of the group, the professing of vows or commitments to the ideal of service. In this more traditional sense of profession, the professions are defined by the ethical requirements of the special relationship of the professional to those served, e.g., 'the doctor–patient relationship'. Though the modern connotation of 'professional' is anyone with advanced training and using those skills in his work, the more traditional sense of paradigm professions is usually reserved for those involving a fiduciary relationship, generally held to be medicine, law, and the clergy.

It is as part of this process of professional definition along ethical lines that the medical and legal professions, along with a number of would-be professional occupational groups such as professional librarians and professional advertisers, have adopted specific written 'codes of ethics'.

'Ethics', then, has at least two distinct levels of meaning, which are often confused. On the one hand, ethics may be understood to be the values or customs which govern the actions of the members of a particular group, tribe, or community. It is into this understanding that considerations of 'professional ethics' falls: professional ethics is reflection upon the moral standards of the professional group, by which its members define their identity as professionals and by which they determine standards for inclusion or exclusion.

The other view of ethics is that which holds that ethics is a discipline which should deduce by reason a truth which could be applied to a particular situation. Such reflection on rules deduced by abstractly formulated questions is often used to criticize the practices of a profession. Thus by 'ethics' one may mean the traditions which define a profession or a methodology for criticizing that tradition. The prevailing account of ethics is that which favors abstract reflection on specific and highly specifiable problems, of which contemporary medicine offers many. The more personal view of ethics asks a rather different sort of question, namely, 'What sort of person do I wish to become in this culture or community or profession?'

THE TASKS OF ETHICS

Upwards perspective	*Downwards perspective*
moral inspiration	regulation of abuse
highest standard	minimal standard
affective	cognitive
teleological	deontological
ultimate principles	etiquette
tacit	explicit

Those tasks associated with the upwards perspective are what one usually thinks of as 'ethics'. What I am calling the 'downwards perspective' are those tasks associated with professional ethics or standards, the tasks associated with maintaining the integrity of the professional group. There is, of course, overlap between the two perspectives. One task is to articulate high ethical standards. Another is getting people to adhere to those standards.

In contemporary American culture, we have come to expect almost instantaneous answers to abstractly formulated problems. We instinctively value ethical behavior, but solid principles carefully nurtured over time are often overlooked. This feature of contemporary culture has been described by the German sociologist Toennies as a breakdown of *Gemeinschaft* and a corresponding shift to *Gesellschaft*, *Gemeinschaft* representing a community of feeling that results from likeness and shared life experience, while *Gesellschaft* represents a more rational, mechanistic way of life, with greater structure and more written and explicit rules and regulations. Broad ethical principles no longer serve as shared values, and there is an attempt to make moral principles explicit as behavioral guidelines.

It is in the setting of a culture of *Gesellschaft* that we see such emphasis on technique and technology. And it is in such a culture that the idea of a profession has come so much to be identified with the expertise of the professional, those aspects of professional life which can be made systematic and explicit.

It is also in the setting of a culture of *Gesellschaft* that there is so much confusion about the appropriate tasks of ethical undertaking, The written codes of ethics, such as the AMA's *Principles of Medical Ethics*, Thomas Percival's *Medical Ethics* or the Hippocratic Oath look deontological in form. They look like a list of rules, which might be applicable in particular situations, but they function more teleologically in the Anglo-American world to suggest the goals or ends which should govern professional conduct. (The Continental European and Latin American codes are in fact deontological, backed by the standing of law.)

Chauncey Leake, American editor of Percival's *Medical Ethics*, suggested that medical ethics was a misnomer, that what Percival was really taking about were matters of etiquette, dealing with the relationships of physicians to one another and to the public, and embodying the tenets of professional courtesy, but not about the ultimate consequences of the conduct of physicians toward their individual patients and toward society as a whole. The point can be granted for Percival's *Medical Ethics* and for the derivative AMA codes without trivializing the undertaking, for the relationship between doctor and patient has traditionally been a central feature of medical practice. One of the central questions facing modern medicine is whether such personal attention is still requisite to quality care or whether it is possible for medicine to be conducted in a more impersonal and technological fashion. I think it is safe to say that in spite of the tendencies for modern medicine to be oriented toward physiological interventions, there remains an expectation on the part of most physicians and patients that the physician should attend to the health needs of the patient as person.

The thrust of Percival's view of medicine is that a physician's effectiveness depends to a large measure on the trust the patient places in the doctor and that certain restraints on the physician are necessary in order to preserve and maintain that trust. The bulk of Percival's trust-building prescriptions are general moral rules of conduct: attention, steadiness, humanity, secrecy, delicacy, and confidence along with qualities of mind: temperance for the sake of clear thought, retirement when senility sets in. Secrecy (confidentiality), compassion and proper care of patients (competence), as well as upholding the dignity of the profession, continued from the ancient oaths.

The code of ethics which the fledgling American Medical Association adopted in 1847 was largely derived from Percival's. A certain protectionism was given emphasis, however, that remained until the 1980 revision, in prescriptions about consultation, solicitation of patients, patient stealing, etc. Though such strictures are often discounted as offering more protection to the practitioner than the patient, it is important to consider the historical roots of that emphasis. In the mid-nineteenth century, licensing of physicians had been discontinued, and quackery by mail-order 'physicians' was rampant. Thus, the apparent 'etiquette' can be argued as a proper ethical principle to protect patients from exploitation and harm.

THE SOCIOLOGICAL DEFINITION OF A PROFESSION

One of the most noticeable features of professional organizations, viewed in sociological perspective, is their attempt to control markets and promote self-interest. The AMA is often seen as simply a trade organization, overlooking the public benefit which follows from a defense of scientific medicine and the promotion of standards. Perhaps the service ideal is less commented upon because it is so inconspicuous, occurring in the privacy of the relationship between doctor and patient and hence not observable by bystanders, but critics of the profession consistently argue that the service ideal is mere lip-service used to obscure professional self-interest.

In 1975 the Federal Trade Commission sued the AMA, arguing that it was in restraint of trade because its code of ethics prohibited advertising. The FTC strategy has been to promote reductions in medical cost through competition. By applying antitrust theories of trade to the professions, the legitimate functions of professional ethics are obscured as 'ethics' is reduced to purely economic considerations.

The theory of professional monopoly was developed by Max Weber, who described the following steps by which the medical profession, as all commercial classes, achieves monopoly power: creation of commodities, separation of the performance of services from the satisfaction of the client's interest (i.e., doctors get paid whether or not their therapies work), creation of scarcity, monopolization of supply, restriction of group membership, elimination of external competition, price fixation above the theoretical competitive market value, unification of suppliers, elimination of internal competition, and development of group solidarity and cooperation.

Weber, however, denied that any form of social activity could be purely economic. He indicated that all activities have an economic aspect insofar as they involve scarcity of resources and thus involve planning, cooperation and competition. But economic considerations alone cannot explain the particular direction taken by any social activity or movement; for this other values have to be taken into consideration. Such careful delineation of an economic component to human motivation was truly radical in its time, but today it has become commonplace to reduce all human motivation to economic considerations. The problem with antitrust analysis of professional monopoly is that it is unidimensional; it considers only economic motives.

The basic issue is whether physicians, or the members of any profession which enjoys the position of monopoly, can place concern for the public good ahead of their own self-interest. Jeffrey Berlant, sociologist and physician, is representative of those who take the economic-reductionist view. Commenting on Percival's *Medical Ethics* (published in 1803 and used as the basis for the AMA's first code of ethics in 1847), Berlant says,

> When Percival declines to call for group sanctions, one wonders how he expects to have virtuous men follow the code without coercion or group controls on behavior. Apparently, he believes that virtuous men of spontaneous good conscience can exist without group regulation and without sanction ([1], p. 83).

It is surprising that a sociologist would overlook the power of unspoken group sanctions among those practicing together in a community, whose work would be well known to one another. Perhaps this is the point that is so hard to appreciate in the latter half of the twentieth century when communities are so vast and social ties so tenuous that it seems that group sanctions must be made behaviorally explicit to have any viability at all.

The doctor–patient relationship is in fact based on trust in the ethics of 'virtuous men (and women) of spontaneous good conscience', a much higher standard than could possibly be achieved by explicit group sanctions.

Trust or trustworthiness is the keystone of medical virtue in the traditional canons of medical ethics from the Hippocratic Oath to Percival's code to the various versions of the AMA codes. Trust is the basis of what it means to be a professional, what it means to be ethical. From the antitrust point of view, any trust, even basic human trust, is suspect as a form of monopolization. Berlant states the case very cogently:

> Such trust-inducing devices of the Percivalian code increase the market value of medical services and help convert them into commodities. Similarly it flatters doctors and helps integrate them into a professional group, thereby furthering group formation. It also creates a paternalistic relationship toward the patient, which may undermine consumer organization for mutual self-protection, thereby maintaining consumer atomization. Essentially it persuades the patient that he need not protect his own interests, either by himself or by organized action. Through atomization of the public into vulnerable patients, paternalism results in the profession's dealing with fragmented individuals rather than bargaining groups. Moreover, by appealing to patient salvation fantasies, trust inducement can stimulate interpatient competition by increasing each patient's desire to see that nothing stand between doctor and himself. Much of the emotional power of the sentiment of the doctor–patient relationship resides in this wish of the patient to save himself at any cost to himself or others ([1], p. 70).

Berlant offers a sharp attack on professional ethics from a particular ideological perspective. But basically this attack is not just on professional ethics, it is also an attack on a kind of community in which people may not be autonomous and independent, but in which people may be dependent and in need of help which they willingly seek. This argument introduces a note of almost cynical suspiciousness into a society which seems almost too willing to trust and to place oneself in the care of another. This is a crisis of confidence – both for medicine and for our civic life in general: to what extent is it possible and necessary to trust and rely on others and to what extent is it possible to remain isolated, self-reliant, and autonomous human beings?

THE IMPORTANCE OF VIRTUE IN PROFESSIONAL GATEKEEPING

A profession, defined by its ethics and the shared commitment of its members to an ideal of service, must attend carefully to what sort of person is allowed to assume the responsibilities and accompanying privileges of professional activity. The determination of who shall be accepted into the profession and the monitoring of the performance of the members of the profession is an important responsibility of professional organizations. The control of entry and exit has been traditionally considered to be an important function for professional self-regulation. Leaving aside for a moment the question of who holds the responsibility for such regulation, I wish to establish that such activities are essential to the idea of a profession and that they are impossible to fulfill without some concept of virtue. If medicine is more than just technical expertise, then some understanding of virtue or 'character' is essential to what we expect of a physician whether we acknowledge it explicitly or not. It is not sufficient that a physician, for example, be knowledgeable about matters of physiology and biochemistry: that physician must also be able to communicate with patients and adhere to certain standards of ethics.

The entry decisions are made quite formally in the United States with medical school admissions (since we graduate almost all the students we admit). Most admissions committees look for students who are not only academically competent, but possess some measure of the requisite virtues. Though the vocabulary for talking about requisite virtues has become largely atrophied and the most emphasis is placed on grades and board scores which can be readily quantified, I believe most admissions committees retain some sense of trying to select candidates for medical education who will make 'good doctors'. I would suggest that there is some concept of virtue at work, even though it usually functions at a tacit level of awareness. Efforts to bring these considerations into explicit focus should only enhance the legitimacy and prominence with which they are taken.

The control of exit from professions is a responsibility variously held by state licensing bodies and professional organizations. The state licensing boards, which have the power to grant licenses, also have the power to rescind them. Professional organizations, such as the AMA and various specialty organizations, can and do expel members from their organizations if they fail to live up to the standards of ethics. There is a widely held perception that professional organizations are reluctant to move against their members, but the past fifteen years have witnessed an increasing awareness on the part of professional organizations of the necessity to 'police their own'. The AMA's Council on Mental Health, for example, introduced a model impaired-physician statute, which has subsequently been adopted by most states, which makes provisions for the licensing boards to review the license of any physician thought to be disabled by virtue of physical or mental illness or substance abuse and to provide legal protection for the whistle-blower. Again, a concept of virtue is in operation in such considerations, for the model statute provides for a humane 'therapeutic' intervention while maintaining the necessity of professional 'discipline'.

Another example of this kind of activity is the increased emphasis of professional organizations on investigating grievances of dissatisfied patients. The activities of the American Psychiatric Association's Ethics Committee is an example of this kind of activity. In 1972 the APA adopted "Annotations Applicable to Psychiatrists" to the AMA's *Principles of Medical Ethics*. Each member of the APA is expected to be familiar with this statement of ethical principles, and it serves as a reference for any complaint of unethical conduct brought against a member.

While sociologists and antitrust theorists might argue that such activities primarily serve the economic interests of the members of the professional group, such gatekeeping activities serve the public interest as well by assuring that certain standards of professional competence and integrity are maintained.

THE CENTRAL VIRTUES OF THE MEDICAL PROFESSION

Thus far I have argued that a concept of virtue is central to the idea of a profession; that moral discourse has been largely handicapped without explicit acknowledgement of the concept of virtue; that the concept is implicit in the activities of professional organizations; and that there has been a potentially destructive aspect to the field of medical ethics (bioethics) operating in a cultural and psychological void, as if there were no such thing as virtue or a moral agent. One might reasonably expect now a catalog of appropriate medical virtues, virtues appropriate to our contemporary culture or to the culture of medicine. I am prepared to offer at least a start on such a catalogue by suggesting two virtues which have been central to the traditional ideals of a profession: trustworthiness and respect for confidentiality.

The notion of trustworthiness is a constant theme throughout the various articulated codes of ethics from the Hippocratic Oath through Percival's *Medical Ethics* to the various revisions of the AMA codes. The physician should be someone whom the patient can trust and someone who inspires confidence in the patient. Perhaps in ancient times even more so than in modern times, this was important because the inspiration of trust in the physician, confidence in the treatment, and hope for improvement were essential to the success of the treatment when the physician had little to offer. In the contemporary era, physiological interventions can be much more specific, but the personal ministrations of the physician to the patient remain, I believe, an indispensable part of medical practice. Knowledge is always incomplete and technical power limited.

There is perhaps an even more basic reason for suggesting that trustworthiness is a basic virtue for the medical profession. That is the psychological primacy of trust as a basic human quality, really the first issue in psychological development with which the newborn and its caretakers must deal. It has been demonstrated that consistency in the nurturing enviroment, i.e., predictability that the parents will meet the infant's needs, has been shown to leave one with a sense of trust in the world and the ability to trust others; absence of such consistency in early childhood leaves one always potentially uncertain and in doubt. It may seem a large jump to draw analogy between the child and the patient (perhaps an adult, autonomous, and rational creature), but there is a fundamental way in which one is quite dependent on the physician when one seeks help. The overlays of adult rationality in no way significantly lessen the psychological significance of that supplication in time of need. The patient needs to know that the physician will not do anything to violate that sense of confidence and expectation that help will indeed be rendered.

The second virtue I would like to suggest is related to the first; respect for confidentiality. Again the tradition of confidentiality has an ancient heritage, going back at least to medieval common law traditions, which protected the confidentiality of communications with a physician, lawyer or clergyman. The psychological reasons for this are evident. One cannot freely disclose to the physician (or other professional) the relevant information (history) that will be needed to make an assessment of the problem and render help unless one knows that that communication is private and not to be disclosed elsewhere. This virtue of respect for confidentiality is particularly relevant in the modern era because so much of medical care is delivered by teams, often including non-professionals, and because of the difficulty of protecting the contents of medical records.

THE VIRTUES OF THE PATIENT

It might seem a bit out of place to consider the virtues of the patient when the popular sentiment seems so much to view the patient as passive in front of the overwhelming power of the professional. Nineteenth century statements such as Percival's code and the early versions of the AMA codes spoke of the responsibilities of the patient to the physician, especially duties such as loyalty, which clearly seem antiquated. If any sense of mutuality of the doctor–patient relationship is to be maintained (or restored), it is reasonable to think of the virtues of the patient.

Again trustworthiness might head the list. Honesty might be a central consideration of such trustworthiness, especially in terms of forthrightness in the disclosure of information which might be relevant to the physician's understanding of the problems for which the patient is seeking help. This is particularly important in sorting out those problems which have a psychosomatic component – and this probably includes a much greater proportion of patient–physician encounters than we generally acknowledge. Furthermore, we are increasingly coming to consider the responsibilities of patients in health matters which involve lifestyle, especially health problems arising from alcohol and tobacco, improper diet, or bad habits (such as insufficient exercise). The physician can treat alcoholic gastritis, for example, but the patient and ultimately the community must accept some responsibility for the treatment of the alcoholic and alcoholism. Still there is a sick role and a social judgment involved in shifting concepts of disease and illness, which are increasingly medicalized or de-medicalized: abortion, child abuse, homosexuality as perversion or alternate sexual preference, hypertension, hysterical conversion, malingering, passive-dependency and other so-called personality disorders. A sophisticated appreciation of all these potentially medical problems is really impossible without recourse to some concept of virtue.

CONCLUSION

Two divergent approaches to ethics, bioethics, and professional ethics, have appeared in recent years to be at cross-purposes. What I am suggesting here is that both have been handicapped without recourse to the concept of virtue. Ultimately what is at stake is a rethinking of cultural values, notions of the responsibility of individuals in a society to one another, the relationships of members of a family and community and relationships of professionals with those served. While a paper of this scope cannot attempt to solve these various cultural issues, I think it can be demonstrated that the concept of virtue in ethics is a legitimate starting point for resolving some of the misunderstandings which have characterized ethics for the past several years.

BIBLIOGRAPHY

[1] Berlant, J.: 1975, *Profession and Monopoly: A Study of Medicine in the United Stated and Great Britain*, University of California Press, Berkeley, California.

[2] Dyer, A. R.: 1980, *Idealism in Medical Ethics, the Problem of the Moral Inversion*, University Microfilms, Ann Arbor, Michigan.
[3] Leake, C. D. (ed.): 1927, *Percival's Medical Ethics*, Williams and Wilkins, Baltimore, Maryland.
[4] Veatch, R. M.: 1973, 'The Generalization of Expertise', *Hastings Center Studies* 1 (2), pp. 29–40.
[5] Weber, M.: 1949, *On the Methodology of the Social Sciences*, The Free Press, Glencoe, Illinois.

Department of Psychiatry,
Duke University Medical Center,
Durham, North Carolina, U.S.A.

EDMUND D. PELLEGRINO

THE VIRTUOUS PHYSICIAN, AND THE ETHICS OF MEDICINE

> Consider from what noble seed you spring: You were created not to live like beasts, but for pursuit of virtue and of knowledge.
>
> Dante, *Inferno* **26**, 118–120

INTRODUCTION

In the opening pages of his *Dominations and Powers*, Santayana asserts that "Human society owes all its warmth and vitality to the intrinsic virtue of its members" and that the virtues therefore are always "hovering silently" over his pages ([32], p. 3). And, indeed, the virtues have always hovered over any theory of morals. They give credibility to the moral life; they assure that it will be something more than a catalogue of rights, duties, and rules. Virtue adds that extra 'cubit' that lifts ethics out of its legalisms to the higher reaches of moral sensitivity.

Yet, as MacIntyre's brilliant treatise so ably attests, virtue-ethics since the Enlightenment has become a dubious enterprise [19]. We have lost consensus on a definition of virtue, and without moral consensus there is no vantage point from which to judge what is right and good. Virtue becomes confused with conformity to the conventions of social and institutional life. The accolades go to those who get along and get ahead. Without agreeing on the nature of the good, moreover, we can hardly know what a 'disposition' to do the right and good may mean.

These uncertainties force us to rely on ethical systems built upon specific rights, duties, and the application of rules and principles. Their concreteness seems to promise protection against capricious and antithetical interpretations of vice and virtue. But that concreteness turns to illusion once we try to agree on what is the right and good thing to do, in a given circumstance.

Despite the erosion of the concept of virtue, it remains an inescapable reality in moral transactions. We know there are people we can trust to temper self-interest, to be honest, truthful, faithful, or just, even in the face of the omnipresence of evil. Sadly, we also know that there are others who

Earl E. Shelp (ed.), Virtue and Medicine, 237–255.

cannot be trusted habitually to act well. We may not be virtuous ourselves, we may even sneer at the folly of the virtuous man in our age, yet we can recognize him nonetheless. It is as Marcus Aurelius said, " . . . no thing delights as much as the examples of the virtues when they are exhibited in the morals of those who live with us and present themselves in abundance as far as possible. Wherefore we must keep these before us" [23].

And, in fact, the virtues are again being brought before us in the resurgence, among moral philosophers, of an interest in virtue-based ethics [9, 11, 13, 17, 21, 29, 33, 34, 35, 38, 39]. For the most part the resurgence is based in a re-examination, clarification, and refurbishment of the classical-medieval concept of virtue. On the whole, the contemporary reappraisal is not an abnegation of rights or duty-based ethics, but a recognition that rights and duties notwithstanding, their moral effectiveness still turns on the disposition and character traits of our fellow men and women.

This is preeminently true in medical ethics where the vulnerability and dependance of the sick person forces him to trust not just in his rights but in the kind of person the physician *is*. The variability with which this trust is honored, and at times its outright violation, account for the current decline in the moral credibililty of the profession. It accounts, too, for the trend toward asserting patients' rights more forcefully in contract as opposed to covenantal models of the physician – patient relationship. In illness, when we are most exploitable, we are most dependent upon the kind of person who will intend, and do, the right and the good thing, because he cannot really do otherwise.

MacIntyre has expressed his pessimism about the possibilities of an ethic of virtue in our times. He ends his tightly reasoned treatise with a surprisingly romantic hope for " . . . a local form of community within which civility and the intellectual and moral life can be sustained in the rough new dark ages which are already upon us" ([19], p. 245). One can accept MacIntyre's astute historical diagnoses without waiting too quietistically for the appearance of his new and 'doubtless very different' St. Benedict to lead us into some new moral consensus. Even in this 'dark age', it may be possible to incorporate the remnants of the classical-medieval idea of virtue into a more satisfactory moral structure that also includes rights and duties.

This essay examines that possibility in the limited domain of professional medical ethics. Using the classical-medieval concept of virtue, it attempts to define the virtuous physician, to relate his virtue to the ends of medicine and to the usual principles applied in medical ethics, as well as to the prevailing rights and duty-based systems of professional ethics.

This essay is part of a larger effort to link a theory of the patient-physician relationship, a theory of patient good, and theory of virtue in a unified conceptual substructure for the professional ethics of medicine [24, 25].

THE CLASSICAL-MEDIEVAL CONCEPT OF VIRTUE

What is virtue? What are the virtues? Are they one or many? Can they be taught? These are still the fundamental first order questions for any virtue-based ethical theory. And, as with so many of the perennial philosophical questions, they were first examined in an orderly way in the Platonic dialogues. In the first dialogues, Socrates raises them with his characteristic to, and fro, probing. Given his circumambulation of definitions, and the difficulties of translating from ancient languages, Socrates' precise meanings remain somewhat problematic.

Most commentators, however, focus on his equilibration in the *Meno*, *Protagoras*, and *Gorgias* of virtue (*aretė*) with knowledge (*epistemė*) [5, 13, 36, 37]. Virtue is here synonymous with excellence in living the good life. This excellence depends upon knowledge of good, evil, and self. It is not specialized knowledge directed to any one activity but rather to living one's whole life well. It must, like an art, be perfected through practice. The individual virtues – courage, justice, temperance, wisdom, and piety – order life towards excellence. But they too depend on knowledge so that, on this view, vice is the result primarily of ignorance.

How much of this intellectualist definition of virtue is Socrates' and how much Plato's is debatable. In the later dialogues, the *Laws* and the *Republic*, more attention is given to justice as a central virtue and to proper ordering of the state [12]. Wisdom is still a virtue but for the statesman it is knowledge of what is good not just for the individual good life but the good of the whole state. The aim of laws should be virtue but the chief of these is wisdom, so that wisdom is the knowledge that 'presides' over justice (*Republic*, Bk. IV).

In the *Republic*, still exploring definitions, Plato draws the analogy of virtue with bodily health – "Virtue then as it sees would be a certain health, beauty and good condition of a soul and vice a sickness, ugliness and weakness" (*Republic*, Bk. IV). He likens virtue to the order and balance between the parts of the body that characterize health. For Socrates that ordering results from intellectual perfection of the art of living a good life. Plato, as Aristotle pointed out, paid more attention than Socrates to the non-rational elements in human life. Virtue, for him is as much a matter of disposition to act in the right way, as it is a desire for the good, arising from knowledge of the good.

Aristotle reacted to Socrates' over-intellectualization of virtue. He devoted large portions of the *Nicomachean Ethics* to his concept of virtue. His own work, Aristotle said, " . . . does not aim at theoretical knowledge like the others (for we are inquiring not in order to know what virtue is, but in order to become good, since otherwise our inquiry would have been of no use), we must examine the nature of actions, namely, how we ought to do them" (*Nicomachean Ethics*, 1099a, 10–11). He does not reject the role of intellect; he judges it a proper part of virtue, but not its whole. Thus he emphasizes that virtue is also concerned with feelings and actions – " . . . just acts are pleasant to the lover of justice and in general virtuous acts to the lover of virtue" (*Nicomachean Ethics*, 1099a, 10–11).

But virtue is not just feeling. It is of two kinds: intellectual and moral – "Intellectual virtue in the main owes both its birth and growth to teaching (for which reasons it requires experience and time) while moral virtue comes about as a result of habit, whence also its name *ethikė* that is formed by a slight variation of the word *ethos* (habit)" (*Nicomachean Ethics*, 1103a, 14–18). Virtue is not simply a passion, or a function, but a " . . . state of character" (*Nicomachean Ethics*, 1106a. 11). Virtue is that state of character "' '. which makes a man good and which makes him do his own work well" (*Nicomachean Ethics*, 1106a, 22–24).

But states of character must be in accord with the 'right rule.' Hence the intellectual virtues play a part in making a man virtuous. Practical wisdom is that aspect of intellect that enables a man to direct his life to its cheif good. The 'right rule' is reached, as Ross says, "by the deliberative analysis of the practically wise man and telling him that the end of human life is to be best attained by certain actions which are intermediate between extremes. Obedience to such a rule is moral virtue" [31].

Aristotle's doctrine of the mean and his divisions and subdivisions of the intellectual and moral virtues are too complex to engage us here [31]. What is significant for the present discussion is the balance Aristotle strikes between moral virtue and reason, feelings, dispositions, and right action. These are Aristotle's modifications and extensions of the Socratic notion of virtue.

Aristotle made more explicit the Socratic orientation of virtue to ends. Thus " . . . every virtue or excellence both brings into good condition the thing of which it is the excellence and makes the work of that thing to be done well . . . " (*Nicomachean Ethics*, 2, 5, 1106a, 15–17). The end of virtue for both Socrates and Aristotle is the good life. What constitutes the good life is for both of them the fulfillment of the potentiality of human nature.

Aristotle summarizes the essential elements which distinguish the virtuous man: " . . . in the first place he must have knowledge, secondly, he must choose the acts and choose them for their own sakes, and thirdly his actions must proceed from a firm and unchangeable character" (*Nicomanchean Ethics*, 11, 4, 1104, 6, 31–34). This combination of character, knowledge, and deliberate choice makes for a virtuous person. Actions themselves are virtuous when they emanate from just such a person (*Nicomachean Ethics*, 1105b, 5–6). For Aristotle, then, virtue resides not only in the act or in the knowledge of virtue, but the act itself must also be done virtuously, as a virtuous man would do it. " . . . But as a condition of the possession of the virtues knowledge has little or no weight, while the other conditions count not for a little, but for everything. Nor does virtue reside in the act itself" (*Nicomachean Ethics*, 1105b, 1–4).

Aquinas' treatment of the virtues is best understood in the context of his total enterprise, to reconcile, emend, and amend the ancient philosophers through the revealed truths of the Christian experience. Precisely to what extent his emphasis on the cardinal virtues coincides with, or departs from, Artistotle is difficult to say. Aquinas too had to surmount the problem of trying to apprehend the nuances of words in an ancient language no longer in daily use. MacIntyre calls him a 'marginal figure' not representing the general opinion of the virtues extant in his time. He does, however, recognize that Aquinas' interpretation of the *Nicomachean Ethics* "has never been better" ([19], p. 166).

For Aristotle, the intellectual and moral virtues were requisite for the full development of man's natural capacities. But for Aquinas, man's natural end was itself insufficient because he had a spiritual destiny that transcends the merely natural [4]. Man is destined to union with God. To this end not only the natural virtues but the supernatural are needed, and these can be known only through the Christian revelation. For the Christian the perfection of the natural virutes is not sufficient [4].

Like Aristotle, Aquinas held virtue to be grounded in good habits, in dispositions to do the right and good thing, but always in association with the right reason. Thus he says, "It belongs to human virtue to make a man good and his work according to reason" [2]. "Through virtue man is ordered to the utmost limit of his capacity" [5]. It is a condition of the perfection of human life. Or said another way, "Virtue is called the limit of potentiality . . . because it causes an inclination to the highest act which a faculty can perform" [1].

Virtues for Aquinas are habits and dispositions that enable a man to reason well (the intellectual virtues) and to act in accordance with a right reason (moral virtues) – *recta ratio agibilium* [20]. These two kinds of virtue can be independent of each other, but in the virtuous man, who strives for the perfection of his human potentiality, they are joined. And they are joined most firmly in the one virtue from which the others derive – prudence, the standard of right willing and acting, the form and mold for the other virtues [29].

Josef Pieper among modern commentators has most clearly expounded Aquinas' teaching on prudence and the other cardinal virtues [29]. Without attempting to restate that teaching here, suffice it to say that for Aquinas all virtue is necessarily prudent, acting in accord with right reason in conformity with the reality of things. This is not prudence in the pejorative modern sense of caution, timidity, rationalized cowardice, cunning, or casuistic deviousness. Rather, it is the analogue of Aristotle's wisdom, the capacity and disposition to will and act rightly in particular, practical, and uncertain circumstances. As such, prudence informs, measures, guides, shapes, and generates the other virtues, by inclining us to choose good means to good ends for man.

The post-medieval transformations of the classical concepts were many: Rousseau opposed virtue to society; for Shaftesbury virtue lay in the pursuit of public interest; Hutcheson and Hume identified it with a 'moral sense'; to Montaigne virtue was 'an innocence, accidental and fortuitous'; for Descartes, strength of soul; for Malebranche, love of order; and for Spinoza, the soul directing itself by a universal and clear idea; Mandeville paradoxically defended the utility of the vices. MacIntyre shows with a wealth of detail the continuing emotivist and intuitionist trends of philosophies of virtue since the Enlightenment.

In the last several years the classical concept has been reexamined by a growing number of moral philosophers. They have underscored such things as the differences in meanings of the word *arete* in Greek, *virtus* in Latin and 'virtue' in English, the distinctions between virtues and skills, the difficulties in defining such words, as 'disposition' and 'habit' as used in the traditional definitions, the relationships between the natural and the supernatural virtues, the relations of virtue, values and concepts of the good, the unity or non-unity of the virtues, whether they are teachable or not, their relevance to health and medical care, and their relationship with duties, rights, and obligations [11, 17, 19, 21, 23, 29].

MacIntyre, for example, extends the Aristotelian concept from qualities internal to practices, to role reference and to community good [19]. Hauerwas relates them to the narrative of a particular people [14]. Some commentators are more critical, some more accepting of the classical concepts. In the end, the great majority say almost the same thing as the ancients but in more modern language. One is tempted to say of virtue, as it has been said of pronography, 'I can't define it but I know it when I see it.'

THE VIRTUOUS PERSON, THE VIRTUOUS PHYSICIAN

Through these multitudinous definitions a set of common themes seems discernible. Virtue implies a character trait, an internal disposition, habitually to seek moral perfection, to live one's life in accord with the moral law, and to attain a balance between noble intention and just action. Perhaps C. S. Lewis has captured the idea best by likening the virtuous man to the good tennis player: "What you mean by a good player is the man whose eye and muscles and nerves have been so trained by making innumerable good shots that they can now be relied upon .... They have a certain tone or quality which is there even when he is not playing .... In the same way a man who perseveres in doing just actions gets in the end a certain quality of character. Now it is that quality rather than the particular actions that we mean when we talk of virtue" [18].

On almost any view, the virtuous person is someone we can trust to act habitually in a 'good' way – courageously, honestly, justly, wisely, and temperately. He is committed to *being* a good person and to the pursuit of perfection in his private, professional and communal life. He is someone who will act well even when there is no one to applaud, simply because to act otherwise is a violation of what it is to be a good person. No civilized society could endure without a significant number of citizens committed to this concept of virtue. Without such persons no system of general ethics could succeed, and no system of professional ethics could transcend the dangers of self interest. That is why, even while rights, duties, obligations may be emphasized, the concept of virtue has 'hovered' so persistently over every system of ethics.

Is the virtuous physician simply the virtuous person practicing medicine? Are there virtues peculiar to medicine as a practice? Are certain of the individual virtues more applicable to medicine than elsewhere in human activities? Is virtue more important in some branches of medicine than others? How do professional skills differ from virtue? These are pertinent questions propadeutic to the later questions of the place of virtue in professional medical ethics.

I believe these questions are best answered by drawing on the Aristotelian-Thomist notion of virtues and its relationship to the ends and purposes of human life. The virtuous physician on this view is defined in terms of the ends of medicine. To be sure, the physician, before he is anything else, must be a virtuous person. To be a virtuous physician he must also be the kind of person we can confidently expect will be disposed to the right and good intrinsic to the practice he professes. What are those dispositions?

To answer this question requires some exposition of what we mean by the good in medicine, or more specifically the good of the patient – for that is the end the patient and the physician ostensibly seek. Any theory of virtue must be linked with a theory of the good because virtue is a disposition habitually to do the good. Must we therefore know the nature of the good the virtuous man is disposed to do? As with the definition of virtue we are caught here in another perennial philosophical question – what is the nature of the Good? Is the good whatever we make it to be or does it have validity independent of our desires or interest? Is the good one, or many? Is it reducible to riches, honors, pleasures, glory, happiness, or something else?

I make no pretense to a discussion of a general theory of the good. But any attempt to define the virtuous physician or a virtue-based ethic for medicine must offer some definition of the good of the patient. The patient's good is the end of medicine, that which shapes the particular virtues required for its attainment. That end is central to any notion of the virtues peculiar to medicine as a practice.

I have argued elsewhere that the architectonic principle of medicine is the good of the patient as expressed in a particular right and good healing action [26]. This is the immediate good end of the clinical encounter. Health, healing, caring, coping are all good ends dependent upon the more immediate end of a right and good decision. On this view, the virtuous physician is one so habitually disposed to act in the patient's good, to place that good in ordinary instances above his own, that he can reliably be expected to do so.

But we must fact the fact that the 'patient's good' is itself a compound notion. Elsewhere I have examined four components of the patient's good: (1) clinical or biomedical good; (2) the good as perceived by the patient; (3) the good of the patient as a human person; and (4) the Good, or ultimate good. Each of these components of patient good must be served. They must also be placed in some hierarchical order when they conflict within the same person, or between persons involved in clinical decisions [27].

Some would consider patient good, so far as the physician is concerned, as limited to what applied medical knowledge can achieve in *this* patient. On this view the virtues specific to medicine would be objectivity, scientific probity, and conscientiousness with regard to professional skill. One could perform the technical tasks of medicine well, be faithful to the skills of good technical medicine per se, but without being a virtuous person. Would one then be a virtuous physician? One would have to answer affirmatively if technical skill were all there is to medicine.

Some of the more expansionist models of medicine – like Engel's biopsychosocial model, or that of the World Health Organization (total well-being) would require compassion, empathy, advocacy, benevolence, and beneficence, i.e., an expanded sense of the affective responses to patient need [8]. Some might argue that what is required, therefore, is not virtue, but simply greater skill in the social and behavioral sciences applied to particular patients. On this view the physician's habitual dispositions might be incidental to his skills in communication or his empathy. He could achieve the ends of medicine without necessarily being a virtuous person in the generic sense.

It is important at this juncture to distinguish the virtues from technical or professional skills, as MacIntyre and, more clearly, Von Wright do. The latter defines a skill as 'technical goodness' – excellence in some particular activity – while virtues are not tied to any one activity but are necessary for "the good of man" ([38], pp. 139–140). The virtues are not "characterized in terms of their results" ([38]), p. 141). On this view, the technical skills of medicine are not virtues and could be practiced by a non-virtuous person. Aristotle held *technè* (technical skills) to be one of the five intellectual virtues but not one of the moral virtues.

The virtues enable the physician to act with regard to things that are good for man, when man is in the specific existential state of illness. They are dispositions always to seek the good intent inherent in healing. Within medicine, the virtues do become in MacIntyre's sense acquired human qualities " . . . the possession and exercise of which tends to enable us to achieve those goods which are internal to practices and the lack of which effectively prevents us from achieving any such goods" ([19], p. 178).

We can come closer to the relationships of virtue to clinical actions if we look to the more immediate ends of medical encounters, to those moments of clinical truth when specific decisions and actions are chosen and carried out. The good the patient seeks is to be healed – to be restored to his prior, or to a better, state of function, to be made 'whole' again. If this is not possible, the patient expects to be helped, to be assisted in coping with the pain, disability or dying that illness may entail. The immediate end of medicine is not simply a technically proficient performance but the use of that performance to attain a good end – the good of the patient – his medical or biomedical good to the extent possible but also his good as he the patient perceives it, his good as a human person who can make his own life plan, and his good as a person with a spiritual destiny if this is his belief [24, 25]. It is the sensitive balancing of these senses of the patient's good which the virtuous physician pursues to perfection.

To achieve the end of medicine thus conceived, to practice medicine virtuously, requires certain dispositions: conscientious attention to technical knowledge and skill to be sure, but also compassion – a capacity to feel something of the patient's experience of illness and his perceptions of what is worthwhile; beneficence and benevolence – doing and wishing to do good for the patient; honesty, fidelity to promises, perhaps at times courage as well – the whole list of virtues spelled out by Aristotle: " . . . justice, courage, temperance, magnificence, magnanimity, liberality, placability, prudence, wisdom" (*Rhetoric*, 1, c, 13666, 1–3). Not every one of these virtues is required in every decision. What we expect of the virtuous physician is that he will exhibit them when they are required and that he will be so habitually disposed to do so that we can depend upon it. He will place the good of the patient above his own and to seek that good unless its pursuit imposes an injustice upon him, or his family, or requires a violation of his own conscience.

While the virtues are necessary to attain the good internal to medicine as a practice, they exist independently of medicine. They are necessary for the practice of a good life, no matter in what activities that life may express itself. Certain of the virtues may become duties in the Stoic sense, duties because of the nature of medicine as a practice. Medicine calls forth benevolence, beneficence, truth telling, honesty, fidelity, and justice more than physical courage, for example. Yet even physical courage may be necessary when caring for the wounded on battlefields, in plagues, earthquakes, or other disasters. On a more ordinary scale courage is necessary in treating contagious diseases, violent patients, or battlefield casualties. Doing the right and good thing in medicine calls for a more regular, intensive, and selective practice of the virtues than many other callings.

A person who is a virtuous person can cultivate the technical skills of medicine for reasons other than the good of the patient – his own pride, profit, prestige, power. Such a physician can make technically right decisions and perform skillfully. He could not be depended upon, however, to act against his own self-interest for the good of his patient.

In the virtuous physician, explicit fulfillment of rights and duties is an outward expression of an inner disposition to do the right and the good. He is virtuous not because he has conformed to the letter of the law, or his moral duties, but because that is what a good person does. He starts always with his commitment to be a certain kind of person, and he approaches clinical quandaries, conflicts of values, and his patient's interests as a good person should.

Some branches of medicine would seem to demand a stricter and broader adherence to virtue than others. Generalists, for example, who deal with the more sensitive facets and nuances of a patient's life and humanity must exercise the virtues more diligently than technique-oriented specialists. The narrower the specialty the more easily the patient's good can be safeguarded by rules, regulations rights and duties; the broader the specialty the more significant are the physician's character traits. No branch of medicine, however, can be practiced without some dedication to some of the virtues [21].

Unfortunately, physicians can compartmentalize their lives. Some practice medicine virtuously, yet are guilty of vice in their private lives. Examples are common of physicians who appear sincerely to seek the good of their patients and neglect obligations to family or friends. Some boast of being 'married' to medicine and use this excuse to justify all sorts of failures in their own human relationships. We could not call such a person virtuous. Nor could we be secure in, or trust, his disposition to act in a right and good way even in medicine. After all, one of the essential virtues is balancing conflicting obligations judiciously.

As Socrates pointed out to Meno, one cannot really be virtuous in part:

> Why did not I ask you to tell me the nature of virtue as a whole? And you are very far from telling me this; but declare every action to be virtue which is done with a part of virtue; as though you had told me and I must already know the whole of virtue, and this too when frittered away into little pieces. And therefore my dear Meno, I fear that I must begin again, and repeat the same question: what is virtue? For otherwise, I can only say that every action done with a part of virtue is virtue; what else is the meaning of saying that every action done with justice is virtue? Ought I not to ask the question over again; for can any one who does not know virtue know a part of virtue? (*Meno*, 79)

VIRTUES, RIGHTS AND DUTIES IN MEDICAL ETHICS

Frankena has neatly summarized the distinctions between virtue-based and rights- and duty-based ethics as follows:

In an ED (ethics of duty) then, the basic concept is that a certain kind of external act (or doing) ought to be done in certain circumstances; and that of a certain disposition being a virtue is a dependent one. In an EV (ethics of virtue) the basic concept is that of a disposition or way of being – something one has, or is not does – as a virtue, as morally good; and that of an action's being virtuous or good or even right, is a dependent one [10].

There are some logical difficulties with a virtue-based ethic. For one thing, there must be some consensus on a definition of virtue. For another there is a circularity in the assertion that virtue is what the good man habitually does, and that at the same time one becomes virtuous by doing good. Virtue and good are defined in terms of each other and the definitions of both may vary among sincere people in actual practice when there is no consensus. A virtue-based ethic is difficult to defend as the sole basis for normative judgments.

But there is a deficiency in rights- and duty-ethics as well. They too must be linked to a theory of the good. In contemporary ethics, theories of good are rarely explicitly linked to theories of the right and good. Von Wright, commendably, is one of the few contemporary authorities who explicitly connects his theory of good with his theory of virtue. This essay, together with three previous ones, is part of the author's effort in that direction [25–27].

In most professional ethical codes, virtue and duty-based ethics are intermingled. The Hippocratic Oath, for example, imposes certain duties like protection of confidentiality, avoiding abortion, not harming the patient. But the Hippocratic physician also pledges: "... in purity and holiness I will guard my life and my art." This is an exhortation to be a good person and a virtuous physician, in order to serve patients in an ethically responsible way.

Likewise, in one of the most humanistic statements in medical literature, the first century A.D. writer, Scribonius Largus, made *humanitas* (compassion) an essential virtue. It is thus really a role-specific duty. In doing so he was applying the Stoic doctrine of virtue to medicine [3, 28].

The latest version (1980) of the AMA 'Principles of Medical Ethics' similarly intermingles duties, rights, and exhortations to virtue. It speaks of 'standards of behavior', 'essentials of honorable behavior', dealing 'honestly' with patients and colleagues and exposing colleagues 'deficient in character'. The *Declaration of Geneva*, which must meet the challenge of the widest array of value systems, nonetheless calls for practice 'with conscience and dignity' in keeping with 'the honor and noble traditions of the profession'. Though their first allegiance must be to the Communist ethos, even the Soviet physician is urged to preserve 'the high title of physician', 'to keep and develop the beneficial traditions of medicine' and to 'dedicate' all his 'knowledge and strength to the care of the sick'.

Those who are cynical of any protestation of virtue on the part of physicians will interpret these excerpts as the last remnants of a dying tradition of altruistic benevolence. But at the very least, they attest to the recognition that the good of the patient cannot be fully protected by rights and duties alone. Some degree of supererogation is built into the nature of the relationship of those who are ill and those who profess to help them.

This too may be why many graduating classes, still idealistic about their calling, choose the Prayer of Maimonides (not by Maimonides at all) over the more deontological Oath of Hippocrates. In that 'prayer' the physician asks: " . . . may neither avarice nor miserliness, nor thirst for glory or for a great reputation engage my mind; for the enemies of truth and philanthropy may easily deceive me and make me forgetful of my lofty aim of doing good to thy children." This is an unequivocal call to virtue and it is hard to imagine even the most cynical graduate failing to comprehend its message.

All professional medical codes, then, are built of a three-tiered system of obligations related to the special roles of physicians in society. In the ascending order of ethical sensitivity they are: observance of the laws of the land, then observance of rights and fulfillment of duties, and finally the practice of virtue.

A legally based ethic concentrates on the minimum requirements – the duties imposed by human laws which protect against the grosser aberrations of personal rights. Licensure, the laws of torts and contracts, prohibitions against discrimination, good Samaritan laws, definitions of death, and the protection of human subjects of experimentation are elements of a legalistic ethic.

At the next level is the ethics of rights and duties which spells out obligations beyond what law defines. Here, benevolence and beneficence take on more than their legal meaning. The ideal of service, of responsiveness to the special needs of those who are ill, some degree of compassion, kindliness, promise-keeping, truth-telling, and non-maleficence and specific obligations like confidentiality and autonomy, are included. How these principles are applied, and conflicts among them resolved in the patient's best interests, are subjects of widely varying interpretation. How sensitively these issues are confronted depends more on the physician's character than his capability at ethical discourse or moral casuistry.

Virtue-based ethics goes beyond these first two levels. We expect the virtuous person to do the right and the good even at the expense of personal sacrifice and legitimate self-interest. Virtue ethics expands the notions of benevolence, beneficence, conscientiousness, compassion, and fidelity well beyond what strict duty might require. It makes some degree of supererogation mandatory because it calls for standards of ethical performance that exceed those prevalent in the rest of society [30].

At each of these three levels there are certain dangers from over-zealous or misguided observance. Legalistic ethical systems tend toward a justification for minimalistic ethics, a narrow definition of benevolence or beneficence, and a contract-minded physician-patient relationship. Duty- and rights-based ethics may be distorted by too strict adherence to the letter of ethical principles without the modulations and nuances the spirit of those principles implies. Virtue-based ethics, being the least specific, can more easily lapse into self-righteous paternalism or an unwelcome over-involvement in the personal life of the patient. Misapplication of any moral system even with good intent converts benevolence into maleficence. The virtuous person might be expected to be more sensitive to these aberrations than someone whose ethics is more deontologically or legally flavored.

The more we yearn for ethical sensitivity the less we lean on rights, duties, rules, and prinicples, and the more we lean on the character traits of the moral agent. Paradoxically, without rules, rights, and duties specifically spelled out, we cannot predict what form a particular person's expression of virtue will take. In a pluralistic society, we need laws, rules, and principles to assure a dependable minimum level of moral conduct. But that minimal level is insufficient in the complex and often unpredictable circumstances of decision-making, where technical and value desiderata intersect so inextricably.

The virtuous physician does not act from unreasoned, uncritical intuitions about what feels good. His dispositions are ordered in accord with that 'right reason' which both Aristotle and Aquinas considered essential to virtue. Medicine is itself ultimately an exercise of practical wisdom – a right way of acting in difficult and uncertain circumstances for a specific end, i.e., the good of a particular person who is ill. It is when the choice of a right and good action becomes more difficult, when the temptations to self interest are most insistent, when unexpected nuances of good and evil arise and no one is looking, that the differences between an ethics based in virtue and an ethics based in law and/or duty can most clearly be distinguished.

Virtue-based professional ethics distinguishes itself, therefore, less in the avoidance of overtly immoral practices than in avoidance of those at the margin of moral responsibility. Physicians are confronted, in today's morally relaxed climate, with an increasing number of new practices that pit altruism against self interest. Most are not illegal, or, strictly speaking, immoral in a rights- or duty-based ethic. But they are not consistent with the higher levels of moral sensitivity that a virtue-ethics demands. These practices usually involve opportunities for profit from the illness of others, narrowing the concept of service for personal convenience, taking a proprietary attitude with respect to medical knowledge, and placing loyalty to the profession above loyalty to patients.

Under the first heading, we might include such things as investment in and ownership of for-profit hospitals, hospital chains, nursing homes, dialysis units, tie-in arrangements with radiological or laboratory services, escalation of fees for repetitive, high-volume procedures, and lax indications for their use, especially when third party payers 'allow' such charges.

The second heading might include the ever decreasing availability and accessibility of physicians, the diffusion of individual patient responsibility in group practice so that the patient never knows whom he will see or who is on call, the itinerant emergency room physician who works two days and skips three with little commitment to hospital or community, and the growing over-indulgence of physicians in vacations, recreation, and 'self-development'.

The third category might include such things as 'selling one's services' for whatever the market will bear, providing what the market demands and not necessarily what the community needs, patenting new procedures or keeping them secret from potential competitor-colleagues, looking at the investment of time, effort, and capital in a medical education as justification for 'making it back', or forgetting that medical knowledge is drawn from the cumulative experience of a multitude of patients, clinicians, and investigators.

Under the last category might be included referrals on the basis of friendship and reciprocity rather than skill, resisting consultations and second opinions as affronts to one's competence, placing the interest of the referring physician above those of the patients, looking the other way in the face of incompetence or even dishonesty in one's professional colleagues.

These and many other practices are defended today by sincere physicians and even encouraged in this era of competition, legalism, and self-indulgence. Some can be rationalized even in a deontological ethic. But it would be impossible to envision the physician committed to the virtues assenting to these practices. A virtue-based ethics simply does not fluctuate with what the dominant social mores will tolerate. It must interpret benevolence, beneficence, and responsibility in a way that reduces self interest and enhances altruism. It is the only convicing answer the profession can give to the growing perception clearly manifest in the legal commentaries in the FTC ruling that medicine is nothing more than business and should be regulated as such.

A virtue-based ethic is inherently elitist, in the best sense, because its adherents demand more of themselves than the prevailing morality. It calls forth that extra measure of dedication that has made the best physicians in every era exemplars of what the human spirit can achieve. No matter to what depths a society may fall, virtuous persons will always be the beacons that light the way back to moral sensitivity; virtuous physicians are the beacons that show the way back to moral credibility for the whole profession.

Albert Jonsen, rightly I believe, diagnoses the central paradox in medicine as the tension between self-interest and altrusim [15]. No amount of deft juggling of rights, duties, or principles will suffice to resolve that tension. We are all too good at rationalizing what we want to do so that personal gain can be converted from vice to virtue. Only a character formed by the virtues can feel the nausea of such intellectual hypocrisy.

To be sure, the twin themes of self-interest and altruism have been inextricably joined in the history of medicine. There have always been physicians who reject the virtues or, more often, claim them falsely. But, in addition, there have been physicians, more often than the critics of medicine would allow, who have been truly virtuous both in intent and act. They have been, and remain, the leaven of the profession and the hope of all who are ill. They form the sea-wall that will not be eroded even by the powerful forces of commercialization, bureaucratization, and mechanization inevitable in modern medicine.

We cannot, need not, and indeed must not, wait for a medical analogue of MacIntyre's 'new St. Benedict' to show us the way. There is no new concept of virtue waiting to be discovered that is peculiarly suited to the dilemmas of our own dark age. We must recapture the courage to speak of character, virtue, and perfection in living a good life. We must encourage those who are willing to dedicate themselves to a "higher standard of self effacement" [6].

We need the courage, too, to accept the obvious split in the profession between those who see and feel the altruistic imperatives in medicine, and those who do not. Those who at heart believe that the pursuit of private self-interest serves the public good are very different from those who believe in the restraint of self-interest. We forget that physicians since the beginnings of the profession have subscribed to different values and virtues. We need only recall that the Hippocratic Oath was the Oath of physicians of the Pythagorean school at a time when most Greek physicians followed essentially a craft ethic [7]. A perusal of the Hippocratic Corpus itself, which intersperses ethics and etiquette, will show how differently its treatises deal with fees, the care of incurable patients, and the business aspects of the craft.

The illusion that all physicians share a common devotion to a high-flown set of ethical principles has done damage to medicine by raising expectations some members of the profession could not, or will not, fulfill. Today, we must be more forthright about the differences in value commitment among physicians. Professional codes must be more explicit about the relationships between duties, rights, and virtues. Such explicitness encourages a more honest relationship between physicians and patients and removes the hypocrisy of verbal assent to a general code, to which an individual physician may not really subscribe. Explicitness enables patients to choose among physicians on the basis of their ethical commitments as well as their reputations for technical expertise.

Conceptual clarity will not assure virtuous behavior. Indeed, virtues are usually distorted if they are the subject of too conscious a design. But conceptual clarity will distinguish between motives and provide criteria for judging the moral commitment one can expect from the profession and from its individual members. It can also inspire those whose virtuous inclinations need re-enforcement in the current climate of commercialization of the healing relationship.

To this end the current resurgence of interest in virtue-based ethics is altogether salubrious. Linked to a theory of patient good and a theory of rights and duties, it could provide the needed groundwork for a reconstruction of professional medical ethics as that work matures. Perhaps even more progress can be made if we take Shakespeare's advice in *Hamlet*: "Assume the virtue if you have it not . . . . For use almost can change the stamp of nature."

BIBLIOGRAPHY

[1] Aquinas, T.: 1875, 'De Virtutibus in Communi', in S. Fretté (ed.), *Opera Omnia*, Louis Virés, Paris, Vol. 14, pp. 178–229.

[2] Aquinas, T.: 1964, *Summa Theologica*, Blackfriars, Cambridge, England, Vol. 23.

[3] Cicero: 1967, *Moral Obligations*, J. Higginbotham (trans.), University of California Press, Berkeley and Los Angeles.

[4] Copleston, F.: 1959, *Aquinas*, Penguin Books, New York.

[5] Cornford, F.: 1932, *Before and After Socrates*, Cambridge University Press, Cambridge, England.

[6] Cushing, H.: 1929, *Consecratio Medici and Other Papers*, Little, Brown and Co., Boston.

[7] Edelstein, L.: 1967, 'The Professional Ethics of the Greek Physician', in O. Temkin (ed.), *Ancient Medicine: Selected Papers of Ludwig Edelstein*, Johns Hopkins University Press, Baltimore.
[8] Engel, G.: 1980, 'The Clinical Application of the Biopsychosocial Model', *American Journal of Psychiatry* **137**: 2, 535–544.
[9] Foot, P.: 1978, *Virtues and Vices*, University of California Press, Berkeley and Los Angeles.
[10] Frankena, W.: 1982, 'Beneficence in an Ethics of Virtue', in E. Shelp (ed.), *Beneficence and Health Care*, D. Reidel, Dordrecht, Holland, pp. 63–81.
[11] Geach, P.: 1977, *The Virtues*, Cambridge University Press, Cambridge, England.
[12] Gould, J.: 1972, *The Development of Plato's Ethics*, Russell and Russell, New York.
[13] Guthrie, W.: 1971, *Socrates*, Cambridge University Press, Cambridge, England.
[14] Hauerwas, S.: 1981, *A Community of Character*, University of Notre Dame Press, Notre Dame, Indiana.
[15] Jonsen, A.: 1983, 'Watching the Doctor', *New England Journal of Medicine* **308**: 25, 1531–1535.
[16] Jowett, B.: 1953, *The Dialogues of Plato*, Random House, New York.
[17] Kenny, A.: 1978, *The Aristotelian Ethics*, Clarendon Press, Oxford, England.
[18] Lewis, C.: 1952, *Mere Christianity*, MacMillan Co., New York.
[19] MacIntyre, A.: 1981, *After Virtue*, University of Notre Dame Press, Notre Dame, Indiana.
[20] Maritain, J.: 1962, *Art and Scholasticism With Other Essays*, Charles Scribner and Sons, New York.
[21] May, W.: Personal communication, 'Virtues in a Professional Setting', unpublished.
[22] McKeon, R. (ed.): 1968, *The Basic Works of Aristotle*, Random House, New York.
[23] Oates, W. (ed.): 1957, *The Stoic and Epicurean Philosophers*, Modern Library, New York.
[24] Pellegrino, E.: 1979, 'The Anatomy of Clinical Judgments: Some Notes on Right Reason and Right Action', in H. T. Engelhardt, Jr., *et al.* (eds.), *Clinical Judgment: A Critical Appraisal*, D. Reidel, Dordrecht, Holland, pp. 169–194.
[25] Pellegrino, E.: 1979, 'Toward a Reconstruction of Medical Morality: The Primacy of the Act of Profession and the Fact of Illness', *Journal of Medicine and Philosophy* **4**: 1, 32–56.
[26] Pellegrino, E.: 1983, 'The Healing Relationship: The Architectonics of Clinical Medicine', in E. Shelp (ed.), *The Clinical Encounter*, D. Reidel, Dordrecht, Holland, pp. 153–172.
[27] Pellegrino, E.: 1983, 'Moral Choice, The Good of the Patient and the Patient Good', in J. Moskop and L. Kopelman (eds.), *Moral Choice and Medical Crisis*, D. Reidel, Dordrecht, Holland.
[28] Pellegrino, E.: 1983, '*Scribonius Largus* and the Origins of Medical Humanism', address to the American Osler Society.
[29] Pieper, J.: 1966, *The Four Cardinal Virtues*, University of Notre Dame Press, Notre Dame, Indiana.
[30] Reader, J.: 1982, 'Beneficence, Supererogation, and Role Duty', in E. Shelp (ed.), *Beneficence and Health Care*, D. Reidel, Dordrecht, Holland, pp. 83–108.

[31] Ross, W.: 1959, *Aristotle: A Complete Exposition of His Works and Thoughts*, Meridian Books, New York.

[32] Santayana, G.: 1951, *Dominations and Powers: Reflections on Liberty, Society and Government*, Charles Scribner and Sons, New York.

[33] Shelp, E. (ed.): 1982, *Beneficence and Health Care*, D. Reidel, Dordrecht, Holland.

[34] Shelp, E. (ed.): 1981, *Justice and Health Care*, D. Reidel, Dordrecht, Holland.

[35] Sokolowski, R.: 1982, *The Good Faith and Reason*, University of Notre Dame Press, Notre Dame, Indiana.

[36] Taylor, A.: 1953, *Socrates: The Man and His Thought*, Doubleday and Co., New York.

[37] Versanyi, L.: 1963, *Socratic Humanism*, Yale University Press, New Haven, Connecticut.

[38] Von Wright, G.: 1965, *The Varieties of Goodness*, The Humanities Press, New York.

[39] Wallace, J.: 1978, *Virtues and Vices*, University of California Press, Berkeley and Los Angeles.

Kennedy Institute of Ethics,
Georgetown University,
Washington, D.C.,
U.S.A.

MARTIN BENJAMIN AND JOY CURTIS

VIRTUE AND THE PRACTICE OF NURSING

> "A woman cannot be a good and intelligent nurse without being a good and intelligent woman."
>
> Florence Nightingale

When, in the mid-nineteenth century, Florence Nightingale revolutionized nursing, practically all nurses were women and what counted as a good or virtuous woman reflected the values and ideals of Victorian England. Thus a question arises as to what extent Nightingale's conception of a 'good and intelligent nurse' can be justified today when health care has been radically transformed by new knowledge and technology and our conception of a good or virtuous woman differs significantly from that which prevailed in her time.

In what follows we begin by sketching Alasdair MacIntyre's account of the relationships among practices, virtues, and the notion of a whole or integrated life [17]. We then show how this framework can be used to illuminate the relationships between Nightingale's conception of nursing and her ideals of Christian service, enlightenment rationalism, and Victorian womanhood. We conclude by considering the extent to which Nightingale's conception of a 'good and inelligent nurse' is compatible with the realities of modern nursing practice and current conceptions of woman's roles.

MACINTYRE'S ACCOUNT OF VIRTUES

Alasdair MacIntyre's recent account of the nature of the virtues provides a useful framework for analyzing virtues in nursing.[1] A virtue, according to MacIntyre's preliminary definition, is "an acquired human quality the possession and exercise of which tends to enable us to achieve those goods which are internal to practices and the lack of which effectively prevents us from achieving any such goods" ([17], p. 178). This definition includes two semi-technical terms, 'practices' and 'internal goods'.

Practices

By a *practice* MacIntyre means

Earl E. Shelp (ed.), Virtue and Medicine, 257–274.

any coherent and complex form of socially established cooperative human activity through which goods internal to that form of activity are realised in the course of trying to achieve those standards of excellence which are appropriate to, and partially definitive of, that form of activity, with the result that human powers to achieve excellence, and human conceptions of the ends and goods involved, are systematically extended ([17], p. 175).

Examples of practices, in this sense, include architecture, agriculture, football, physics, literary criticism, music making, the making and sustaining of family life, democratic citizenship, medicine, and nursing. Each has a history, a tradition, and standards of excellence; and each requires that participants cultivate certain virtues.

What MacIntyre terms 'goods internal to a practice' are those human goods that can be achieved only by engaging in a practice and submitting to its constraints and standards of excellence. They are called 'internal' goods, first, because they can be specified only in terms of a practice and, second, because experiencing them as goods depends on participating in a practice. External goods, on the other hand, though achievable by submitting to the constraints and standards of certain practices, are only contingently related to these practices.

Consider, for example, the practice of nursing. A number of important internal goods are associated with being a competent nurse. Among these are the satisfaction of using one's knowledge, talent, and skills to care for the sick, to help administer various treatments, and to teach clients how to maintain health. To enjoy these goods one must submit to the rules and standards of excellence for good nursing. These rules and standards partially define the practice and, although not immune to criticism and revision, their historical development is partly definitive of nursing as a practice. The satisfaction of self-respect obtained by competently participating in, and furthering the development of, nursing practice are goods internal to it.

In addition to such internal goods, the practice of nursing is usually accompanied by certain external goods such as a salary, job security, and occasionally public recognition in the form of promotions, raises, awards, etc. Such goods are not internal to nursing, for they can be achieved in a variety of ways. Moreover, one who participates in nursing *only* for the sake of such external goods will be tempted to violate the internal constraints and standards of the practice for the sake of such goods. Thus, for example, nurses who refrain from filling out incident reports about their errors or mistakes in the hope of being externally rewarded for 'error free nursing' are at the same time corrupting the practice of nursing and sacrificing the goods internal to it.

In filling out his account of a practice, MacIntyre draws two important contrasts, each of which will be important for our subsequent discussion. First, he emphasizes that "a practice, in the sense intended, is never just a set of technical skills" ([17], p. 180). Although a practice may include technical skills, these skills do not define the practice. Rather, practices are identified, transformed, and enriched by the practitioner's regard for those internal goods that are determined in part by the history of the practice. Thus, "to enter into a practice is to enter into a relationship not only with its present practitioners, but also with those who have preceded us in the practice, particularly those whose achievements extended the reach of the practice to its present point" ([17], p. 181). It is for this reason that, despite the differences between the technical skills required by contemporary nursing and those required in Florence Nightingale's time, an understanding of Nightingale's conception of virtue and nursing practice is helpful for understanding contemporary conceptions of virtue and nursing practice.

The second important contrast is between practices and institutions.

> Institutions are characteristically and necessarily concerned with ... external goods. They are involved in acquiring money and other material goods; they are structured in terms of power and status, and they distribute money, power and status as rewards. Nor could they do otherwise if they are to sustain not only themselves, but also the practices of which they are the bearers. For no practices can survive for any length of time unsustained by institutions ([17], p. 181).

Thus, a hospital is an institution, while nursing is a practice. Although institutions are essential for sustaining practices, their competitive preoccupation with the acquisition of external goods also poses a constant threat to the integrity of these practices. For example, a hospital's competitive preoccupation with the acquisition of money, power, and status can threaten as well as sustain the practice of nursing. "In this context," MacIntyre maintains, "the essential function of the virtues is clear. Without them ... practices could not resist the corrupting power of institutions" ([17], p. 181).

Virtues

Having sketched the notion of a practice and having distinguished internal from external goods, we may better understand MacIntyre's definition of a virtue: "A virtue is an acquired human quality the possession or exercise of which tends to enable us to achieve those goods which are internal to practices and the lack of which effectively prevents us from achieving any such goods." Although different practices may require one to cultivate different manual, technical, and intellectual skills and to submit to different standards of excellence, they all, according to MacIntyre, require roughly the same core virtues:

> Every practice requires a certain kind of relationship between those who participate in it. Now the virtues are those goods by reference to which, whether we like it or not, we define our relationships to those other people with whom we share the kind of purposes and standards which inform practices ([17], p. 178).

Three particular virtues that MacIntyre maintains are necessary for the integrity of any practice are honesty, courage, and justice.[2]

Honesty is important because without truthfulness the trust that is required if practitioners are cooperatively to pursue complex (internal) goods is undermined. Justice – that is, treating others in respect of merit or desert according to impersonal standards – is essential if a practice is to be free of corruption. And courage, the capacity to risk harm or danger to oneself, is important because genuine care and concern for the goods internal to a practice may require those who participate in it to take certain self-endangering risks.

Thus a virtuous nurse will be truthful with colleagues, including physicians where the practices of nursing and medicine overlap. Insofar as clients and their families are also conceived to be participants in the practice, she will be truthful with them. A virtuous nurse will avoid various forms of favoritism, and instead treat other nurses, physicians, and clients as they deserve to be treated. This is not, however, to say that she treats all members of each class equally, but rather that where she treats individuals differently she does so in accord with well-grounded, impersonal criteria. Finally, insofar as some degree of self-endangering risk is required to protect clients or to maintain the integrity of her practice, a virtuous nurse will be courageous enough to undertake it.

Integrity

To conclude, we must briefly connect this account of practices and virtues with the notion of a whole life; that is, we must say a word about the relationship between a virtuous practitioner and a virtuous person.

A person occupies a number of different roles in his or her life. A nurse, for example, is also likely to engage in other practices; for example, making and sustaining family life and participating in democratic politics. What does she do, then, when the requirements of one practice conflict with those of another – when, for example, the claims of family life conflict with those of her nursing practice? The fact that such conflicts can arise shows that the account of virtues in terms of practices is partial and incomplete and must be supplemented with "an account of a whole human life conceived as a unity" ([17], p. 188). At the center of such an account will be the virtue of integrity; that is, the notion of a whole or integrated life in which the various roles or components are more or less whole or unified. Such a conception of a human life is necessary if we are to talk of a life's being meaningful and if we are systematically to be able to resolve conflicts for an individual among the requirements of different practices.

When Florence Nightingale wrote of a 'good and intelligent nurse', she referred in part to virtues and technical skills related to the practice of nursing. And when she wrote of a 'good and intelligent woman', she referred in part to a woman who is trying to live a whole, integrated, or meaningful life. Thus we turn now to an account of how she saw the relationship between the two.

A GOOD NURSE AS A GOOD WOMAN

Nightingale's View of a Good Nurse

Florence Nightingale's vision of a good nurse can be traced in part to the history of Christian ideals and to Enlightenment rationalism's stress on clear-headed thinking.

Early Christians, believing that the rich were bound by faith to share their wealth with the poor, practiced with religious fervor the age-old custom of hospitality. Church deacons turned private homes into hospitals which served the sick, the insane, the destitute, travelers, lepers, foundlings and orphans, aged men and women ([9], pp. 51–52). These early Christians served God by serving others – especially the sick and the poor – and considered self-sacrifice a noble act. Thus, in early Christianity the tradition was established that a commitment to nursing meant a life of charity, martyrdom, and penance ([20], pp. 134–143).

Monastic orders and the great medieval military nursing orders, such as the Knights Hospitallers of St. John, kept Christian virtues central to nursing. The Hospitallers' white Maltese cross, a sign of purity worn upon their habit, had eight points representing the eight virtues or beatitudes that Jesus proclaimed in the Sermon on the Mount ([20], p. 186).

During the 12th and 13th centuries, secular forms of organized care developed that were based upon Christian values. But those religious nursing orders which offered little or no resistance to the Church's 1545 decree that nuns should live in strict enclosures, saw nursing care in the community deteriorate. Luther's emphasis on faith rather than works as a means of salvation also resulted in fewer religiously motivated women entering nursing. And Henry VIII's suppression of monastic orders in England resulted in nurses being drawn increasingly from the illiterate classes ([9], pp. 93–95). Thus, deterioration in hospital nursing spread throughout Europe and England. In the mid-19th century Charles Dickens captured the low level to which nursing had fallen in his well known character, Mrs. Sarah Gamp. According to Dickens, nurse Gamp was a coarse, vulgar person whose nose "was somewhat red and swollen, and it was difficult to enjoy her society without becoming conscious of a smell of spirits" ([12], p. 29).

Given the very low level of nursing practice at the time, it is no surprise that Florence Nightingale became a legend. She had, after all, reduced mortality at Barrack Hospital during the Crimean War from 42% to a fraction over 2% ([24], p. 1022). Nightingale had overcome horrible conditions at that hospital, which before she came had no sewage system, no laundry, no supplies, and no fit food for the sick ([12], pp. 37–38).

Nightingale's *Notes on Nursing: What It Is and What It Is Not*, with its statement of sanitary principles and assertions that patients must be treated humanely, quickly became a classic guide to nursing practice. And Nightingale's methods of nursing practice spread rapidly throughout the world. The school she founded trained nurses to organize and teach other nurses. Her 'nursing missioners' revolutionized nursing practice by introducing educated, trained, higher class women into hospital nursing in England and other countries, including the United States ([9], pp. 126–128).

Nightingale claimed that she used the word "nursing for want of a better." In her view nursing previously had "been limited to signify little more than the administration of medicines and the application of poultices." She taught that, "It ought to signify the proper use of fresh air, light, warmth, cleanliness, quiet, and the proper selection and administration of diet – all at the least expense of vital power to the patient" ([21], p. 209). She defined nursing as having "charge of the personal health of somebody . . . and what nursing has to do . . . is to put the patient in the best condition for nature to act upon him" ([4], p. 9).

Although Nightingale warned against nurses becoming dependent upon literary rather than clinical education, she argued that a nurse must use her mind. Nightingale stressed that a good nurse must think independently. She wrote, for example, that "It is as impossible in a book to teach a person in charge of sick how to *manage* as it is to teach her how to nurse . . . but it *is* possible to press upon her to think for herself" ([21], p. 215). She also stressed the importance of both making accurate observations and reflecting upon them: "Observation tells *how* the patient is; reflection tells *what* is to be done . . . . Observation tells us the fact, reflection the meaning of the fact" ([21], p. 255). Thus, Nightingale's view of nursing reflects the high value she placed upon clear-headed reasoning.

Finally, and most importantly, Nightingale continued the nursing tradition of Christian service and wrote that a good nurse responded to a calling. In an address to American nurses she warned against certain dangers, one of which was making nursing a profession and not a calling, and she asked, "What is it to feel a *calling* for anything? Is it not to do our work in it to satisfy the high ideal of what is the right, the best, and not because we shall be found out if we don't do it?" ([21], p. 272). Through such writings and directives to her nursing school, Nightingale kept Christian duty central to her conception of a good nurse.

Nineteenth Century Nursing Practice, Virtues, and Integrity

As MacIntyre explains, three particular virtues – honesty, courage, and justice – are necessary components of any practice, which includes nursing practice after Nightingale's 19th century reforms. Honesty, of course, was an important virtue, since relationships of trust define the intimacy of nursing. In recognition of the necessity of honesty, Nightingale specifically required that a good nurse be honest, truthful, and trustworthy ([21], p. 258). Courage, too, was obviously a component of late-19th century nursing practice, since to care for patients with contagious diseases, knowing full well the dangers, demanded of a nurse great personal courage. For example, Agnes Jones, Nightingale's "dearest and best pupil," an exceptionally beautiful and gifted woman, engaged in poorhouse reform and as a result died of typhus ([6], pp. 78–79). As Nightingale pointed out, a nurse must have "the courage, the coolness of the soldier" ([21], p. 275). Justice was another component of late-19th century nursing. Trained nurses served in a variety of settings – private homes, infirmaries for the poor, well established teaching hospitals, colonial hospitals – and they offered health teaching and care to the sick, to persons of all ages, occupations, and social classes.

A virtuous nurse – honest, courageous, and just – achieved certain internal goods through the practice of nursing. These included 'discipline', which in Nightingale's terms 'is the essence of moral training'. Nineteenth century nursing practice in the Nightingale tradition required discipline which fostered the full development of a nurse's potential both as a practicing nurse and as a spiritual person. According to Nightingale, who claimed to be quoting the best "lady trainer of probationer nurses" whom she knew, "It [discipline] is education, instruction, training – all that, in fact, goes to the full development of our faculties, moral, physical and spiritual, not only for this life, but looking on this life as the training ground for the future and higher life . . ." ([21], pp. 264–265). Nightingale's fight against licensure for nurses and her remarks against the movement to make nursing a profession rather than a calling, relate in part to the internal goods of nursing practice, that is, to the spiritual growth of a nurse as that nurse first learned and then practiced nursing.

Internal goods of nursing practice also included the opportunity to be of 'real usefulness'. The practice of nursing demanded that the nurse, like Agnes Jones, bring help to those around her through her own physical labor, executive skills, and mental acumen ([21], pp. 295–296). The usefulness of nursing practice could easily be seen when the nurse gave medicines, stimulants, or treatments as prescribed and provided fresh air, light, warmth, cleanliness, quiet, and diet. Since "health-at-home nursing" meant "exactly the same proper use of the same natural elements with as much life-giving power as possible to the healthy" ([21], p. 267), a nurse could also be of practical use in bringing health teaching to the public. This often meant bringing it to the very poor.

But nursing practice offered a nurse more than internal goods such as personal spiritual growth and opportunities to use intelligence, executive skills, and knowledge of nursing. Nineteenth century nursing practice also included certain external goods, especially if a nurse remained head of a ward for her entire lifetime. A nurse, as female head of a hospital family, could work "as contentedly as a mother at the head of her family," her duties, privileges, and authority clearly recognized ([21], p. 301). Nursing positions offered security and public recognition. Even though nursing in England during this period was more steady and settled than in the United States, where an entire hospital nursing force might change within two or three years ([21], p. 301), the hospital system in the United States, like that in England, was based upon the model of a family ([5], pp. 17–18). As such, it offered unmarried nurses opportunities for a respectable, secure life.

In terms of our framework designed to examine nursing practice in relation to the virtue of integrity, that is, to a whole or integrated life, 19th century nursing practice allowed women to combine a whole or integrated *woman's* life with a new and exciting career. Late 19th century nursing practice offered women opportunities to engage in meaningful, useful work at a time when opportunities for work in the world of any kind were strictly limited, especially for well-educated women. The view of nursing after Nightingale's reforms were instituted allowed educated women of social standing to practice nursing ([6], pp. 131–132). Respectable women who engaged in nursing practice were expected to continue to uphold Victorian values. Women's virtues, especially subordination and chastity, were expected of a nurse if she were to be accepted as a respectable person, especially if she were to be accepted as a respectable sister in charge of a poorhouse or men's ward or a respectable nurse engaged in district nursing.

Subordination and obedience to men were central to a woman's (and thereby a nurse's) character in the 19th century. Florence Nightingale's view of women, and consequently her views of nursing, embodied this position. She thereby reflected the belief in women's natural inferiority and rightful subordination that male religious and intellectual leaders had held for centuries.

Historical View of a Good Woman

Historically, where virtue has been concerned, the crucial fact for our discussion is that women have not been equal to men. Even the word 'virtue' originally excluded women altogether, since it derived from the Latin *virtus*, meaning 'manliness', and *vir*, meaning 'man' ([19], p. 1432). In ancient Greece women were unable to be virtuous in the same sense as men; but women could be virtuous in one special respect – in controlling their passion ([17], p. 128). Aristotle argued that women were defective males who were less able than men to think rationally and who were thus naturally subordinate to them ([1], pp. 41–54).

Christianity did not fundamentally change this emphasis on women's inferiority and subordination. By the end of the 1st century, interpretations of the Apostle Paul's teachings hardened attitudes about women's subordination ([22], pp. 136–137). In the 5th century St. Augustine affirmed woman's inferiority and emphasized her sinfulness ([23], pp. 156–163), and in the 13th century St. Thomas Aquinas, drawing upon Aristotle and Augustine, firmly grounded woman's subordination and inferiority in Roman Catholic Church doctrine. His writings claim that because of woman's sexuality, she is inferior to men physically, intellectually, and morally ([18], pp. 213–221). In European society, obedience in women remained a paramount virtue, even though Luther's doctrine of Christian vocation in the world, as well as other changes in religious doctrine during the Reformation, ultimately resulted in greater personal freedom and responsibility for women ([10], pp. 292–314).

Even as new and radical ideas gained acceptance in the 18th century, intellectually influential men continued to characterize women's virtues in a way that ensured their continued subordination and inferiority. While discussing the ideal woman – the lady – Kant praised her virtues, which derived from the gentlewoman's feminine sensibilities and refined taste rather than from her rational faculties. Accompanying the virtue of feminine charm – a most important virtue – were virtues such as neatness, cleanliness, delicacy, modesty, and taste ([1], pp. 127–144).

Nightingale did not challenge this conception of a good woman when she showed how women could create a respected and valued practice. In the late 19th century she wrote: "A really good nurse needs to be of the highest class of character. It needs hardly be said that she must be (1) Chaste . . . (2) Sober, in spirit as well as in drink . . . (3) Honest . . . (4) Truthful . . . (5) Trustworthy . . . (6) Punctual . . . (7) Quiet, yet quick . . . (8) Cheerful, hopeful . . . (9) Cleanly . . . (10) Thinking of her patient and not of herself" ([21], pp. 257–258). That Nightingale headed her list with 'chaste' signaled her combination of two sets of virtues: those associated with being a nurse and those associated more specifically with being a good woman as traditionally conceived.

Nightingale explained that a nurse must be chaste "in the sense of the Sermon on the Mount – a good nurse should be the Sermon on the Mount herself." Thus, she indicated that a nurse should be morally pure, decent, and modest. But she also meant that a chaste nurse – implicitly a woman – should be a good woman; for, she continued, "It should naturally seem impossible to the most unchaste to utter even an immodest jest in her presence" ([21], pp. 257–258). Thus, the virtues embedded in the traditional conception of a 'good woman', as well as those of the practice of nursing itself, are central to Nightingale's conception of nursing and to her influence on modern nursing.

An Emerging Tension

Nightingale's conception of a good nurse as a good woman contained the seeds of an internal conflict. The problem was that the virtues of Enlightenment rationalism built into her conception of a good nurse were incompatible with her traditional conception of a good woman; the clear-headed, independent thought and action required by the former would not always square with the subordination to (male) physicians required by the latter.

On the one hand, for example, Nightingale emphasized the importance of a nurse understanding "all her business" ([21], p. 256) – since she believed that nearly all physicians' orders were conditional ([21], p. 254). For example, she taught that a nurse must be trained to "know her business," and even though loyal to medical orders, she must not be servile: "True loyalty to orders cannot be without the independent sense or energy of responsibility, which alone secures real trustworthiness" ([21], p. 264). For Nightingale a nurse must have, in addition to a hearty interest in a patient and a technical interest in his care and cure, an intellectual interest in the case ([21], p. 275). On the other hand, she did not teach her nurses to act independently either in relation to nursing superiors or to physicians to spite of her caustic remark that obedience was "suitable praise for a horse" ([6], p. 133). For she taught that, "To obey *is* to understand orders, and to understand orders is really to obey. A nurse does not know how to do what she is told without such 'training' as enables her to understand what she is told, or without such moral disciplinary 'training' as enables her to give her whole self to obey. A woman cannot be a good and intelligent nurse without being a good and intelligent woman" ([21], p. 255). And by a 'good and intelligent woman', she meant a woman in the traditional sense.

This implicit tension between Nightingale's notion of a virtuous nurse and her conception of a virtuous woman was to become more pronounced in the twentieth century. Nonetheless, we must recognize and appreciate the differences between Nightingale's time and our own. For, as Nancy Boyd has pointed out, Florence Nightingale showed that for a 19th century woman the practice of nursing could be an exciting adventure without challenging Victorian relationships between men and women ([7], p. 187).

A GOOD NURSE AS A GOOD PERSON

Decline of the Subordinate Woman in Nursing

Florence Nightingale's conception of nursing took root in American nursing, with a heavy emphasis on woman's virtues rather than on virtues associated with independent thought. The Florence Nightingale Pledge, written not in England but by a nurse at Harper Hospital in Detroit, Michigan, in the 1890s, reflects American nursing practice at the turn of the century ([12], p. 141):

> I solemnly pledge myself before God, and in the presence of this assembly: *to pass my life in purity* and to practice my profession faithfully. I will abstain from whatever is deleterious and mischievous, and will not take or knowingly administer any harmful drug. I will do all in my power to maintain and elevate the standard of my profession and will hold in confidence all personal matters committed to my keeping and all family affairs coming to my knowledge in the practice of my calling. *With loyalty will I endeavor to aid the physician in his work* and devote myself to the welfare of those committed to my care ([12], p. 142, our emphasis).

As the lines we have emphasized in the pledge indicate, a nurse is supposed to be a good woman who is loyal to the (male) physician. No mention is made of initiative or self-reliance and little of accountability. In the mid-1950s one of the co-authors of this chapter (JC), like hundreds of student nurses for over a half century, recited the Nightingale Pledge at graduation from nursing school. Obedience to supervisors and physicians remained a main tenet of nursing ethics until the rebirth of feminism in the 1970s [11].

But calls to be obedient have never been absolute. Nurses have never said that they should obey physicians who make obvious mistakes. Moreover, we can find in nursing history opposition to obedience as a major nursing obligation. As early as 1916 Isabel Stewart wrote: "The traditional virtues of the good nurse are: obedience, the spirit of self-sacrifice, courage, patience, conscientiousness, and discretion. These are good, but under the newer conditions they are not alone sufficient. I think we have not placed enough emphasis on the more positive and vigorous qualities, such as self-reliance, the power of leadership, and initiative" [11].

Isabel Stewart was suggesting that nursing should change as the place of women in American life changed. The twentieth century was to bring critical political, economic, and sexual changes in the role and conception of women. For example, in 1920 the suffrage movement culminated in the vote for women; in the 1940s World War II was a turning point in women's occupational opportunities; and in the 1960s the 'pill' revolutionized women's ability to control conception. Some nurses recognized the need to reflect the implications of these changes both in their practice and in their codes of ethics. The following brief survey traces such changes in nursing codes.

In 1897 American nurses first tried to develop a code of ethics, but according to Lavinia Dock, it was "another piece of useless trouble" since instead of writing an original code, they tried rewriting the Code of Ethics of the American Medical Association ([8], p. 129). Efforts to create a code continued for more than fifty years. But not until 1950 did the American Nurses' Association first adopt its Code for Professional Nurses. Continued efforts led to revision of the 1950 code and to final adoption of the present Code for Nurses in 1976 ([26], pp. 1–5).

A partial review of the 1950 and the 1976 codes points to a definite change in what nurses consider their important professional ideals. Values stemming from nursing's heritage of woman's virtues, especially those relating to obligations to physicians, linger in the 1950 code but are eliminated in 1976. For example, the 1950 code asserts that "The nurse is obligated to carry out the physician's orders intelligently" and that "The nurse sustains confidence in the physician and other members of the health team" ([16], p. 165). In the 1976 code such statements are eliminated and the ethical nurse is one who "collaborates with members of the health professions" and who "safeguards the client from incompetent, unethical, or illegal practice" [3].

Another obvious change in the codes in the last twenty-five years is from a more or less determinate set of 'do's' and 'don't's' to a set of guidelines that require more independent thought and judgment. The 1950 code, for example, states that "The fundamental responsibility of the nurse is to conserve life and to promote health" ([16], p. 165). But advances in knowledge and technology, together with various social changes, have made such a conception of a nurse's fundamental responsibility overly restrictive. For nurses as well as for physicians, the emphasis now is on the rights of patients – an emphasis reflected in the Patient's Bill of Rights of the American Hospital Association [2] as well as recent revisions of the codes of ethics of both the American Nurses' Association and the American Medical Association [27]. In a rights-based conception of ethics the patient's situation and his or her values and life plan determine the specific form of a health care professional's duties rather than a specific list of 'do's' and 'don't's'. Thus the 1950 statement that 'The fundamental responsibility of the nurse is to conserve life and to promote health" is replaced as the first item in the 1976 Code for Nurses with: "The nurse provides services with respect for human dignity and the uniqueness of the client . . ." [3]. A nurse who practices according to the 1976 code must employ independent thought and judgment in order to determine which nursing action would be most respectful of human dignity and the client's uniqueness.

Also in the 1976 code those ideal qualities that Isabel Stewart wished could be more emphasized in 1913 – self-reliance, the power of leadership, and initiative – are clearly emphasized. After 60 years these are seen, for example, in the statements that "The nurse assumes responsibility and accountability for individual nursing judgments and actions" and that "The nurse exercises informed judgment and uses individual competence and qualifications as criteria in seeking consultation, accepting responsibilities, and delegating nursing activities to others" [3].

Thus, corresponding to social changes that have occurred with regard to the role and virtues of women in the twentieth century we can identify changes in codes of nursing. As the virtues of a good woman become less identified with the traditional stereotype, so too do the virtues of a nurse.

Twentieth Century Nursing Practice, Virtues, and Integrity

Returning once more to the framework for analyzing virtues and MacIntyre's account of the three virtues which are necessary components of any practice, we can see that honesty, courage, and justice are as important today in nursing practice as they were in previous times. Certainly a nurse of the 1980s is expected to be a trustworthy person. The ANA 'Code for Nurses' is based upon a trusting relationship with clients (as, for example, a nurse safeguards the client's right to privacy) and with co-workers (as, for example, a nurse collaborates with members of the health professions to meet the health needs of the public). A nurse must also be courageous, not only in caring for persons with infectious diseases or in handling potentially dangerous drugs and materials, but in speaking out to safeguard clients from incompetent, unethical, or illegal practices of other health care workers. A nurse must also be courageous in working, as the code directs, "to establish and maintain conditions of employment conducive to high quality nursing care." In some instances, for example, a nurse will be confronted by what appears to be an irreconcilable conflict between the integrity of her practice and the demands of a hospital. In this event, if all less drastic measures fail, she may have to engage in conscientious refusal, whistle-blowing, or a strike if nursing practice is to maintain minimally decent standards. Finally, justice is a critical component in current nursing practice; the Code, reflecting this importance, clearly states: "The nurse provides services with respect for human dignity and the uniqueness of the client unrestricted by considerations of social or economic status, personal attributes, or the nature of health problems" [3].

Like her 19th century nursing predecessors, a 20th century virtuous nurse achieves certain internal goods. These include using skills and knowledge to help persons care for themselves, grow, gain strength, and at times accept loss or death. Nursing practice offers an unusual opportunity to learn first-hand about certain features of the human condition. In observing the responses of clients and families *in extremis*, in being privileged to learn some of their innermost thoughts, wishes, hopes, and fears, and in listening to their life stories, the thoughtful nurse develops a deeper understanding of their lives and of her life as well. Such intrinsic rewards continue to be an important part of a nursing practice. For example, Margretta Styles cites a study in Wales in which practicing nurses were inclined toward traditional rewards of an intrinsic nature such as self-esteem that they derive from their work rather than those of an external nature such as professional advancement ([25], p. 102).

External goods to current nursing practice, however, differ from those of the past century. Public recognition is less an external reward for most nurses today than it was for pioneer nursing leaders. Indeed, research indicates that contemporary public recognition of the value of nursing practice is problematic since nurses are currently often portrayed through books, movies, plays, and television as empty-headed sexual playmates [13–15]. Furthermore, a young unmarried woman is no longer expected to move from her Victorian family into a hospital, which served as a surrogate family. Nurses, approximately 98% of whom are women, participate in American political, educational, social, and economic life with many fewer restrictions than those imposed on women in the nineteenth century. Nurses today are salaried and live outside of the hospital or nurses' home and work for the same kind of rewards as do most persons in our society – recognition for a job well done, promotions, a good salary, a secure home, a retirement fund, etc.

CONCLUSION

The tension implicit in Nightingale's conception of a nurse between the virtues of a good nurse and the virtues of a good woman is gradually being resolved as we move away from arbitrary and restrictive sexual stereotypes. The virtues of honesty, courage, and justice remain vital to the practice of nursing. But no longer must a nurse be quite so concerned about whether exercising these virtues is incompatible with being a good or virtuous woman as traditionally conceived. For what counts as a good or virtuous woman today differs little from what counts as a good or virtuous man. Thus we might consider revising the quotation from Florence Nightingale with which we began as follows:

> One cannot be a good and intelligent nurse without being a good and intelligent person.

Not only does this reformulation acknowledge that the goods of nursing, both internal and external, are accessible to men, but it also indicates that women seeking these goods must cultivate human virtues generally, rather than those traditionally and arbitrarily reserved for them.

NOTES

[1] Although we have reservations about much in MacIntyre's book, we largely accept his account of the relationships between certain virtues, practices, and the goods relevant to a practice. MacIntyre's presentation is richer and more detailed than our brief summary may suggest, Readers interested in the full account are advised to consult ([17], pp. 174–189).

[2] Honesty, courage, and justice are what might be called 'sustaining' virtues – they are necessary to sustain practices. The present discussion of nursing *as a practice* is basically limited to the way nursing instantiates these virtues and how this instantiation, both in Florence Nightingale's time and our own, is related to the notion of a whole or integrated life.

BIBLIOGRAPHY

[1] Agonito, R.: 1977, *History of Ideas on Woman: A Source Book*, G. P. Putnam's Sons, New York.

[2] American Hospital Association: 1973, 'Statement on a Patient's Bill of Rights', *Hospitals* 47, 41.

[3] American Nurses' Association: 1976, *Code for Nurses with Interpretive Statements*, American Nurses' Association, Kansas City, Missouri.

[4] American Nurses' Association: 1980, *Nursing: A Social Policy Statement*, American Nurses' Association, Kansas City, Missouri.

[5] Ashley, J. A.: 1976, *Hospitals, Paternalism, and the Role of the Nurse*, Teachers College Press, Columbia University, New York.

[6] Baly, M. E.: 1980, *Nursing and Social Change*, 2nd ed., William Heinemann Medical Books Ltd., London.

[7] Boyd, N.: 1982, *Josephine Butler, Octavia Hill, Florence Nightingale: Three Victorian Women Who Changed Their World*, MacMillan, New York.

[8] Dock, L. L.: 1912, *A History of Nursing*, Vol. 3, G. P. Putnam's Sons, New York.

[9] Dock, L. L. and Stewart, I. M.: 1938, *A Short History of Nursing*, 4th ed., G. P. Putnam's Sons, New York.

[10] Douglass, J. D.: 1974, 'Women and the Continental Reformation', in R. R. Reuther (ed.), *Religion and Sexism: Images of Women in the Jewish and Christian Traditions*, Simon and Schuster, New York, pp. 292–318.

[11] Jameton, A. L.: 1983, *Nursing Practice: The Ethical Issues*, Prentice-Hall, Englewood Cliffs, New Jersey.
[12] Kalisch, P. A. and Kalisch, B. J.: 1978, *The Advance of Modern Nursing*, Little, Brown and Company, Boston.
[13] Kalisch, P. A. and Kalisch, B. J.: 1982, 'The Image of Nurses in Novels', *American Journal of Nursing* **82**, 1220–1224.
[14] Kalisch, P. A. and Kalisch, B. J.: 1982, 'The Nurse as a Sex Object in Motion Pictures 1930–1980', *Research in Nursing and Health* **5**, 147–154.
[15] Kalisch, P. A. and Kalisch, B. J.: 1982, 'The World of Nursing in Prime Time Television, 1950–1980', *Nursing Research* **31**, 358–363.
[16] Kempf, F. C.: 1957, *The Person as a Nurse*, 2nd ed., MacMillan Company, New York.
[17] MacIntyre, A.: 1981, *After Virtue*, University of Notre Dame Press, Notre Dame, Indiana.
[18] McLaughlin, E. C.: 1974, 'Equality of Souls, Inequality of Sexes: Woman in Medieval Theology', in R. R. Ruether (ed.), *Religion and Sexism: Images of Women in the Jewish and Christian Traditions*, Simon and Schuster, New York, pp. 213–266.
[19] Morris, W. (ed.): 1969, *The American Heritage Dictionary of the English Language*, American Heritage Publishing Co., Inc., and Houghton Mifflin Company, Boston.
[20] Nutting, M. A. and Dock, L. L.: 1907, *A History of Nursing*, Vol. 1, G. P. Putnam's Sons, New York.
[21] Nutting, M. A. and Dock, L. L.: 1907, *A History of Nursing*, Vol. 2, G. P. Putnam's Sons, New York.
[22] Parvey, C. F.: 1974, 'The Theology and Leadership of Women in the New Testament', in R. R. Ruether (ed.), *Religion and Sexism: Images of Women in the Jewish and Christian Traditions*, Simon and Schuster, New York, pp. 117–149.
[23] Ruether, R. R.: 1974, 'Misogynism and Virginal Feminism in the Fathers of the Church', in R. R. Ruether (ed.), *Religion and Sexism: Images of Women in the Jewish and Christian Traditions*, Simon and Schuster, New York, pp. 150–183.
[24] Smith, F. T.: 1981, 'Florence Nightingale: Early Feminist', *American Journal of Nursing* **81**, 1021–1024.
[25] Styles, M. M.: 1982, *On Nursing: Toward a New Endowment*, C. V. Mosby Company, St. Louis.
[26] Sward, K. M.: 1978, 'The Code for Nurses: An Historical Perspective', in American Nurses' Association, *Perspectives on the Code for Nurses*, American Nurses' Association, Kansas City, Missouri, pp. 1–9.
[27] Veatch, R. M.: 1980, 'Professional Ethics: New Principles for Physicians', *Hastings Center Report* **10**, 16–19.

Michigan State University, Department of Philosophy,
College of Nursing, East Lansing, Michigan,
U.S.A.

KAREN LEBACQZ

THE VIRTUOUS PATIENT

In Solzhenitsyn's powerful novel *Cancer Ward*, we first meet Ludmila Dontsova in her role as doctor ([25], p. 1). For thirty years, she has developed medical theory and gathered empirical evidence about cancer. Later, her body racked with unmistakable signs of disease, she must make the difficult transition from physician to patient:

> . . . suddenly, within a few days, her own body had fallen out of this great, orderly system. It had struck the hard earth and was now like a helpless sack crammed with organs – organs which might at any moment be seized with pain and cry out . . . . As from today she ceased to be a rational guiding force . . . , she had become an unreasoning, resistant lump of matter . . . . Her world had capsized, the entire arrangement of her existence was disrupted" ([25], p. 446).

Thus does Dontsova experience the terrifying transition from the "noble estate" of medicine to the "taxpaying, dependent estate of patients" ([25], p. 419).

Most of us will never be of the 'noble estate' of medicine; but all of us will experience at some time the 'taxpaying dependent estate of patients'. As Susan Sontag puts it, "Everyone who is born holds dual citizenship, in the kingdom of the well and in the kingdom of the sick. Although we all prefer to use only the good passport, sooner or later each of us is obliged . . . to identify ourselves as citizens of that other place" ([26], p. 3).

What does it mean to be a 'good' patient? Are there virtues appropriate to this 'estate' or 'citizenship'? Are there vices to be avoided? These are the questions to be addressed in this essay.

"THE PERFECTION OF A THING"

The term 'virtue' can have numerous connotations. For our purposes, it suffices to adopt a broad definition – ". . . the excellence or perfection of a thing" ([3], p. 472). A virtue is a kind of excellence. To exhibit virtue is to exhibit excellence.

Four assumptions about virtue inform the following analysis. First, virtue is a mean. Extremes in either direction tend not to be virtues. Thus, 'patience'

Earl E. Shelp (ed.), Virtue and Medicine, 275–288.

may be a virtue, but too much patience is not. Some extremes constitute vices; others are simply inappropriate responses or lack of 'excellence'.

Second, virtues are not simply character traits but responses to situations. Bearing pain without comment might indicate courage, but it might also indicate fear or masochism. Courage can be assessed only in terms of an appropriate – or, indeed, excellent – response to a situation [15].

Third, there is an element of cultural conditioning in the definition and valuation of virtues [14]. A Stoic acceptance of pain might be considered 'virtuous' in one culture or time, but not in another. This is particularly true in the area of pain and illness, since the very experience of pain or illness is itself culturally conditioned [11, 26]. The view of 'virtues' given here is culturally limited and suggestive, not exhaustive.

Finally, 'virtue' may be different from 'the virtues'. 'The virtues' are qualities of character in response to situations. 'Virtue' implies an integrity or cohesiveness of character that goes beyond and is different from those individual qualities [10]. It may be easier to assess individual virtues than to locate any single pattern of life that constitutes integrity or 'virtue' in this broader sense.

"A MORE ONEROUS CITIZENSHIP"

Since virtues are in part responses to circumstance, to know the virtues of patients we must examine the circumstance of the patient. Dontsova's collapse mirrors three morally relevant aspects of that circumstance: the biological fact of illness ("organs which might at any moment be seized with pain") and its social and psychological concomitants ("her world had capsized, the entire arrangement of her existence was disrupted").

To be a patient is, first, to experience bodily (psycho-somatic) change. Sometimes there is pain: "I was lost in pain and drugs and . . . I thought I was screaming . . . " ([24], p. 92). It closes the mind and heart: " . . . a kind of invisible but thick and heavy fog invades the heart, envelops the body, constricting its very core" ([25], p. 54; see also [24]). It enervates: " . . . I came home, and just petered out . . . . I had severe headaches . . . and I was spent, spent" ([24], p. 116). It drags: "The acute headache gone, a middle-sized one is always there. This is what is called taking things slowly" ([24], p. 117). And physical appearance changes: " . . . my hair fell out in large clumps" ([1]. p. 208).

So to be a patient is in the first place to deal with illness itself. But illness is not just biological: "(P)ain commandeered, engaged and exhausted all my faculties, energies, and talents; pain tested all my weaknesses and shortcomings: pain filled most of the time and dictated the use of all the time . . . " ([27], p. 55). Pain can change perceptions of time, dilute energy, and generally render a person 'not himself': "I felt dissociated, estranged from myself – my flesh, my future" ([1], p. 5). "As a patient I lived with a useless body in a disconnected present" ([28], p. 28). It erodes typical response: "I am weaker and hence not able to explain or even make really rational decisions about big matters . . . " ([1], pp. 202–203). "One suddenly becomes uncertain about things taken most for granted . . . " ([28], p. 38).

In short, illness represents a threat to the *self* and requires changes in self-concept. Among the threats noted by commentators are these: (1) loss of one's life story [5], (2) erosion of self-image [17], (3) extreme vulnerability [29], (4) rupture between the self and the body [9, 17], (50) loss of ability to predict or take things for granted – hence, loss of worldview [5, 29], and (6) confrontation with mortality [17, 29].[1]

Furthermore, illness is experienced as something that 'happened to me' or 'befell me'. It is not chosen. It is an 'unwelcome intruder' ([20], p. 76). It seems 'unfair' and evokes a sense of injustice [29]. It thus raises questions of theodicy – "why did this happen to me?" ([6], p. 255). It threatens autonomy: "It was no longer he, but the tumor that was in charge" ([25], p. 274); " . . . I now possessed it, or it me . . . " ([1], p. 5). As Pellegrino argues, in illness we lose most of the freedoms we normally associate with being fully human. We are therefore "condemned to a relationship of inequality with the professed healer" ([17], p. 159; see also [29]).

And so illness brings with it a new social role. It brings expectations – 'right' or 'wrong' ways of acting:

> Being ill in a nursing home became my next task, a sombre dance in which I knew some of the steps. I must conform. I must be correct. I must be meek, obedient and grateful, on no account must I be surprising ([24], p. 91).

Indeed, we cannot speak of a virtuous 'patient' until we speak of the 'sick role'. 'Patient' is a role, and the virtues of a patient have to do with excellence in the role as well as responses to pain and personal change.

In our culture, " . . . the sick person is expected to submit to the techniques and facilities of medical science . . . (,) to cooperate with his physician, and in this way do everything he can to facilitate his recovery" ([7], p. 116). As one commentator puts it, "The patient's role was to accept treatment, make payments, and not to ask questions" ([21], p. 84).

The sick role is a task, requiring some acumen to perform:

I must do my work of being a patient with care. This was work that one did by lying still, remembering, judging. Deciding when your discomfort justified asking for help . . . . I made some mistakes and then I was contrite and very reasonable. Patients must like and dislike as little as possible ([24], pp. 92–93).

As difficult as coping with physical symptoms and pain may be, for many patients the change in social role is far worse – and hence, far more threatening to their sense of 'virtue'. Ironically, it may also be necessary for the healing process: Engelhardt argues that patient compliance (hence, medical success) depends on the establishment of a good patient–physician relationship, including the acceptance of the sick role [5].

The role changes an entire world: "This damned illness had cut right across his life, mowing him down at the crucial moment" ([25], p. 246). It reduces the most important person to impotence and insignificance: "He was no longer a vital cog in a large, important mechanism" ([25], p. 396); "In the clinic and at home she was unique and irreplaceable. Now she was being replaced" ([25], p. 449); "What I am hearing is the beginning of my daily existence, with this difference, though, that now I have no function in it" ([28], p. 25). It sanctions violations of privacy: " . . . they did all the things it is my precious privacy and independence to do for myself" ([24], p. 94). It changes the very structures of life: " . . . time in a Nursing Home was different from other time" ([24], p. 95).

To be a 'patient', therefore, is to confront physical distress, personal threat, and a new social role.[2] It is, indeed, a "more onerous citizenship" ([26], p. 3). The 'virtues' of the patient are qualities of excellence in response to the stresses of pain, discomfort, physical limitation, loss of autonomy, violation of privacy, vulnerability, and loss of self.

"I HAVEN'T BEEN A VERY GOOD PATIENT"

Patients internalize expectations not only about what they ought to do (e.g., take their medicine), but also about who they ought to be. Sontag suggests that fatal illness " . . . has always been viewed as a test of moral character . . . " ([26], p. 41). Any illness can be seen as such. And given the stresses of illness, it is no wonder that many patients feel they fail the test: "I haven't been a very good patient compared to the others on the ward . . . . I've probably been the worst patient on this ward . . . " ([7], p.114).

And what made this patient 'the worst' on the ward? – he was not able to act 'happy' and to be constantly 'making the best of everything'. His lament expresses poignantly the bind created for patients by a (stereo) typical image of the ideal patient that persists in contemporary Western culture.

In this image, the good patient is 'long suffering'. She does not complain, but endures – indeed, with a cheerful countenance. She obeys – not making too many demands, being cheerful and friendly to support staff, following doctors' orders. The acute patient tries to get well, accepts advice, eats chicken-noodle soup, swallows pride and pills, and does as requested. The chronic patient adjusts to limitations, forges a new life based upon them, and bears pain with good will. Finally, she does not burden her family with all her anxieties. In short, she is 'happy' and 'makes the best of everything'.[3]

Living up to this stereotypical image is difficult for many patients. But should they even try? Is this the 'virtuous patient'? Is this the most appropriate response to the patient's situation described above?

I think not. Nor do I think that Renee Fox's patient was indeed 'the worst' on the ward. A more accurate understanding of three important virtues suggests a different view.

These three virtues are *fortitude, prudence,* and *hope*.[4] A careful examination of each of them suggests a range of expression of 'virtue' for those who must cope with being a 'patient'.

"I HAVE TO BE A MIRACLE OF QUIET"

Daniel Foster suggests that patients facing serious illness go through three phases of questioning: (1) informational – "What's wrong? Is there treatment?" (2) behavioral – "Can I do it? Do I have enough courage?" and (3) religious – "Why did this happen to me?" ([6], pp. 254–255). The second stage is the one we are concerned with here. Many patients take courage, or fortitude, to be the most important task. They want to know, 'do I have the courage to do it?' In fact, Zaner suggests that much of the 'medical talk' related to the informational stage really amounts to urging courage on patients [29]. Care providers as well as patients are likely to think of courage as the first virtue of the patient.

But what is courage? What does fortitude mean in this situation? Like the patient quoted above, many take courage to be synonymous with long-suffering – being 'happy' and 'making the best' of things. A closer look at the classical virtue of fortitude will help to show that this is not the only possible manifestation of this virtue. The patient who is cheerfully resigned is not necessarily the one who shows most virtue.

According to Pieper, the classical virtue of fortitude involves bravery for the sake of the good ([18], p. 122). Its essence is not lack of fear, for without fear there is no bravery. Courage means facing fear, and not allowing oneself to be forced into evil by fear. For example, to be courageous might mean accepting a life-saving operation in spite of fear of the pain involved.

There are two aspects of this bravery: endurance and attack. Contemporary Western culture prizes endurance and tends to equate it with fortitude. And some patients do indeed live this aspect of fortitude. Renee Fox describes patient Leo Angelico as one who accepted his situation and endured in it ([7], p. 187). In *Cancer Ward*, when fiesty Oleg is finally released from the ward, he watches a goat stand absolutely still for minutes on end and reflects: " . . . That was the sort of character a man needed to get through life" ([25], p. 503). Coping with the limitations of age and illness, Florida Scott-Maxwell also evidences a desire for the endurance aspect of fortitude:

> A garden, a cat . . . , to walk in woods and fields, even to look at them, but these would take strength I have not got . . . . So . . . I must make the round of the day pleasant, getting up, going to bed, meals . . . . I have to be a miracle of quiet to make the flame in my heart burn low . . . ([24], p. 131).

To accept limitation, to endure in the face of the unchangeable, to quiet the heart – these are aspects of fortitude.

There is a place, then, for *patience* among the virtues, though patience should not be mistaken for "indiscriminate, self-immolating, crabbed, joyless, and spineless submission to whatever evil is met with . . . " ([18], p. 129). To be patient is to preserve cheerfulness and serenity of mind in spite of injuries that result from the attempt to realize the good or to live humanly. The virtue of fortitude in its aspect as endurance or patience has to do with keeping one's spirit from being broken by fear, grief, or sadness. This is clearly one virtuous response to the situation of the patient.

"I HEARD THE ANIMAL GROWL IN ME"

But endurance is not the only aspect of fortitude, and patient acceptance is not the only appropriate response to illness. As Illich puts it, "A myriad of different virtues express the different aspects of fortitude that traditionally enabled people to recognize painful sensations as a challenge and to shape their own experience accordingly" ([11], p. 93). Indeed, Stringfellow goes so far as to assert that the endurance of pain, far from being a virtue, may be a form of 'vanity' – a vice – for it tempts patients to seek justification in their own actions ([27], p. 52).

Wrath and attack – refusal to accept and endure – are also part of fortitude. Wrath is that part of fortitude that gives strength to attack evil. "The power of anger is actually the power of resistance in the soul", declears Pieper ([18], p. 193). Hence, Florida Scott Maxwell exhibits the virtue of fortitude not only by seeking to be a 'miracle of quiet' but by railing against invasions of privacy:

Then the rage I knew so well rose in me and threatened all. I heard the animal growl in me when they did all the things it is my precious privacy and independence to do for myself. I hated them while I breathed, 'Thank you, nurse' ([24], p. 94).

And so does Morris Abram when he expresses outrage at violations of hygiene in his care and declares "I never surrendered to the system" ([1], p. 205). To rail both against illness and against the limitations of the sick role can be an expression of the virtue of fortitude. In a situation characterized by so much loss of autonomy and control, the anger that expresses one's *self* can be a reassertion of autonomy – the 'power of resistance in the soul'.

And autonomy is clearly an issue for patients. Abram asserts, "I felt I could only survive it by insisting on control" ([1], p. 209). Throughout *Cancer Ward*, the fictional patients struggle with their control or lack of it: "Anything so long as he didn't have to sit passively waiting for death . . . (I)t was only when they took the initiative that they found inner calm" ([25], p. 177). "Vadim still took pleasure in the way he kept himself under control . . . " ([25], p. 378). "The tumor was trying hard to sink him, but he was still steering his course" ([25], p. 389).

Control is hard-won in the face of fear and the onslaught of disease and deterioration: "The tumor, which had begun by annoying him, then frightening him, had now acquired rights of its own. It was no longer he but the tumor that was in charge" ([25], p. 274). "He began to feel the terrifying absurdity of these paltry surroundings and idiotic conversations, and the urge to rip apart his polished self-control and howl as a wild animals howls at its snare . . . " ([25], p. 300). Like Florida Scott Maxwell, Solzhenitsyn's fictional patients experience rage at loss of control and they wanted to 'howl' or 'growl' – "like an animal" – in the face of the evils upon them. Such rage can be itself the expression of the virtue of fortitude in its aspect as wrath and attack.

If the first behavioral concern of patients is courage – 'Can I do it?', 'Do I have the strength?' – then a core virtue of the patient is fortitude. Fortitude can be expressed in acceptance or in wrath and anger. There is no single 'virtuous' response.

"I COULD SEE THE BIRDS"

But fortitude presupposes another virtue: *Prudence*. For if fortitude is bravery in the face of evil, it requires the ability to perceive evil and good, to know what is required. And this ability to perceive the real and to discern what is required as a 'fitting' response, is what is understood by the classical virtue of prudence.

For some, this may be a surprising statement. Prudence has come to mean largely "a due regard for one's own welfare" ([21], p. 84) or "timorous, small-minded self-preservation" ([18], p. 4). Indeed, the original meaning of the virtue has become so distorted in modern times that Dom Helder Camara cries out: "Prudence, you have been so disfigured that . . . I almost feel the need to shout, 'Be rash!' Above all, don't forget that boldness is a virtue . . ." ([2], p. 111). However, as Camara implies and Pieper makes clear, an interpretation of prudence in terms of conformism, compromise, and fear is a 'disfigurement' of the virtue.

In its classic sense, prudence has not to do with self-protectiveness, but with acting in accord with the real. Bravery that does not accurately perceive danger is false courage, bravado or foolhardiness, and it borders on vice. Hence, the exercise of fortitude or other virtues depends on the perfected ability to perceive the situation accurately and the willingness to act on what one perceives. This is what prudence brings.

If prudence means acting in accord with the real, or letting the truth of real things determine our actions, then we can fall short of this virtue in two ways: (1) by failing to perceive accurately, or (2) by being unwilling to act.

The strength of the Patients' Rights movement today may lie in its urging of patients to *act* on their own behalf. Thus, it falls under the second aspect of prudence – the willingness to be responsible and to act. When Florida Scott Maxwell railed against the invasion of her privacy, the nurse responded, "You're the kind that get well quickly. Some want everything done for them, just won't take themselves on at all" ([24], p. 95). 'Taking oneself on' and being responsible is part of prudence. It puts flesh on the bones of the 'covenant' or 'partnership' between patient and physician [12, 19]. Fox's subjects responded to their situations with many devices for taking initiative and being responsible: they became experts on their disease, kept meticulous medical records on their progress, and generally had a 'take up your bed and walk' attitude. In this, they exhibited part of the virtue of prudence.

However, prudence would not always mean 'taking up your bed' and walking. It depends on the nature of the illness. How do we know whether to 'be a miracle of quiet' or to 'let the animal growl'? The answer lies in the first aspect of prudence – the aspect of perception. Perception is key.

Before acting, we must listen. "'Prudence as cognition' . . . includes above all the ability to be still in order to attain objective perception of reality" ([18], p. 13). Talcott Parsons pointed out long ago that we are an activist society (cf. [8]). We expect patients to *do* things in order to become well. 'Taking up one's bed' and 'walking' are approved in this culture because they fit this activist model. But what we do must be in accord with the real if it is to be virtuous (prudent). It is possible to be so busy 'doing' that we forget to listen. Action that excludes contemplation rejects a part of prudence.

Hence, Pieper stresses the role of silence, of listening, of contemplation as key to prudence. Prudence requires both docility (in its true sense of teachableness, or openness) and trust, necessary for "receptive silence before the truth of real things" ([18], p. 21).

Ironically, illness and the accompanying role of patient can themselves be the path to prudence. This occurs in two arenas. First, patients who learn to 'listen' to their bodies learn about themselves. Illness may force an examined life [16]; changes in the body can have positive meaning for the self [9]. "If a man never became ill he would never get to know his own limitations" ([25], p. 148).

But it is not only limitations that are learned in illness. Illness cries out for meaning, and can even produce a story through which meaning is now given to life [5]. For instance, Abram reflects:

> My first thought was that my illness was the dramatization of a metaphor . . . . My spiritual malaise was deep – to the very marrow of my bones, and now in the marrow of my bones, a disease was threatening to snuff me out ([1], p. 271).

Patients learn about themselves and about the meaning of their lives from their confrontations with illness.

Second, "illness tends to remove the hardness. The experience of suffering somehow makes it easier to see the suffering of others and to want to help" ([6], p. 267). A subject of Renee Fox described it this way:

> After you've been in for a while, your outlook on the hospital and your illness changes . . . . Instead of just lying in bed and thinking about your troubles, you get interested in the people around you and what's happening to them . . . ([7], p. 135).

Fox notes that a common method of 'coping' with illness on Ward F-Second was learning to reach out in love to others. This reaching out follows upon a change in perspective that is part of the virtue of prudence.

Indeed, van den Berg goes so far as to suggest that "The illness of the body can be the condition for a soundness of mind which the healthy person misses easily" ([28], p. 73). Sick people who do not cling stubbornly to memories of health may discover "a new life of surprising intensity" ([28], p. 68). Abram notes that his bout with cancer brought an 'urgency' to his life and that his values changed:

> I have tried to focus on what is important .... The list of acquaintances called friends has been trimmed while real friendships have been intensified. I give more of myself to fewer .... Cancer can bring balance and its own kind of wisdom ([1], p. 272).

The perspective that mystics have often urged – the ability to take joy in little things, to seek nothing, to be attentive to minute changes in the day – these are sometimes the gifts of illness. As Ted Rosenthal put it when he was dying, "I could see the birds for the first time in my life" [22].

"WHAT I NEED IS AN I.V. OF SANI-FLUSH"

To be able to learn from illness requires a kind of spiritual or emotional health. Indeed, both patients and physicians argue that the connection between physical and emotional health is 'indivisible' [1, 5]. Hence, among the virtues necessary for the patient will be those related to emotional – and spiritual – health.

Key among these is *hope*: hope for improvement, hope for the courage to withstand it all, hope for meaning to emerge out of the chaos, pain, and sense of injustice. This theological virtue therefore takes its place with prudence and fortitude as central among the virtues of patients.

Hope involves perfecting the will to trust in the attainment of the end [3]. The key is trust – letting go of control, and depending on fate, or God, or something outside oneself. Hope means that when we perceive the real and it turns out to be terrifying, we do not despair. Hope has to do with transcendence – with not being limited by what appears to be necessary.

Central to the development and display of hope for most patients is a sense of humor. Humor keeps hope alive amd makes transcendence possible. Gloomy Pavel Nikolayevich is stunned by the appearance of Chaly on the cancer ward. "There was nothing of the exhaustion of cancer on his face. His smile twinkled with confidence and joie de vivre ... " ([25], p. 304). Chaly laughs, jokes with the nurses – "A woman a day keeps the doctor away" – and plays cards. Pavel reflects, "Maybe Chaly was right, maybe if you tackled your illness the way he did it would slip away of its own accord. Why mope?" ([25], p. 307).

Fox's patients also used humor as a way to deal with problems and relieve stress. Fox suggests that humor was "the inner emotional 'language' of the ward" ([7], p. 175). It was the way patients dealt with the stresses of everyday life, their fears of death and disability, and their deepest concerns. One patient joked:

> I think the reason I'm so fouled up is because of all the gooey stuff they're always sticking in my veins. My veins must be coated an inch thick with all that sticky albumin and resin. What I need is an I.V. of Sani-Flush . . . ([7], p. 171).

Similarly, Abram reports the following exchange between a formerly depressed patient and his nurse, after the patient has begun to rise above depression:

> 'This therapy has aged me, you know.'
> 'We can't make you any younger . . . . '
> 'I'm not here to get younger . . . I'm here to get older' ([1], p. 224).

The joke shows clearly the acceptance of fate (he can not look younger again) and yet its transcendence (he can refuse to give up, and still hope for years ahead). Humor is a way of refusing to capitulate to fate, even when prudence dictates a sorry end. Humor gives perspective and the ability to transcend the real even as one accepts it.

Of course, humor can also be a way of avoiding the obvious or refusing to come to terms with reality. It is possible that Fox's patients and the fictional patients of *Cancer Ward*, almost all of whom are men, use humor because they are denied a more direct 'emotional language'. Rather than expressing their fears directly, they joke about potential loss: " . . . my wife is thirty-six. Well, if she doesn't want to take me back, I'll trade her in for two eighteen-year-olds" ([7], p. 171).

But humor is clearly also a way of making sense – "or, at least mock sense", as Fox puts it – of their predicament. It is therefore the link to the virtue of hope. Hope makes sense of things that are senseless, trusts that there is meaning in the midst of chaos, and finds the courage to endure or attact in situations where prudence alone might fail. Hope is the response to the third stage of questioning: 'Why did this happen to me?' "For what may I hope?" ([6], p. 255). As a sign of hope, humor keeps us human in the midst of tragedy.

"YOU MUST BARE YOUR HEART AND EXPECT NOTHING IN RETURN"

These three virtues, then – fortitude, prudence, and hope – are central to the task of being a patient. They are among the 'excellent' responses to pain, physical limitation, bodily change, loss of autonomy, violation of privacy, vulnerability, and the 'sick role'.

Yet though we can describe these virtues, there is clearly no one way of expressing them and no single pattern that 'should' be followed by patients. To some extent, the way we experience and express being a patient will be the result of all that we have been before [27]. Yet where courage seemed non-existent before, it can emerge [6]. As Renee Fox concluded after her study of the men on Ward F-Second, "(p)at answers and personal judgments now come harder and seem less satisfactory" ([7], p. 207). Both health and illness are tasks that must be learned. Each culture will judge some paths more 'excellent' than others, but the integrity to shape our experience, to endure and transcend suffering, to perceive and love the real – there remains some mystery here. As Ted Rosenthal describes it, "You will feel so all alone, abandoned, come to see that life is brief .... You must bare your heart and expect nothing in return" ([22], pp. 81–82). Perhaps in the long run, the 'virtuous patient' is the one who bares her heart, becomes vulnerable to the experience, and expects nothing in return ....

NOTES

[1] In view of these changes experienced in the 'self' – particularly in the case of serious illness – one might ask whether the concept of 'virtue' can even be applied to patients. If the development of virtue requires decision, rationality, and coherence of character ([10, 18]), then perhaps it does not apply to instances where coherence is interrupted and rationality undermined.

[2] Indeed, there may be several new roles or changes in several old ones (e.g., one's role as spouse or citizen may change). Furthermore, 'inpatient' is a different role than 'out-patient' and 'chronic patient' is a different role than 'acute patient'. Thus Engelhardt speaks of "... sick roles as expected ways to act, given a particular disease" ([5], p. 148).

[3] I use 'she' to describe the patient here because this (stereo) typical image is akin to the 'virtues' assigned to women during the last part of the 19th century (cf. [4]).

[4] In classical theology, the first two of these are 'cardinal' virtues (fortitude a 'moral' and prudence an 'intellectual' virtue) and the last is a 'theological' virtue [3]. From the patient's perspective, such niceties of academic distinction fade, and I therefore ignore them here. For the patient's perspective here I have utilized not only first-person accounts of illness, but also some fictionalized accounts, in the conviction that fiction often permits honed confrontation with issues of virtue.

BIBLIOGRAPHY

[1] Abram, M. B.: 1982, *The Day Is Short*, Harcourt, Brace, Javonovich, New York.

[2] Camara, D. H.: 1981, *A Thousand Reasons for Living*, Fortress, Philadelphia.

[3] *The Catholic Encyclopedia*: 1922, Vol. 15, Encyclopedia Press, New York.

[4] Douglas, A.: 1977, *The Feminization of American Culture*, Avon Books, New York.

[5] Engelhardt, H. T.: 1982, 'Illness, Diseases, and Sicknesses', in V. Kestenbaum (ed.), *The Humanity of the Ill*, University of Tennessee Press, Knoxville, pp. 142–156.

[6] Foster, D. W.: 1982, 'Religion and Medicine: The Physician's Perspective', in M. E. Marty and K. L. Vaux, *Health/Medicine and the Faith Traditions*, Fortress, Philadelphia, pp. 245–270.

[7] Fox, R. C.: 1959, *Experiment Perilous*, University of Pennsylvania Press, Philadelphia.

[8] Friedson, E.: 1973, *Profession of Medicine: A Study of the Sociology of Applied Knowledge*, Dodd, Mead and Co., New York.

[9] Gadow, S.: 1982, 'Body and Self: A Dialectic', in V. Kestenbaum (ed.), *The Humanity of the Ill*, University of Tennessee Press, Knoxville, pp. 86–100.

[10] Hauerwas, S.: 1974, *Vision and Virtue*, Fides Press, Notre Dame.

[11] Illich, I.: 1975, *Medical Nemesis: The Expropriation of Health*, Caldor and Boyers, London.

[12] Jonas, H.: 1969, 'Philosophical Reflections on Experimenting with Human Subjects', in P. Freund, *Experimentation with Human Subjects*, George Braziller, New York, pp. 1–31.

[13] Kestenbaum, V. (ed.): 1982, *The Humanity of the Ill*, University of Tennessee Press, Knoxville.

[14] MacIntyre, A.: 1981, *After Virtue*, University of Notre Dame Press, South Bend.

[15] Mandelbaum, M.: 1969, *The Phenomenology of Moral Experience*, Johns Hopkins Press, Baltimore.

[16] O'Nell, J.: 1982, 'Essaying Illness', in V. Kestenbaum (ed.), *The Humanity of the Ill*, University of Tennessee Press, Knoxville.

[17] Pellegrino, E. D.: 1982, 'Being Ill and Being Healed: Some Reflections on the Grounding of Medical Morality', in V. Kestenbaum (ed.), *The Humanity of the Ill*, University of Tennessee Press, Knoxville.

[18] Pieper, J.: 1966, *The Four Cardinal Virtues*, University of Notre Dame Press, Notre Dame.

[19] Ramsey, P.: 1970, *The Patient as Person*, Yale University Press, New Haven.

[20] Rawlinson, M. C.: 1982, 'Medicine's Discourse and the Practice of Medicine', in V. Kestenbaum (ed.), *The Humanity of the Ill*, University of Tennessee Press, Knoxville, pp. 19–85.

[21] Reeck, D.: 1982, *Ethics for the Professions: A Christian Perspective*, Augsburg Press, Minneapolis.

[22] Rosenthal, T.: 1973, *How Could I Not Be Among You?*, Avon Books, New York.

[23] Sarton, M.: 1978, *A Reckoning*, W. W. Norton, New York.

[24] Scott-Maxwell, F.: 1968, *The Measure of My Days*, Alfred A. Knopf, New York.

[25] Solzhenitsyn, A.: 1969, *Cancer Ward*, Bantam Books, New York.

[26] Sontag, S.: 1977, *Illness as Metaphor*, Farrar, Strauss, and Giroux, New York.
[27] Stringfellow, W.: 1970, *A Second Birthday*, Doubleday, New York.
[28] Van den Berg, J. H.: 1966, *The Psychology of the Sickbed*, Duquesne University Press, Pittsburgh.
[29] Zaner, R. M.: 1982, 'Chance and Morality', in V. Kestenbaum (ed.), *The Humanity of the Ill*, University of Tennessee Press, Knoxville, pp. 39–68.

Pacific School of Religion,
Berkeley, California,
U.S.A.

MARC LAPPÉ [1]

VIRTUE AND PUBLIC HEALTH: SOCIETAL OBLIGATION AND INDIVIDUAL NEED

INTRODUCTION: PERSONAL AND INSTITUTIONAL VIRTUE

To the extent that virtue requires human excellence, it is an attribute of persons whose actions exceed the norm established by customary standards, mores, or laws. Where the health of individuals is involved, virtue has traditionally been discussed in terms of the activities of physicians towards their patients. Albert Schweitzer's activities in the Gabon are considered virtuous because he put his patients' needs above his own. McCullough describes the conditions under which virtuous behavior may be seen in the doctor–patient relationship [20]. According to McCullough, attributes like humanity, decency, and altruism are components of virtuous physician conduct.

For health as it pertains to groups of individuals, i.e., *public* health, virtue has been all but neglected as a concept. Ethical aspects of societal obligations for rendering health care have only recently received attention in the United States. In its report, 'Securing Access to Health Care', the President's Commission for the Study of Ethical Problems in Medicine and Biomedical and Behavioral Research described the need for ensuring that the population as a whole be provided a 'decent minimum' of health care [1]. The Commission concluded that society has an ethical obligation to ensure equitable access to health care for everyone. But, in acknowledging this moral requirement, the Commission underscored that societal obligations should be balanced with individual obligations.

The report thus fell short of advocating health care as an unlimited right. Instead, the Commission found that it was appropriate to ask individuals to pay their 'fair share' of their own health costs, without 'excessive burden'. While recognizing that health care has a special place in relieving suffering, preventing premature death, restoring function, increasing opportunity, providing health related information, and serving as a vehicle for mutual empathy and compassion, it stopped short of recognizing any obligations on the part of institutions to be virtuous in their health-supporting functions. The furthest the Commission appeared to be willing to go was to acknowledge that "health care needs are too complex . . . to permit care to be regularly

Earl E. Shelp (ed.), Virtue and Medicine, 289–303.

denied on the grounds that individuals are solely responsible for their own health" [1]. Such a view narrows the scope of societal obligations and reduces the role of virtue in public health systems.

The Commission thus established a middle ground in health care policy. In effect, their recommendations endorsed the concept of a 'health floor' beneath which no person should be permitted to fall, but denined that the obligation to provide health care was solely a societal one. Holding that health care is an essential commodity, as the Commission does, creates a necessary but not a sufficent basis for developing a theory of virtue and health.

A theory of virtue in public health care can use the Commission's findings as entry-level considerations but must recognize that its conclusions may be inadequate. A virtuous theory of public health would transcend cost/effectiveness or questions of locus of support for various health services. Virtue in public health, as we will use it here, requires recognition of individual needs and vulnerabilities, as well as extraordinary effort, generosity, or moral excellence.

At the root of this essay lies the conviction that health care – and health protection – cannot simply be apportioned according to some formula of distributive justice or equality, because health is not a commodity. Nor can the conduct of public health be resolved into behavioral rules. The provision of public health cannot be reduced to the same conduct rules which apply to physicians (cf. [29]), in the main because groups have extremely heterogeneous needs. Virtuous health policies can only be seen in the context of an appropriate conception of the goals of public health – and their lexical ordering according to ethical criteria. These criteria, in turn, must incorporate a theory of health as well as a theory of the moral obligations entailed in its delivery.

It is the purpose of this article to establish the boundaries of public responsibility and obligation in promoting and protecting health as they would be set in a society which regarded the protection of health as a virtue rather than a necessity. In this sense, this paper breaks new ground in applying ethics to the classic elements of public health. But before this concept can be developed, the limitations of a theory of virtue as it applies to whole populations must be explored.

DIFFICULTIES IMPLICIT IN THE CONCEPT OF VIRTUE AS APPLIED TO PUBLIC HEALTH

The concept of a virtue in health policy is problematic on two counts. First, because virtue is primarily applied to the conduct of individuals, describing the actions of institutions which comprise the public health system in comparable moral terms is potentially flawed. Second, public health has traditionally been conducted on the basis of purely pragmatic considerations of efficacy, efficiency, and cost/benefit considerations, while virtue nominally applies to abstract attributes such as beneficence, sacrifice, compassion, and altruism – all characteristics which defy mensurate analysis [13].

One way to resolve these potential contradictions is to recognize that conduct directed towards specific ends, whether of individuals or collections of individuals, can be morally assessed in terms of the appropriateness of their means. Public health per se can be seen as having at least three interdependent goals which can be achieved through human action, both individual and collective:

(1) the furtherance or promotion of intrinsic factors requisite for health (health promotion);

(2) the minimization of adverse health impacts and sustenance of well-being (health maintenance); and

(3) the protection of individuals and elimination of factors which can diminish well-being (health protection).

Each of these objectives can be seen along a continuum of human actions, beginning with those steps which can be taken by individuals acting alone (health promotion) and extending to those which can only be taken with the involvement of institutions (health protection). Each of these objectives can be achieved by means which can be considered virtuous: and, each objective can itself be enhanced or amplified in a manner which likewise embodies the precepts of virtue. To establish the limits of virtue in public health, each objective will be examined in terms of health policies which encourage their achievement, and compared for possible lexical ordering of their ethical acceptability.

For the purpose of this essay, the description of health and its prerequisites which I previously developed [18] will be used. The key elements of health are affected both by the acts of individuals and by extraindividual factors found in the genome and the environment. (For the purposes of this paper, health per se may be defined as a sustained state of well-being specific to a given age.) The key point of my previous essay was that whatever elements of health might be attainable through individual action could likewise be compromised by extrinsic factors, e.g., exposure to environmental pollutants. Thus, while health can be promoted by dint of individual effort, it can be lost through the actions of others over which individuals have little or no control [18].

The moral significance of this observation is that health cannot properly be described as solely within the ambit of individual choice. Additionally, some individuals will be more vulnerable to extrinsic, health-limiting factors than will others, because of physiological maturity, genetics, or pre-existing chronic health problems. An example would be the dramatically different drug-metabolizing abilities of genetically dissimilar individuals and their impact on medical treatment demands [14]. Establishing hospital-wide policies which recognize and compensate such vulnerabilities to protect the at-risk patient, can be seen as part of a virtuous policy.

A specific value judgment which I have previously derived (see [18]) is that because of the lexical ordering of the obligation to prevent harms over that obligation to do good [11], it will be more important to derive a theory of virtuous public health behaviors as it applies to minimizing harms than one which centers on promoting good. Each of the possible models of health-related behaviors to be discussed potentially embodies elements of virtue. Three such models follow.

THREE MODELS OF HEALTH ATTAINMENT

Health Promotion

A wide constellation of human activities – largely attainable through individual action – has been shown to reinforce or promote well-being. Several specific activities have recently been given broad empirical support. Among these are dietary habits which reduce the risk of cancer [25]; improved cardiovascular function through aerobic exercise; and the reduction of lung and liver disease through the elimination of smoking and alcohol. The beneficial effects of these behaviors have been widely heralded in publications like the Lalonde report [16], and the U.S. Surgeon General's publication, *Healthy Americans* [30].

While novel in their newly found empirical basis, there is really nothing strikingly new about health promotional activities. More relevant to this discussion is the recognition that health promotion per se has been – and continues to be – deemed a virtue.

Attainment of personal excellence in health through individual action was recognized by the Greeks as a component of *arete* or virtue. Perfection of human abilities by dint of individual effort was deemed to be a moral obligation. Excellence in terms of grace, beauty, or physical prowess was widely esteemed during the Classic Period in ancient Greece. At the root of this

Greek concept of virtue is the belief that excellence is the highest form of human attainment.

The Greeks (particularly Aristotle) believed that each person has a basic biological makeup which could be perfected by his own actions. Excellence in health presupposed that persons have an intrinsic propensity to be healthy. Maximal strength, prowess, and achievement in battle were considered virtuous in men, while women were expected to have more 'quiet' virtues of cooperation, support, and love (cf. [15]). Significantly, the Greeks gave little consideration to inborn deficiencies or vulnerabilities. Individuals like Demosthenes, who overcame a speech impediment, were recognized as exceptional. One city-state in particular, Sparta, elevated human physical beauty to such a level that they were willing to destroy infants who were born with physical deformities.

Contemporary protagonists of health promotion, while unwilling to take their beliefs to this Spartan extreme, nonetheless advocate self-directed behaviors that emulate the Greek ideal. In its most characteristic form, health promotion consists of encouraging individuals to undertake behaviors which support or augment their well-being, and to avoid practices which potentially undermine it. Various institutions can encourage these goals.

An example would be the universal provision of exercise facilities and on-the-job exercise time at corporate expense. Several West Coast based institutions like the Shakle Corporation and several Silicon Valley firms have implemented just such programs.[2] From an institutional perspective, the example of corporate-funded exercise programs has a minimal claim to virtue, since they enhance the attractiveness of the company to prospective employees at relatively minimal corporate expense. The extent of virtue present can be said to be proportional to the degree to which participants benefit from the corporate policy, independent of secondary benefits which accrue to the company itself in terms of greater employee loyalty, reduction of sick leave, minimization of work days lost, etc.

It is difficult to assess the value to be ascribed to even the most virtuous of such institutional policies, because of the blurring of institutional and individual objectives. Similarly, even the most enlightened form of health promotion, e.g., the provision of essential health-related information to illiterate or otherwise impaired groups for whom this data would otherwise be inaccessible, can fall short of the condition that virtuous behavior be self-less. This in large part due to the fact that provision of health education subtly – and sometimes overtly – conveys a sense of responsibility for self-care to the individual while simultaneously discharging some degree of obligation on the state or other proferring agency from more direct forms of aid. Some programs, e.g., hypertension screening, would be deemed virtuous when coupled with follow-up and free provision of treatment services.

Health Maintenance

To understand the complexities of this model, it is useful to consider health as being largely assured for most individuals during their formative stages once past their post-newborn development (i.e., through adulthood). Coupled with the observation that the curve of infant mortality drops off precipitously after the perinatal period, the model is premised on the fact that post-natal survival after the age of 5 years is largely assured for most individuals in the industrialized countries where the major causes of death in adolescence and young adulthood are accidents, injuries, or suicide [3]. According the some Eastern philosophical schools, this period is one in which individuals should live in relative harmony with their surroundings.

Virtuous health policies are not merely those which disrupt this natural harmony the least, but which ensure that individuals have the fewest possible departures from a 'natural relationship' to their environment. The closest philosophical model of this ideal is the Taoist notion of virtue or *te*. As developed by Lao Tzu, acting in accord with nature is the ideal. Health and well-being, according to this Chinese model, are assured by following the path of 'non-action' or *wu wei*, a concept that entails 'acting with nature'.[3]

It is possible to construe that sustaining an optimal level of health is a virtuous activity, when that activity is carried out by individuals. Thus, one can argue that virtue consists of keeping healthy to the point that basic parameters of fitness are constantly kept above the norm of the population. Excellence here consists of going beyond the obligation to maintain only the minimum of health needed to ensure discharge of societal, familial, or religious duties. To the extent that keeping health at such a level transcends the normal prerequisities of individual responsibility, it may be construed as virtuous. But such an assessment is problematic, particulary where individual achievement becomes narcissistic, as can be the case among some Hollywood actors or actresses.

Virtuous health-respecting behavior can also be ascribed to those groups of individuals who live with the minimum requirements of well-being, while causing the least radical disruption of the physical environment. A typical case would be Mahatma Ghandi and his followers for whom health maintenance was a personal ideal. But in Western societies, maintenance of health is more commonly a co-dependent activity. Thus, various countries have developed differing policies on the degree to which purveyors of potentially health impairing foods, alcohol, or cigarettes may be constrained. Public virtue, as contrasted to individual virtue in this model, would consist of adopting policies which encourage the greatest likelihood of health maintenance for a given population. A case study can be found in the history of the development of an international code to limit the promotion of infant formula.

A CASE STUDY: THE WORLD HEALTH ORGANIZATION'S INTERNATIONAL CODE OF MARKETING BREAST MILK SUBSTITUTES

The most striking example of the application of virtue in this model may be the acceptance by the Nestlé Corporation of the World Health Organization's recommendations for limiting the promotion of infant formula among new mothers. this policy evolved from an intense debate which took place in the late 1960s over the rectitude of institutional behavior in promoting the use of breast milk substitutes, particularly in the Third World. The continued promotion of formula in the face of substantial scientific information which indicated the primacy of breast-feeding over bottle-feeding was deemed particularly culpable because of the increased likelihood of bottle-fed newborn deaths from diarrheal diseases – deaths which have now been shown to be preventable through breast milk feeding [12].

In light of the demonstrable impairments of infant health and survival which accompanied an early post-natal shift to breast milk substitutes (largely through the increased incidence of diarrheal diseases), the World Health Organization formally issued guidelines for limiting the promotion of infant formula in developing countries on May 21, 1981. The purpose of the Code was to "contribute to the provision of safe and adequate nutrition for infants, by the protection and promotion of breast feeding . . . " [4].

The Code specified in part that the content of any informational or educational materials be directed at informing the reader about the superiority of breast feeding; the nutrituonal needs of the mother; the negative impact of introducing partial bottle supplemental feedings; the likely irreversibility of the decision not to breast feed; and, the social and financial implications of electing to use infant formula or other breast milk substitutes. It also required companies which marketed infant formula to curtail drastically advertising and promotion, to limit the distribution of free samples or equipment, and to curtail previously effective marketing strategies such as mass media promotions.

By 1972, Nestlé, which holds the largest market share of Third World formula and milk supplemental products, had begun to modify its marketing strategies to limit the amount of direct advertising and promotion of their product to ensure that a maximum of women had a degree of freedom of choice in electing to breast feed their babies, a freedom which had been severely compromised earlier by more aggressive company policies. Over the next four years, Nestlé began to carry out the intent of the WHO guidelines by strengthening the emphasis on breast feeding in its promotional materials and restricting the distribution of advertising or promotional materials for its product. This phase reached a milestone in 1978 when Nestlé withdrew mass media advertising for infant formula in all developing countries [23]. On March 16, 1982, Nestlé unilaterally adopted the provisions of the WHO Code as its corporate policy. Since then, the degree and extent of their actual compliance in the field with the WHO Code, particularly as it applies to limiting sample-distribution, has been challenged by church groups [22]. The relative success and compliance of Nestlé with these international guidelines is being monitored by an ad hoc committee chaired by Senator Muskie (D-Maine).

To the extent that Nestlé actively follows its own polices in this regard, the corporation may be said to be exercising a virtuous health policy. Virtue here consists of taking a policy position on an essential, health-maintaining behavior at some, perhaps substantial, corporate expense. While Nestlé has not documented the extent of the reduction in its sales of breast milk substitutes, it has been largely limited to pre-existing markets by self-imposed restrictions on promotion and mass media advertising. The degree to which Nestlé's other-respecting activities have actually reduced corporate profits is also largely unknown, since infant formula makes up only a small percentage of the company's gross income. Nonetheless, the elements, of generosity, self-sacrificing behavior, and respect for indigent cultural values – however belatedly shown – establish a broad model for virtuous health policies at a corporate level.

Health Maintenance Organizations

In contrast to corporate behavior which results in sacrifice, contemporary health maintenance organizations (HMOs) often conduct their activities with an eye towards assuring the maximal return to their stock-holders (cf. [28]). While HMOs certainly do considerably more than merely advocate health protective strategies or monitor the inevitable physiological decline of their clients through periodic testing, it is questionable whether in practice they carry out the full intent of their mission. Virtuous policies for ensuring health maintenance would in principle include the most aggressive health screening programs at the margin of cost-effectiveness (e.g., cervical and breast cancer screening; hypertension testing and followup; glaucoma screening before the age of 40 in high-risk populations, etc.).

To the extent that many HMOs limit their screening to the most cost-effective test, or restrict the use of highly expensive, but essential, diagnostic tests (e.g., isotope liver scans), their public health policies lack virtue. While some seek to keep clients in as good health as possible through the judicious use of appropriate screening, monitoring and routine physical examinations (e.g., blood pressure testing, urine tests, etc.), it is the exception rather than the rule to find HMOs which advocate the use of the most definitive and expensive diagnostic tests when less expensive ones will serve as well.

Another institutional practice which must be considered from the vantage point of virtue is the Holistic Health movement (cf. [17]). This movement encourages lifestyles which are most congruent with maintaining health and well-being. Representative practices include ingestion of unprocessed oils and whole grains, naturopathic practices, exercise systems which are designed to augment natural body rhythms or energies (e.g., *Tai Chi*), and general health practices which reinforce natural functions (e.g., colonic irrigation). Rather than being construed as extra-ordinary or self-sacrificing and hence virtuous, these behaviors are best seen as choices available to groups who elect alternative life styles.

Virtue in activities which simply promote or maintain health is thus by definition difficult to assign, since the activities tend to be self-serving and lacking generosity in that the techniques or skills are rarely freely extended to others or given to whole communities. Rather, they depend on individual choice and volition, and where more generalized (as with the Nestlé Corporation), are limited to charitable or altruistic acts which restore the possibility of well-being, rather than actively protecting or ensuring the preservation of health.

Thus, to the extent that both of the first two models deal with conduct which is largely if not exclusively within the province of individuals and not institutions, they largely entail individual virtue and not the collective or institutional virtue we are seeking from public health agencies. This element of virtue may be found in the collective governmental health policies of nations like Canada, Sweden, or Denmark, which have a socialized system of medical care delivery. These and other Health Policies generally have been extensively compared in the United States, the United Kingdom, and the

Federal Republic of Germany (see [19]), and are outside the scope of this article.

Health Protection

The third model of health to which virtue applies is that of health protection. This concept places the greatest emphasis on steps which can be taken to prevent hazardous or noxious factors from impinging on vulnerable persons. Virtue here would consist of efforts which would minimize the likelihood of occurrence of adverse environmental or other deleterious events outside the control of the individual who are not otherwise capable of avoiding exposure themselves.

A CASE STUDY: REDUCING ENVIRONMENTAL LEVELS OF LEAD

Based on the second National Health and Nutrition Examination survey conducted by the National Center for Health Statistice, four percent of U.S. children six months to five years old have elevated levels of lead in their blood [24]. The limit set for designating and 'elevated' blood reading is now thought to be within the values associated with mental impairment and behavioral disorders [26]. Based on findings both in the United States and Great Britain, the clearest and most compelling environmental association with blood lead levels in children is the lead level in gasoline and the amount of leaded gasoline sold per capita [2, 5]. The most compelling data suggesting a causal relationship between the use of leaded gas and human body burdens of lead is the association between the time course of the significant drop in blood levels between 1976 and 1980 and the reduction in sales of leaded gas in the U.S. during the same period [2].

Voluntary corporate reductions in the total sales of leaded gasoline beyond those set by the Clean Air Act would further improve the disturbing picture of lead contamination still extant. In Britain, reducing lead in gasoline to 0.15 grams per liter would lower lead in exhaust emissions by more than half. Such a move would be virtuous because it would reduce an otherwise unavoidable source of contamination and resulting health burden on the most vulnerable population (children less than 5 years old), and because it would require some sacrifice of corporate profits. Perhaps because of this economic cost, the Royal Commission study in the lead problem in England has recommended a step-wise phase-out of lead by 1990 rather than an outright ban [5].

Any voluntary program on the part of a corporate entity that affords health protection for those in the community generally at corporate expense can be said to be virtuous.

EXPOSITION OF ELEMENTS OF INSTITUTIONAL VIRTUE

As applied to public health, actions of institutions which can be said to be virtuous are thus seen to be those which strongly reinforce the goals of one or more of the above models. Because virtuous behavior at an institutional level involves conduct that transcends legal requirements, virtuous institutional policies are most likely to involve elements of altruism, generosity, or justice. Choices which substitute cultural or moral norms (such as philanthrophy) for institutional ones (e.g., profit-making), or which entail sacrifice of purely institutional goals, such as the voluntary restriction of markets for a given product (cf. Nestlé's action, above), are not in themselves virtuous unless they are linked to positive health outcomes, And even then, we might say that the outcome in question does not qualify the act as virtuous unless it required that institutional practice for its success. Thus, the voluntary restriction of infant formula promotion is only virtuous to the extent that without such restriction, little or no breast feeding would occur in the native population. Similarly, the enjoyment of a modicum of health and reasonable prospects for full mental capacities among children would have to be linked to the proposed reduction of environmental lead levels for lead reduction policies to be virtuous. Neither of these associations, however likely, has been scientifically proven.[4]

The borderline case for virtuous behavior would be in those circumstances where institutional choices are only indirectly linked to a public health goal. Virtue in public health policy should also be considered in relation to the degree to which the health need in question is beyond individual attainment, and extent to which the persons affected by the policy are dependent or vulnerable. Policies directed towards the newborn age, or infirm – to the extent that they cannot be reciprocated by these same persons – are potentially virtuous to a greater degree than the same policies directed towards fully competent adults. The ethically relevant aspect of virtue as it pertains to health is thus the degree of generosity, altruism, or sacrifice (beneficence) of the provider coupled with the vulnerability and need of the groups likely to benefit by the policy.

DISTINCTIONS AMONG THE MODELS

To the extent to which behaviors of individuals or groups within the context of these models further commonly recognized public health ends, they potentially embody elements of virtuous public health behaviors. But in practice, groups of whole populations must sometimes participate in the acts necessary to achieve a specific health goal, for instance, in population-wide hypertension screening. To the extent that individual acts serve only that individual (i.e., are self-serving), the goal achieved, no matter how virtuous, cannot establish the virtue of the policy in question. Thus, life extension techniques, recently popularized by two university scientists (cf. [27]), are in principle virtuous modalities according to the first model of virtue outlined above. But the fact that the programs to be followed require substantial expenditures of capital and benefit only the individual who undertakes them renders the program more narcissistic than virtuous.

Both of the first two models thus depend on the actualization of individual choices regarding health. Any resulting health improvement must be institutionalized through implementation of policies which encourage universal participation in health promotion or maintenance; thus, it is unlikely that the same degree of collective good can be achieved as through the third model – implementation of health protection strategies, where a single regulatory act can passively protect a whole population. By a purely utilitarian test, strategies directed at promoting or maintaining health may not be as likely to be virtuous as those directed at protecting health; conversely, health protective strategies are more likely to be paternalistic and hence less virtuous because they are less respecting of autonomy than either of the other two.

Thus, health promotion is achieved in the main through the delivery of effective health education, and only marginally through group inducements (e.g., reductions in workmen's compensation insurance premiums for industies which have gymnasia). In places other than the U.S., this may not be true, as pre-work time exercise or other health-promoting activities can be mandatory (e.g., in Japanese automotive plants). Health maintenance relies similarly on voluntary compliance with guidelines for periodic check-ups and examinations, and rarely entails mandatory participation in programs designed to ensure that basic physiological parameters are maintained at an appropriate point.

Health protection strategies are by their definition almost always achievable only through collective or institutional action. Thus, reductions in the heavy metal content of sewage effluents requires the participation of all municipalities in a given water district; immunization programs the cooperation of whole populations; and effective fluoridation the compliance of whole communities served by a given watershed. In the sense that collective action is needed to ensure the maximal effectiveness of public interventions under the third model, it is simultaneously the most appropriate and the most significant focus for virtuous activities according to the communitarian model of public health (cf. [10]).

CONCLUSION

To achieve the public health objective of maximizing health protective strategies, Western democracies, particularly the U.S., rely almost exclusively on the regulatory approach. Compliance, surveillance, and enforcement activities for many of the key environmental, occupational, and public health laws implemented in recent years have been less than optimal [7]. A case in point is the spate of regulations which followed implementation of the various acts which were disigned to assure equality of access of handicapped persons to institutions and buildings.

Beginning with the Architectual Barriers Act of 1968, which required that all buildings which were built with federal funds or leased by the Federal Government be made accessible to the handicapped, and culminating with the Social Secruity Disability Amendments of 1980, which removed certain disincentives to work from the handicapped, enforcement and implementation have been spotty and less than optimal [9].

According to DeJong and Lifchez, the regulatory approach to ensuring accessibility to the handicapped is inherently adversarial – and hence least likely to generate the type of virtuous institutional behavior which would maximize the likelihood of the most salubrious effect. What is needed, according to these two authors, are incentives and inducements to those institutions which stand in the way of full implementation of the act [8]. Two such vehicles, notable technical assistance and economic incentives are seen as prerequisties to the optimal implementation of these acts. Virtue in the absence of supporting institutions (here defined as the provision of special architectural and structural services to the handicapped) is thus highly improbable in the absence of institutional structures which provide incentives towards its achievement.

But by definition, to be virtuous, actions should be taken with the minimum of inducement or reward. Actions taken for their own sake are much more worthy of recognition than those taken for reward or profit. Thus, virtue in public health will likely continue to be achieved through the selfless acts of those who define and establish the conditions which best meet the needs of the most needy – and provide the necessary institutional supports and inducements to involve those who are less virtuous in the implementation of their programs.

NOTES

[1] Supported by NSF/NEH Grant No. 1SP–8114338. The excellent secretarial service of Alice Murata is acknowledged.

[2] Some 78 percent of responding California employers report health promotional activities at their sites, including accident prevention, alcohol/drug abuse abatement, mental health counseling, stress management, fitness, hypertension screening, and smoking cessation [9].
[3] An exposition of this concept can be found in the collection of Eastern Philosophy and photographic essays by John Chang McCurdy, David Brower and Marc Lappé [21].
[4] This is especially true for promoting breast milk substitutes. Although a recent study has shown a modest negative effect on breast feeding from provision of free samples in a Western society (Canada), the authors are quick to point out the limited nature of their conclusions for less-developed countries (cf. [6]).

BIBLIOGRAPHY

[1] Abram, M.: 1983, *Summing Up Final Report on Studies of the Ethical and Legal Problems in Medicine and Biomedical and Behavioral Research*, U.S. Government Printing Office, Washington, D.C.

[2] Annest, J. L., *et al.*: 1983, 'Chronological Trend in Blood Lead Levels Between 1976 and 1980', *New England Journal of Medicine* **308**, 1373–1377.

[3] Anonymous: 1980, *Vital Statistics of the United States*, Vol. 2, Part A, DHHS Publishing Co., (PHS) 81–1101, Hyattsville, MD.

[4] Anonymous: 1981, *International Code of Marketing Breast-Milk Substitutes*, 34th World Health Assembly, Geneva, Switzerland.

[5] Anonymous: 1983, 'Lead in Petrol: A Long Farewell', *Lancet* **1**, 912.

[6] Bergevin, Y., *et al.*: 1983, 'Do Infant Formula Samples Shorten the Duration of Breast-Feeding?', *Lancet* **1**, 1148–1151.

[7] Bogen, K.: 1983, 'Quantitative Risk-Benefit Analysis in Regulatory Decision-Making: A Fundamental Problem and an Alternative Proposal', *Journal of Health Politics, Policy and Law* **8**, 120–143.

[8] DeJong, G. and Lifchez, R.: 'Physical Disability and Public Policy', *Scientific American* **248**, 40–49.

[9] Fielding, J. and Breslow, L.: 1983, 'Health Promotion Programs Sponsored by California Employers', *American Journal of Public Health* **73**, 538–542.

[10] Forster, J.: 1982, 'A Communitarian Ethical Model for Public Health Interventions', *Journal of Public Health Policy* **3**, 150–163.

[11] Frankena, W.: 1973, *Ethics*, Prentice-Hall, Englewood Cliffs, NJ.

[12] Glass, R., *et al.*: 1983, 'Protection Against Cholera in Breast-Fed Children by Antibodies in Breast Milk', *New England Journal of Medicine* **308**, 1389–1392.

[13] Green, R.: 1982, 'Altruism in Health Care', in E. Shelp (ed.), *Beneficence and Health Care*, D. Reidel, Dordrecht, Holland, pp. 239–254.

[14] Idle, J., *et al.*: 1983, 'Protecting Poor Metabolisers, A Group at High Risk of Adverse Drug Reactions', *Lancet* **1**, 1388.

[15] Kerferd, G.: 1967, 'Arete', in P. Edwards (ed.), *Encyclopedia of Philosophy*, MacMillan Publishing Co., New York, pp. 147–148.

[16] LaLonde, M.: 1974, *A New Perspective on the Health of Canadians: A Working Document*, Department of National Health and Welfare, Ottawa.

[17] Lappé, M.: 1979, 'Holistic Health: A Valuable Approach to Medical Care', *Western Journal of Medicine* **131**, 475–477.

[18] Lappé, M.: 1983, 'Values and Public Health: Value Considerations in Setting Health Policy', *Theoretical Medicine* **4**, 71–92.

[19] Lockhart, C.: 1979, 'Values and Policy Conception of Health Policy Elites in the United States, the United Kingdom, and the Federal Republic of Germany', *Journal of Health Politics, Policy and Law* **6**, 98–119.

[20] McCullough, L.: 1981, 'Justice and Health Care: Historical Perspectives and Precedents', in E. Shelp (ed.), *Justice and Health Care*, D. Reidel, Dordrecht, Holland, pp. 51–71.

[21] McCurdy, J., *et al.*: 1974, *Of All Things Most Yielding*, McGraw Hill, New York.

[22] Methodist Federation for Social Action: 1983, 'Why Still Boycott Nestlé – A "Sample" Lesson', *American Journal of Public Health* **73**, 769.

[23] Muskie, E.: 1982, *First Quarterly Report of the Nestlé Infant Formula Audit Commission*, Washington, D.C.

[24] National Center for Health Statistics: 1982, 'Blood Lead Levels for Persons 6 Months – 74 Years of Age: United States, 1976–80', Public Health Service, Hyattsville, MD (DHHS Pub. No. (PHS) 82–1250).

[25] National Research Council: 1982, *Diet, Nutrition and Cancer*, National Academy Press, Washington, D.C.

[26] Needleman, H., *et al.*: 1979, 'Deficits in Psychologic and Classroom Performance of Children With Elevated Dentine Lead Levels', *New England Journal of Medicine* **300**, 689–695.

[27] Pearson, D. and Shaw, S.: 1983, *Life Extension: A Practical Scientific Approach*, Warner Books, New York.

[28] Shroeder, S. and Showstack, J.: 1979, 'The Dynamics of Medical Technology Use: Analyses and Policy Options', in S. Altman and R. Blendon (eds.), *Medical Technology: The Culprit Behind Health Care Costs?*, DHEW Publication No. (PHS) 79–3216, Hyattsville, MD, National Center for Health Services and Bureau of Health Planning, pp. 178–212.

[29] Siegler, M.: 1979, 'A Right to Health Care: Ambiguity, Professional Responsibility and Patient Liberty', *Journal of Medicine and Philosophy* **4**, 148–157.

[30] U.S. Surgeon General: 1979, *Healthy People: The Surgeon General's Report on Health Promotion and Disease Protection*, U.S. Department of Health, Education and Welfare, U.S. Government Printing Office, Washington, D.C.

Social and Administrative Health Sciences,
University of California School of Public Health,
Berkeley, California,
U.S.A.

SECTION IV

CRITIQUE

TOM L. BEAUCHAMP

WHAT'S SO SPECIAL ABOUT THE VIRTUES?

As several historical essays in this volume attest, virtue theories have had a history of popularity in both philosophy and medicine that conforms to the statistical concept of regression to the mean. From past highs of excessive adulation, to lows of fallow neglect, the virtues seem to rise and fall in value with the irregularity of the stock market. We now seem headed on the upside, and this collection of essays is one testimony to that trend. But these trends should be met with the virtue of equanimity: There will be a downside, and the reactions found in this volume to Alasdair MacIntyre's *After Virtue* [15] are expressions of the downside potential.

A rush to the virtues has always been, and always will be, followed by a rush away from them. There is nothing wrong with the virtues; indeed, there is something eminently right about them. But, contra MacIntyre, rights-based and duty-based moral theories are no less promising than virtue-based accounts. Rights, principles, duties, and the like should not be *replaced* by the virtues, as MacIntyre, Hauerwas, and others seem to suggest. Each of these categories is serviceable, I shall argue in this essay, because they are our means to the ends of morality. This goal-oriented conception is a consequentialist theory, but a generous and tolerant one. It finds value in all these categories, without overvaluing any one of them. The constant tendency to overvalue a particular approach to ethics continues to afflict moral theory, and in my judgment impairs several essays in this volume. In this critique I develop a programmatic approach to moral theory that has the objective of placing these categories in proper perspective, without over- or under-evaluation.

Gert and Pincoffs offer in their essays some important descriptions and definitional claims about the virtues, and I shall not challenge or expand on their analyses. I note, however, that it is vital to distinguish between a virtue as a *trait or quality* of a person and a *statement* that some trait or quality is virtuous. The latter is a linguistic formulation of a standard, in some cases a moral standard. Just as we distinguish between an *act* of a person that conforms to a moral rule and the *moral rule* itself, so we can distinguish normative formulations of virtue – such as descriptive models of character – from possession of the trait referred to in the statement or incorporated

Earl E. Shelp (ed.), Virtue and Medicine, 307–327.

in the model. I shall use 'virtue' to refer to traits of character and 'virtue standard' to refer to general formulations of virtue, models of virtue, and the like.

WHAT VALUE SHALL WE PLACE ON THE VIRTUES?

Al Jonsen and Andre Hellegers some years ago argued that "exhortations to virtue constitute the heart of code ethics" [14]. They maintained that an overemphasis on duties and rights has led to a neglect of the virtues, and therefore to the loss of something essential to a well-developed medical ethics. MacIntyre has, of course, vigorously defended virtue, over duty and rights, as the preferable moral category. Both of the physicians writing in this volume have continued this denigration of duties and rights, while treating the virtues as deserving profound respect and dignity. I shall begin with their contentions.

Allen R. Dyer argues in his essay that biomedical ethics has been "severely handicapped' without the concept of virtue, a concept that he believes 'restores a lost dimension' that 'moral rules' and 'utilities' fail to provide (pp. 223–235). Edmund Pellegrino – quoting Santayana – suggests that the virtues "give credibility to the moral life; they assure that it will be something more than a catalogue of rights, duties and rules" (p. 237). He thus finds virtue systems superior to ethical systems developed from rights and principles. He notes, however, that the contemporary reappraisal in philosophy is "not an abnegation of rights, or duty-based ethics, but a recognition that rights and duties notwithstanding, their moral effectiveness still *turns on* the disposition and character traits of our fellow men and women" (p. 338, italics added). He quotes Frankena's competing claim that from the perspective of an ethics of duty, it is *virtue* that is *dependent* on a prior duty-based account of what *ought* to be done (p. 247). Thus, Frankena gives priority to duty, while Pellegrino, MacIntyre, and many others now give priority to virtue. The latter ordering is rapidly being elevated to status as the received view among those who are 'reappraising' and promoting the virtues.

This problem about whether duty depends on virtue, or virtue on duty, raises profoundly important questions that deserve sustained reflection in contemporary philosophy. There is much to command both viewpoints, as Jim Childress and I have argued elsewhere [3]. Yet I believe there is more to be said *against* both viewpoints, because each overvalues its preferred basic category. One of my purposes is to show how misleading it can be to make any such category 'fundamental to' or 'prior in' a moral theory. The virtues serve an important function in the moral life, but they do not serve either as primary action guides or as primary sources of moral appraisal. I shall also argue for a consequentialist view that integrates these moral notions in a single system, without the fragmentation that MacIntyre and others have proclaimed. In the end, the sober appraisal of the value of the virtues is that they do not deserve the savior role awarded them by MacIntyre or the wave of the hand given them in this volume by Nielsen.

I shall defend this claim by showing that the virtues are of a less unique importance than many believe, precisely because they *correspond to* or are *correlative to* other forms of moral action guides, including moral principles and moral rights. If I am right, then the virtues should be accorded no higher priority status than 'competing' categories – which, on my analysis, turn out not to be in serious competition at all.

VIRTUE OVER DUTY?

The moral virtues seem profoundly important because they are integral to a person's moral character. Physicians, patients, and all persons have characters that exhibit various levels of the virtues and vices, where virtues are understood as established dispositions or habits to do that which is morally commendable. Virtuous traits of character seem to be of special significance in crisis environments in medicine that are too hurried to permit reflection on duty or on the justification of conduct by appeal to a rule. One important objective of the cultivation of virtue is to render fulfillment of duty a routine matter, rather than a continuous struggle to do one's duty, and this is one reason why Aristotle's ethics rightly has a greater appeal to some – myself included – than Kant's. Aristotle's is a more realistic and appealing picture of the human condition, because morality flows from cultivated or natural habit rather than a powerful struggle against inclinations in the attempt to emulate the habits of a holy will – a goal one can approximate but never achieve.

The person who acts charitably because duty demands it seems less admirable than the person who contributes spontaneously from well-formed habits of virtue, and without need of a strict moral law. As Philippa Foot rightly says, "The man who acts charitably out of a sense of duty is not to be undervalued, but it is the [spontaneous contributor] who most shows virtue and therefore to the other that most moral worth is attributed" ([9], pp. 12–14). Not surprisingly, Foot finds Aristotle's theory more prescient and congruent with our intuitions about the moral life than Kant's, particularly because the Kantian quasi-legal metaphors of law, duty, and command seem disembodied and inadequate to the moral life. Is it not better to be generous-spirited than to act from the prod of duty?

The merits of Foot's view are considerable, but there is an important respect in which a sweeping dismissal of Kant and other theories that appeal to principles of duty (including virtually all prominent deontological and utilitarian theories) is too harsh and one-sided. No principle-of-duty-based – hereafter simply duty-based – theory need deny the importance of virtues, and any viable theory of principles of duty, in my judgment, will include an account of virtue. If I am right, then we should not set up stereotypes of theories – as do some writers in this volume – such that one must be either a defender of duty-based *or* virtue-based – *or* rights-based – theories. A morality of principles of duty should enthusiastically recommend settled dispositions to act in accordance with that which is morally required, and a proponent of virtue ethics should encourage the development of principles that express how one ought to act. It is a defect in any theory to overlook all of these ways of expressing what is important in the moral life.

As noted above, some writers in ethical theory suggest that we need only some one or the other of these categories; but the more prudent course is to view these approaches as the reverse sides of a nickel, and neither as worth less than 5¢. Let me now proceed to an argument for this view.

THE CORRESPONDENCE BETWEEN VIRTUE STANDARDS AND PRINCIPLES OF DUTY

I am suggesting that principles of duty correspond to virtue standards. Yet virtue standards such as kindness, generosity, and affection often function to express ideals of conduct rather than duties – and so do not cleanly correspond to duties. I shall argue momentarily that the problem is that these virtues function, in part, as the complement of supererogatory actions in duty-based theories. A principle like beneficence is a better starting point because it directly corresponds to the virtue standard of benevolence: They function both as supererogatory moral ideals *and* as basic standards in a shared morality. In the shared – or what I shall call the 'ordinary' – morality, for every moral principle, such as nonmaleficence and respect for persons, there corresponds or can be made to correspond one or more virtue standards. Virtues such as nonmalevolence and respectfulness, then, are dispositions to act in accordance with duties.

I am not contending that actions are virtues or that virtue standards are logically equivalent to moral principles. But I am maintaining the following: Principles and virtues standards are *both* alike general action guides; virtues in the context of ordinary morality are dispositions to do what persons ought to do as a matter of duty; and principles of duty express our convictions about the proper character that persons should cultivate. It will not do to object to this account that it is *duty*-based because *action*-based, while virtue has to do with character ('being') rather than action ('doing'). As Aristotle repeatedly says, virtue is integrally tied to *actions* that *ought* to be performed. Aristotle constantly speaks of 'virtuous acts'; and, as Ferngren and Amundsen correctly note in their insightful essay, *arete* has been used at least since Socrates to refer to 'right *conduct*,' the source of our idea of 'moral *behavior*' (p. 9).

The two categories of virtue standards and principles of duty can in principle be conceived as in a relation of *perfect* correspondence: For *every* principle of duty there is a corresponding trait of character or virtue, which is simply a disposition to act as specified in the principle; and for *every* virtue of character, there is a corresponding action that conforms to a principle of duty. This may be diagrammed as follows (cf. [2], pp. 165–166):

	Duty-Based Theories [Correspond to]	Virtue-Based Theories
Ordinary Moral Guides	Principles of Duty as Guides of Conduct	Virtue Standards as Guides of Conduct
Extraordinary Moral Guides	Supererogatory Principles as Guides of Conduct	Supererogatory Virtue Standards as Guides of Conduct

'Ordinary moral guides' refers to (1) those (prima facie) moral principles shared in common by all persons and (2) those principles that are shared in common by persons in special roles or professions – such as physicians, nurses, public health officials, and the like. Both (1) and (2) are principles of duty. I shall take '*duties*' to be synonymous with '*obligations*', because both function to assert what *must* be done by all moral agents. This form of assertion covers only part of the moral ground of what *should* or *ought* to be done. The 'extraordinary moral guides' designated above function as ideal standards of what should or ought to be done, not as a matter of the common shared morality but as a matter of what uncommonly exceeds that morality by doing more than it demands. Supererogatory conduct, then, is neither morally required nor morally wrong, is morally praiseworthy, is not morally blameworthy, and voluntarily exceeds what is morally required (cf. [12], p. 115). In the case of supererogatory acts that exceed dutiful acts, 'principles of duty' are *extended as* moral *ideals* and the *duty* part fades from significance. Those who fail at the level of self-imposed, exceptional moral 'oughts', obviously need not also fail at the level of execution of moral duties.

This general category of supererogation need not entail the heroic, saintly, or extraordinary in the sense of exceptionally excellent. If you give aid to a tourist who stops you on the street and asks for directions, or if you invite a blind man to move ahead of you in line, nothing in morality requires this behavior, but you do not so far exceed the demands of morality that you require exceptional praise either. Here we find something of a no man's land in the moral terrain: Morality does not require the action, but a failure to be of aid is nonetheless a moral defect. For those in the role of parents, morality tends to be more demanding. A parent is not required to donate a needed organ to a child, but is required in a wide variety of circumstances to volunteer aid to a child in danger or need. This indicates that we should not divide the moral life too *bluntly* into ordinary moral guides and extraordinary moral guides, as the above chart suggests. The imperatives of the moral life rest on an ill-formed continuum of stringency, and there are many points on the continuum.

The program of correspondence suggested in the above chart can be outlined by listing a few select principles and virtues:

The Principle or Duty of	corresponds to	*The Virtue Standard of*
Beneficence		Benevolence
Truth-telling		Truthfulness
Gratitude		Gratefulness
Fidelity		Faithfulness
Confidentiality		Confidentialness

Several authors in this volume seem implicitly to require such a conception in order to reason as they do.

Critics of this conception will argue that virtues such as sincerity, integrity, fortitude, rectitude, conscientiousness, hopefulness, and perseverance have no clear correspondending duty. The 'long sufferingness' of which Karen Lebacqz writes in her essay seems altogether beyond any role that an ethics of *duty* could provide. This criticism is partially right, but misleading and not damaging to the case I have made. I believe that many duties *do* correspond to such virtues. For example, I think there are ordinary moral requirements of thoroughness, scrupulousness, objectivity, and fairness. More important is that there is no reason to think that each virtue singled out in the English language has a corresponding principle that is likewise singled out in the language. Many virtues are not moral virtues, and some virtues – As Michael Slote has recently shown – are contra-moral in nature ([18], pp. 77–107). But even in the case of exclusively moral virtues, our moral language is untidy: One virtue may encompass many duties, and one duty may correspond to many virtues. Moreover, one reason why the virtue of integrity and the principle of human dignity have proved elusive in moral theory is that they presuppose or organize a vast array of duties and virtues that fall under them.

The most pressing problem, however, is not that of a one-to-one correspondence between virtues and dutiful acts. The major problem is that virtues standards, and virtue theory generally, permit no clear distinction to be drawn between (1) ideals that surpass, supersede or rise above the ordinary morality and (2) virtues that everyone is expected to manifest as a matter of shared morality. By contrast, an ethics of principles that specifies duties seems to admit more readily of a division into required or dutiful actions (those required by principles of duty) and morally praiseworthy but nonetheless nonrequired actions, generally designated as supererogatory. The duty-based theory, by its very use of duty as a central category, requires that a principle specify both what is dutiful and what exceeds duty. But the virtue-based theory has no obvious need for it, because the virtues do not inherently require, command, or compel actions. However, this difference is not as distinct as it at first appears.

This problem can be examined further by reference to Lappé's and Pellegrino's essays in this volume. Lappé opens with a strikingly obvious but significant point extracted from the roots of Greek culture: "*To the extent* that virtue *requires* human *excellence*, it is an attribute of persons whose *actions exceed* the norm established by customary standards, mores, or laws" (p. 289, italics added). Pellegrino writes similarly that "Virtue ethics expands the notions of benevolence, beneficence, conscientiousness, compassion, and fidelity well beyond what strict duty might require. It makes some degree of supererogation mandatory because it calls for standards of ethical performance that exceed those prevalent in the rest of society" (pp. 246–247 and p. 250 for the foundation in Greek philosophy).

The long association through Plato and Aristotle of virtue with *excellence* (*arete*) has led Lappé, Pellegrino, and many others to the confusion that virtue entails a level of excellence that *exceeds* the established norm, and yet is *required*. So understood, virtuous conduct is both supererogatory and required; and, as Ferngren and Amundsen note, "the ordinary person has no *arete*" (p. 4). But a good-samaritan conception of virtue in morals is misleading. A moral virtue is a disposition to do what is morally good and not merely what is morally super good. A virtue standard is a norm of the moral life, not merely a target for the person of saintly, heroic, or excellent character. The association with excellence deriving from the Greek conception of proper functioning need not spill over to the supererogatory. Virtues of respectfulness, benevolence, and fairness, for example, are not exclusively for the morally excellent, meaning those who 'exceed the norm' in Lappé's sense. Such virtues establish our shared moral expectations for good and decent human relationships. Persons who violate those standards violate ordinary canons of morality, not canons of excellence that exceed the norm.

Lappé's and Pellegrino's mistakes arise because 'virtue' is an equivocal term in English, and one ripped from different eras of history and culture, as MacIntyre quite correctly argues, and as Ferngren and Amundsen nicely display. One definition of 'virtue' in the *Oxford English Dictionary* is "conformity of life and conduct with the principles of morality." A second definition in the same source is "a particular moral excellence; a special manifestation of the influence of moral principles . . . . Superiority or excellence; . . . unusual . . . distinction." The first definition situates virtue in the context of ordinary morality, although not necessarily of excellence in ordinary morality. The second situates virtue in the context of extraordinary excellence in a morality of ideals (although perhaps ideals that are *extensions* of the principles that form the ordinary morality).

Lappé and many writers in ethics present a confused picture of the role of the virtues by treating 'excellence' as if these two senses were one. Lappé writes that virtuous behavior "*transcends* legal requirements and involves *altruism* and justice" (p. 299, italics added). Pellegrino similarly treats virtue-based ethics as inherently "altruistic" and "elitist" because of its adherents' high demands (p. 252). Altruism, elitism, and transcendence of legal requirements entail supererogation – as Pellegrino notes – while justice is among the most fundamental of those moral categories that we use to fix the terms of duties and rights. Again, Lappé writes that "Virtue in public health . . . requires recognition of individual needs and vulnerabilities, as well as *extraordinary* effort, generosity, or moral excellence" (p. 290). The first part of this assertion ('requires recognition of individual needs and vulnerabilities') need not involve supererogation in the promotion of the public health. A morally sound and 'virtuous' public health policy establishes the norms, standards, and laws of the activity, to use Lappé's terms. But the second half of the assertion is conceived entirely in terms of extraordinary moral excellence, which surpasses basic standards and laws. 'Sacrifice' is a category fitted for supererogatory action; 'insightfulness' into public health vulnerabilities is not in this category; and 'beneficence', in my judgment, belongs in both. Yet Lappé treats sacrifice, insightfulness, and beneficence as if there were no such distinctions to be made (p. 299). We had best keep these categories apart, or we will go far astray of the mark in writing about virtue and about the moral life generally.

Lappé's example of the Nestlé Corporation's acceptance of the WHO's recommendations for controlling the marketing of infant formula serves to illustrate this problem. Lappé writes that Nestlé is exercising a virtuous health policy if it follows the WHO's proposal: "Virtue here consists of taking a policy position on an essential, health-maintaining behavior at some, perhaps substantial, corporate expense" (p. 296). He contrasts this behavior with the behavior of HMOs that set maximal return to stockholders as their major policy objective. Let us concede (what is not obvious) that Nestlé is acting virtuously in the basic sense of virtue I have adopted in this paper. That is, let us concede that the company has a disposition to do what is morally commendable.

It does not follow, and nothing in Lappé's essay shows, that Nestlé is exercising a virtuous health policy in his sense of 'exceeding the norm'. Perhaps Nestlé 'transcends legal requirements', but *this* transcendence will not amount to supererogation or 'extraordinary moral excellence'. Even if *substantial* corporate expense is involved, the action may not be supererogatory. In historical fact, Nestlé's policy has served two functions for the corporation. It has led to the development of a basic set of badly needed moral standards for corporate responsibility (nothing supererogatory about this) and has corrected (insofar as possible) its tarnished public image. (The HMOs cited by Lappé, by contrast, may not even be living up to basic moral requirements. Alternatively, these HMOs may not have suffered the moral problems that Nestlé has encountered because of Nestlé's own history of corporate activities, which were very profitable, at the time, for stockholders.)

From these arguments I conclude as follows: Moral principles of duty correspond to moral standards of virtue. This is not always a one-to-one correspondence, but we could reconstruct principles and virtue standards so that there would be a perfect one-to-one correspondence. Thus, benevolence would correspond to beneficence, fairness to justice, truthfulness to truth-telling, etc. In cases like friendliness, gratitude, responsiveness, and the like, we would have to be careful not to require – through a principle of *duty* – that which is supererogatory. Many but not all virtues do fall into the category of supererogation, just as many actions but not all actions do.

I shall return later to some theoretical problems that this account must face. Then, in the final section, I shall discuss reasons why the correspondence holds by trying to situate my claims in a broader moral theory. However, before treating these problems, I want to show that the category of moral rights can also be fitted into the same program of correspondence that I have outlined for principles and virtues.

THE CORRELATIVITY OF RIGHTS, DUTIES, AND VIRTUES

A similar correspondence theory can be developed for rights-based approaches to ethics, which are also often said to rest on a different 'foundation' than either virtue-based or duty-based theories. According to the correlativity theory of rights and duties – which I have elsewhere defended ([2], pp. 201–205) – one person's right entails the duty of another to refrain from interfering or to provide some benefit, and any duty similarly entails a right. (A weaker version is that rights entail duties, although not all duties entail rights, but I shall not discuss this version.) The language of rights is thus always translatable into the language of duties.

By adopting this correlativity thesis, we can construct the following abstract schema of the relationship between virtue standards, principles of duty, and rights:

	Duty-Based Theories [Correspond to]	Virtue-Based Theories [and are both correlative to]	Rights-Based Theories
Ordinary Moral Guides	Principles of Duty as Guides of Conduct	Virtue Standards as Guides of Conduct	Rights Claims as Guides of Conduct
Extra-ordinary Moral Guides	Supererogatory Principles as Guides of Conduct	Supererogatory Virtue Standards as Guides of Conduct	X

Under this conception, there can be fundamental moral virtues that correspond to fundamental principles and to fundamental rights, and there can be derivative rules in each category that are derived from the fundamental standard(s).

If this program were carried out by extensively listing duties, virtues, and rights, the result would be more expansive lists than we are accustomed to seeing in moral theories, and this proliferation of duties and rights, unlike virtues, may seem objectionable. However, I am inclined to think that this outcome would be salutary for moral theory, rather than a defect, because clarity would be enhanced. Consider the right to privacy in light of the correspondence and correlation theses:

respect for privacy	respectfulness for privacy	right of privacy
(Principle of Duty)	(Virtue Standard)	(Right)

Although respect for *autonomy* and respect for *persons* have been widely discussed in contemporary moral theory, 'respect for privacy' may seem odd and, by comparison, nonfundamental, even gratuitous. However, there has been discussion both as to whether privacy is a fundamental value [8, 17], and whether it should be distinguished from many other concepts with which it has been associated, such as autonomy [17, 19]. The analysis I am suggesting requires that we more clearly distinguish duties, virtues, and rules that have often been conflated and treated as identical. The current confusion in moral theory over the differences between privacy and autonomy is a prime example.

The correlativity thesis has been challenged on grounds that several classes of duties do not entail rights, and that certain rights do not entail duties. Duties of charity, kindness to animals, and duties of conscience are the staple of such objections. However, this problem can be handled by distinguishing between duty required by a shared principle of ordinary morality and 'duty' required by a self-imposed principle of extraordinary morality. For example, duties of charity and conscience are often supererogatory, and duties of kindness to animals occupy can uncertain status. Thus, the same distinction between supererogatory or extraordinary morality and ordinary (nonsupererogatory) morality discussed earlier again rescues us from theoretical problems: The correlativity thesis holds for all and only duties of ordinary, nonsupererogatory morality. It will of course not always be easy to determine the class in which a particular obligation (or virtue) properly belongs, and there may be several obligations and several persons having obligations that correspond to a single right. But this untidiness does not impair the correlativity thesis; it only shows that correlativity is complex.

I conclude that the correspondence and correlativity theses developed above provide the outlines of an integrated theory of morality according to which virtue standards, principles of duty, and valid claims of right express different emphases in the moral life and are complementary categories.

Rights, for example, are particularly appropriate for contexts in which a good, service, or liberty can rightfully, and if necessary forcefully, be demanded as one's due; the bearer of a right can validly make demands, and others are validly constrained from interferences with the exercise of that right and from failures to provide that which the person is owed. Protection-by-rights is a highly suitable model for certain contexts, while generally unsuitable for those in which virtue is most suitable. It does not follow that one does not have rights that correspond to virtue standards, but only that the distinct emphases attached to our various moral standards are fashioned for and peculiarly suited for certain purposes. For example, we have a right to be treated kindly by others, but contexts in which kindness is appropriate are not usually contexts suited for forcefully demanding something as one's due.

Although the broad program outlined in this section for an integrated theory requires more elaborate argument than I have supplied, or can supply in this essay, a few gaps can be filled by considering one potential objection to this approach.

PROBLEMS WITH THE CORRESPONDENCE AND CORRELATIVITY THESES

My integrated account of duties, virtues, and rights faces several difficulties. We have mentioned the problem that innumerable virtues have no one-to-one corresponding duties or rights, and many rights have no one-to-one correlative duties or virtues. But the problem is broader than an absence of one-to-one relationships. There are many virtues that seem to have no connection whatever to principles of duty, let alone a one-to-one connection. Typical examples are modesty, cautiousness, integrity, loyalty, cheerfulness, sincerity, appreciativeness, cooperativeness, and commitment. None of these virtues is easily translated into a duty, and they seem even more distant from the language of rights. The list could be greatly expanded: What corresponds to sobriety, friendship, courteousness, promptness, and prudence? Or compassion, courage, conscientiousness, devotion, civility, and caring? Can these parts of the moral life be reigned in under a conception of duty? To lovers of virtue like Hauerwas and MacIntyre, the program of reduction I have suggested will seem ludicrous, for they will think it misses what is most important about the virtues: They reflect a communal moral life and the integrity of persons for which principles of duty are simply unsuited. At best, then, it might be argued, there is only a partial and not a complete correspondence across these categories.

The way to bridge this chasm of correspondence is to resort again to the distinction in the above tables between extraordinary moral standards and ordinary moral standards, together with an account of the distinct emphases that are placed on these categories in specific contexts: Principles serve to specify duties; rights to protect interests and warrant forceful claims; and virtues to establish models of conduct.

Consider compassion. Not everyone is equally suited to be compassionate, and some therefore possess the virtue more than others. But does not the moral life suggest that we *ought* to be compassionate, i.e., that we have duties of compassion? Whether or not a requirement to be compassionate derives from some more *fundamental* principle(s) such as beneficence or respect for persons, I find it unthinkable that a comprehensive catalogue of the moral life would omit a duty of compassion from its contents. It may be that we do not have to be heroically compassionate – for that would be supererogatory – but surely we can violate a duty to be compassionate. The physician who fails of compassion in handling the questions of a pain-wracked patient fails in his or her duty to the patient. This is a moral, not merely a professional duty. The failure is a moral failure that may also be directly traced to a failure of virtue, a malformation of character. Just as we do not have to be heroically compassionate, so the person who displays the virtue of compassion is not expected (I would prefer to say 'required') to be virtuous to heroic proportions. We have little difficulty – at least in the paradigm cases – in distinguishing (1) actions that are morally required by principle P and (2) actions that fall under principle P but surpass its requirements. Similarly we need to distinguish (1) virtues that ought to be cultivated and (2) virtues that exceed the standards of ordinary morality.

In his book *Lest Innocent Blood Be Shed*, Philip Hallie writes about the vice of *cruelty* and the 'opposite' virtue of *hospitality*. The latter he analyzes, for his purposes, as unsentimental, efficacious love. He places this analysis in the context of a French village that defied the Nazi conquerors of France at great personal risk to individual citizens and saved the lives of approximately 6,000 people [10]. This virtue of hospitality is difficult to analyze in conventional principle-based systems, and it might seem that any translation would be highly artificial and unsuccessful. But the problem here, I suggest, is not over the correspondence thesis I have developed, but rather over the proper description and status of such actions. Does hospitality exceed the demands of ordinary morality? Would a failure of hospitality amount to neglect? cowardice? fearfulness? Are these always vices? Is it morally required to assume the risks entailed by pursuing Hallie's vision of hospitality? These are not easy questions, because, while heroism and supererogation flourished at this tiny French village, our moral sense is not so keen or our moral system so well sculpted that we can speak without hesitation about moral requirements in such circumstances.

Many writers in moral theory seem confused as to whether their arguments should be expressed in terms of virtues or duties, and also over whether various virtues cited are moral or nonmoral. An interesting example is the virtue of autonomy – or shall we say the principle of (respect for) autonomy? While the latter is the more typical use of terms in contemporary philosophy, consider Stanley Benn's account of autonomy:

> The autonomous man is the one who, in Rousseau's phrase, "is obedient to a law that he prescribes to himself," whose life has a consistency that derives from a coherent set of beliefs, values, and principles, by which his actions are governed. Moreover, these are not supplied to him ready-made as are those of the heteronomous man; they are *his*, because the outcome of a still-continuing process of criticism and re-evaluation . . . .
>
> The principles by which the autonomous man governs his life make his decisions consistent and intelligible to him as his own; for they *consitute* the personality he recognizes as the one he has made his own ([6], pp. 123f).

Benn's theory, as he puts it, is of a person or self-made 'personality' whose 'life has a consistency' of autonomous actions across time – a '*kind* of agent'. The autonomous person thus is consistent, independent, in command, resistant to control by authorities, and the source of his or her basic values and beliefs. The person's life is ordered by a set of rules expressive of the person's will. The life as a whole expresses a coherence and directedness that may break down in particular cases or be frustrated by circumstances, but which generally remains intact through strength of will, fortitude, and resolve to see to it that his or her conception is consistently carried out.

Benn is interested in the character of an individual. His theory of autonomy could easily be analyzed as a normative theory of human virtues that emphasizes commendable dispositions and habits such as independence and consistency in making evaluations. Such traits need not be *moral* virtues, but they do establish a person's character – much as certain traits of actors and actresses mark them as having a distinct character or quality in acting, including their virtues as performers. John Benson has suggested that autonomy is *best* analyzed as a human virtue essential to well-being:

> The virtue of autonomy is a mean state of character with regard to reliance on one's own powers in acting, choosing, and forming opinions. The deficiency [involves] vice: [being] credulous, gullible, compliant, passive, submissive, overdependent, servile . . . .
>
> To be autonomous is to trust one's own powers and to have a disposition to use them, to be able to resist the fear of failure, ridicule or disapproval that threatens to drive one into reliance on the guidance of others . . . . Autonomy, like courage – to which it is closely allied – is an essential virtue that everyone needs . . . because without it one cannot live effectively the life of a member of society ([7], pp. 5, 8).

This analysis turns on strength and stability of character – traits rather than various doings. While we may need to know nothing about the quality of autonomy in a person's general character in order to know whether particular actions are autonomous, we might be at least as much interested in the *person* as in the person's *actions*. This is in fact Benn's interest, although he couches his arguments less in categories of virtue than principle. Thus, resistance to social conformity, reflectiveness, understanding, insightfulness, and resistance to manipulation are the tools of his analysis, not principles of duty or even extensions of those principles.

A CONSEQUENTIALIST CONCEPTION OF MORAL STANDARDS

Bernard Gert argues in his essay that "Some moral virtues are going to be rule dependent: to have the virtue involves following the appropriate moral rule. . . . [But] it may even be possible to eliminate any mention of the rules and to describe the moral virtues as those virtues that an impartial rational person would want both for himself and for all others" (p. 19). Gert then ties *both* his account of the moral rules and the moral virtues to a conception of rationality and to what all rational and impartial persons would want. This approach seems to me correct insofar as it supports the correspondence and correlativity theses defended previously. When Gert writes about kindness, he describes it as the virtue that "involves *acting* so as to prevent or relieve the suffering of evils by others." His description may be insufficient as an analysis of kindness, but that problem may be set aside. His conception of the moral life and of constituents such as kindness can easily be turned in the direction of a duty-based theory *or* in the direction of a virtue-based theory. In both cases, the objective behind our interest in categories like duty and virtue is the goal of preventing or relieving suffering and evil. And this is why Gert can claim, rightly, that kindness is "directly related to the *aim* of morality" (p. 105). I believe that Aristotle's theory should be similarly interpreted. This problem of the aim of morality leads to a final set of comments about aims, goals, objectives, consequences, and the like.

John Stuart Mill held that the 'object of virtue' is the 'multiplication of happiness' [16]. He viewed the principle of utility and the virtues and rules derived from it as instruments of the happy life. This great exponent of the principle of utility, and the duties derived from it, had no trouble accommodating virtues in his system. Mill knew that the purpose of general moral standards – whether in the form of virtues, duties, or rights – is to achieve certain desirable outcomes and to avoid certain undesirable outcomes. This is the reason, and ultimately the justification of our scheme of moral duties, virtues, and rights. Contemporary consequentialists have not abandoned this perspective as a relic of 19th century philosophy. For example, in his recent book, *Moral Thinking*, R. M. Hare relies on a strikingly similar approach. He holds that "intrinsic moral virtues" have "corresponding principles" and that in the training of children we want them most to learn not merely to adopt a policy or practice of obeying principles but rather to possess "firm dispositions of character" – i.e., virtues – that accord with the principles. As with Mill, the justification of both training in duty and virtue is human happiness ([11], pp. 194–197, 103–105).

The Greek tradition too has emphasized the connection between the virtues and a life of happiness or human flourishing. As Pellegrino notes, "the *end* of virtue for both Socrates and Aristotle is the good life" (p. 240). Ethics has a use and objective for Aristotle: to order individual and (as a branch of politics) communal life so that we can live as well as possible through conventions and practices specifically arranged to promote well-being (*eudaimonia*). Human excellences, including moral virtues, are capacities for living according to the goals and purposes of human activities. The virtues are general *dispositions* leading to *states* of well-being or of preventing the reverse. Aristotle's whole moral theory in this interpretation is teleological ([1], Book 2).

Properly qualified, I accept such a consequentialist approach to justification. Following both Mill and Aristotle, I take the object of morality to be closely tied to the creation and maintenance of conditions that allow the pursuit of a well-structured life of well-being, including conditions that minimize the threat of injury and pain. And, with a debt to both Hume [13] and G. J. Warnock [20], I believe that the specific purpose of moral principles and of standards such as virtue is to minimize the tendency for things to go wrong or badly in interpersonal human relationships. Because humans are not characterized by what Hume calls 'extensive benevolence', conditions can seriously deteriorate in human affairs as a result of limited resources, limited time, limited information, and, most importantly, sharply limited sympathies with the plight or needs of others. Moral action guides are needed in a culture to counter the limited sympathies of the human condition, because human motivations can lead to unfortunate and even tragic circumstances, especially if self-interest is allowed to be the overriding factor.

Medical ethics and other professional ethics serve such a function in their restricted professional settings: Their moral principles and virtues serve to counter those limited sympathies that hamper the encounter between health professional and patient, or investigator and subject. Duties, virtues, and rights prevent things from going wrong, in ways they otherwise might, by directing people toward the protection of the interests of others rather than becoming mired in their self-centered interests. Moreover, as Marx Wartofsky argues in his essay in this volume (p. 180), virtuous actions should be understood as those which serve to realize the goods built into practices such as the institution of medicine.

Warnock has shown that the point of moral standards for action and character is to make things better than they otherwise might be. Ethical principles, moral rights, and dispositions of virtue function to countervail limited sympathies. Good principles, good dispositions, good protections, and the like are our tools to the countervailing of many limited human tendencies, such as the infliction of damage, discrimination, deception, and indifference. This conception, which I believe in all essentials to be correct, leads Warnock to the following claims about the relationship between virtues and principles:

> We seem to be led to four (at least) general types of *good disposition* .... Somewhat crudely named, those of (1) non-maleficence, (2) fairness, (3) beneficence, and (4) non-deception. We venture the hypothesis that these (at least) are fundamental *moral virtues*.
>
> But we can now manipulate this conclusion about "good [moral] dispositions" ... to derive from this the proposition that we have here, by the same token, four fundamental moral *standards*, or moral principles. To have and to display, say, the moral virtue of non-deception could be said to be to regulate one's conduct in conformity to a *principle* of non-deception [20].

My claim is that virtues, principles, and rights are all instruments, or means, to these ends of the moral life. These objectives determine our choices of means, and in some contexts the subtle differences of emphasis appropriate to virtues, principles, and rights motivate us to select one category as more suitable than another for the context at hand. Consider as an example of an argument that emphasizes virtue Henry Beecher's pioneering monograph, *Experimentation in Man*. Beecher was worried about the consequences of rule-based or regulatory approaches to the control of experimentation involving human subjects. He held that rules and regulations are "more likely to do harm than good," and will utterly fail to "curb the unscrupulous". Beecher argued that the Nuremberg Code's unqualified insistence on the consent of all subjects "would effectively cripple if not eliminate most research in the field of mental disease," as well as throw out the use of placebos. "Even the 'obvious' matter of consent," he held, "is not so easy to live up to as it sounds" ([4], pp. 52–58). The thrust of his essay was to heighten awareness of the complex character of the problem, to insist on the overwhelming importance of sound training in scientific methodology, to make the scientific practices consonant with moral sensitivity, and to suggest that we cultivate virtue in physicians ([14], pp. 15–17, 43–44, 50). He went on in his subsequent writings to propose that we stem the tide of "thoughtlessness" and "carelessness" through the most "reliable safeguard" against abuses in research involving human subjects, viz. "the presence of an intelligent, informed, conscientious, compassionate, responsible researcher" ([5], pp. 1354–1355). Beecher's approach relies on an argument that the best outcome will be achieved by educating physicians through a virtue-based ethic rather than a rule- or duty-based one. There is a notable absence in his work of any approach based on the *rights* of subjects – an absence that led to subsequent criticism of Beecher's suggestions.

Beecher's approach resembles the views of Allen Dyer in this volume, who argues that "The doctor–patient relationship is in fact based on trust in the ethics of 'virtuous men (and women) of spontaneous good conscience', a much higher standard than could possibly be achieved by explicit group sanctions. Trust or trustworthiness is the keystone of medical virtue in the traditional canons . . . " (p. 230). Like Beecher – and MacIntyre and Hauerwas – Dyer is worried that duties, sanctions, rights, and the like are not as well *suited* to the medical context as the virtues. Placed in the perspective of appropriate emphases, as discussed previously, this claim seems unproblematic. An emphasis on character can serve clinical medical ethics well, especially if we have the objective of minimizing the line between the required and the supererogatory. It would not serve well to protect the interests of citizens who have a right to food under a government-sponsored food-stamp program.

In many contexts, principles that permit 'ought' statements and rigidly formulated duties serve our ends better than the language of virtue. For example, in the years following publication of Beecher's monograph, the very people that it influenced the most – officials at N.I.H. – adapted his ideas to schemes of government regulation governing research with human subjects.

Shortly therefore, we saw the apperance of a long chain of moral rules and regulations – e.g., The Yellow Book (1971) as well as the moral guidelines proposed by the National Commission for the Protection of Human Subjects. N.I.H. officials were convinced – after years of experience, and rightly in my judgment – that an approach based on the virtues-of-the-investigator had insufficient teeth and bite. Moral imperatives – like statements of rights – are blunter and more effective instruments for guiding conduct where exhortations to virtue are hard to transmit or need a disinterested audience. One chooses an instrument or tool for the job at hand – which is why rights are so often listed in political and adversarial contexts: They are our most direct and forceful instruments, because fashioned to allow us to demand what is our due.

To summarize: In those contexts in which reliance on a person of good character is most likely to achieve the moral ends we desire, then a theory of the virtues may be the superior account. This belief underlies Beecher's proposal. On the other hand, the person of virtue may often be perplexed about what should be done, or which course of action is the right one. Indeed, the person of good character may be the first to know that his or her character is insufficient to yield the answer. Hence, a discussion of duty, right, or the morality of actions may seem in some contexts to be more important than a doctrine of the virtues for achieving our ends. In still other contexts where the prod of duty or the protection of rights best achieves our objectives, these moral standards will surpass appeals to virtue.

In his essay, Gert takes up some of these considerations about consequences and argues as follows:

> There may have been, and there still may be, some philosophers who are pure deontologists or pure consequentialists, but any plausible moral theory involves elements of both. That is, all plausible moral theories will involve *both rules and consequences*. Any stable moral rule will have exceptions and it will sometimes be necessary to consider the consequences of violating the rule in order to determine what exceptions are justifiable. Further, even the justification of the rules themselves will involve consequences, either directly or indirectly. However, consequences alone will not be sufficient, for without rules there will be no way for that kind of impartiality which is recognized by everyone to be an essential feature of morality, to be adequately incorporated into the theory (italics added).

I agree that a rule (or a meta-rule) of impartiality is required for morality, but this rule is itself justified by appeal to consequences that we seek to promote or to avoid. Gert presents us with a false dichotomy between consequences and rules. This seems particulary odd since he admits that the justification of the rules will (always, and not just often, in my view) require appeal to the consequences. I do not see why anyone would promote a moral rule or a conception of virtue without some goal in mind such that the consequences of having the rule or promoting the virtue will serve as the most fitting means to that end. I may be a 'pure consequentialist' in ethics, but I hope one who has a fair appreciation of rules, and indeed an appreciation that does not too far deviate from Gert's when it comes to impartiality.

I have tried to show in this essay that philosophers such as MacIntyre who hold out a profound and dazzling promise for virtue ethics, while lamenting the history of duty-based and rights-based ethics, overvalue the virtues and undervalue duties and rights. I think they are also deficient in the virtues of tolerance and appreciativeness. While I would agree that discussion now occurring about the importance of virtue will make a significant contribution to moral theory, it is easy to overestimate its potential. If my analysis in this paper is correct, then MacIntyre's critique of the last 300 years (and beyond) of moral theory can be turned on his theory of virtue with the same ferocity that he has turned it on others. I believe that every criticism of modern ethical theories that holds for duty-based and rights-based theories holds no less for virtue-based theories. Remember, too, that regression to the mean is a good statistical model for placing your bets on the future history of moral theories.

BIBLIOGRAPHY

[1] Aristotle: 1963, *Nichomachean Ethics*, in *The Philosophy of Aristotle*, trans. A. E. Wardman, Mentor Books, New York.

[2] Beauchamp, T. L.: 1982, *Philosophical Ethics*, McGraw-Hill, New York.

[3] Beauchamp, T. L. and Childress, J. F.: 1983, *Principles of Biomedical Ethics*, Oxford University Press, New York.

[4] Beecher, H.: 1959, *Experimentation in Man*, Charles C. Thomas, Springfield, Ill.

[5] Beecher, H.: 1966, 'Ethics and Clinical Research', *New England Journal of Medicine* **274**, 1354–1360.

[6] Benn, S.: 1976, 'Freedom, Autonomy, and the Concept of Person', *Proceedings of the Aristotelian Society* **LXVI**, 113–128.

[7] Benson, J.: 1983, 'Who is the Autonomous Man?', *Philosophy* **58**, 1–15.

[8] Caplan, A.: 1982, 'On Privacy and Confidentiality in Social Science Research', in T. Beauchamp, *et al.*, (eds.), *Ethical Issues in Social Science Research*, The Johns Hopkins University Press, Baltimore, pp. 315–325.

[9] Foot, P.: 1978, *Virtues and Vices*, Basil Blackwell, Oxford.

[10] Hallie, P.: 1979, *Lest Innocent Blood Be Shed*, Harper & Row, New York. See also 1981, 'From Cruelty to Goodness', *The Hastings Center Report* **11**, 23–28.

[11] Hare, R. M.: 1981, *Moral Thinking: Its Levels, Method and Point*, Oxford University Press, Oxford.

[12] Heyd, D.: 1982, *Supererogation: Its Status in Ethical Theory*, Cambridge University Press, Cambridge.
[13] Hume, D.: 1894/1975, *An Enquiry Concerning the Principles of Morals*, ed. L. A. Selby-Bigge, 3rd edition, Ph.D. Nidditch, Oxford University Press, Oxford.
[14] Jonsen, A. and Hellegers, A. E.: 1974, 'Conceptual Foundations for an Ethics of Medical Care', in L. Tancredi (ed.), *Ethics of Health Care: Papers of the Conference on Health Care and Changing Values*, National Academy of Sciences, Washington, pp. 3–20.
[15] MacIntyre, A.: 1981, *After Virtue*, University of Notre Dame Press, Notre Dame.
[16] Mill, J. S.: 1974, *Utilitarianism*, in *Utilitarianism, On Liberty, and Essay on Bentham*, ed. with an Introduction by Mary Warnock, New American Library, New York.
[17] Parent, W.: 1983, 'Recent Work on the Concept of Privacy', *American Philosophical Quarterly* **20**, 341–355.
[18] Slote, M.: 1983, *Goods and Virtues*, Clarendon Press, Oxford.
[19] Thomson, J.: 1975, 'The Right to Privacy', *Philosophy and Public Affairs* **4**, 295–315.
[20] Warnock, G. J.: 1971, *The Object of Morality*, Methuen & Co., London.

Department of Philosophy and
Kennedy Institute of Ethics,
Georgetown University,
Washington, D.C. 20057,
U.S.A.

ROBERT M. VEATCH

AGAINST VIRTUE: A DEONTOLOGICAL CRITIQUE OF VIRTUE THEORY IN MEDICAL ETHICS

The essays in this volume provide many lists of virtues in different cultures and different times. Those lists include the 'virtues' of cunning, hatred of enemies, condescension, skill, obedience to husbands, whimsicality, and filial piety as well as gentleness, meekness, compassion, and humaneness. That in itself should suggest that there are serious problems to be solved in developing a full and sound theory of virtue in medical ethics. Ferngren and Amundsen, in the first essay in this volume, make clear the dramatic contrast between the classical Homeric virtues in Greek thought and the major gentle virtues of the Christian tradition. The Homeric hero possessed the character of cunning, courage, self-reliance, loyalty, love of friends, hatred of enemies (as well as courtesy, generosity, and hospitality) while the Pauline virtues included much gentler fare: love, joy, peace, longsuffering, gentleness, goodness, faith, meekness, and temperance. It might, in fact, be argued that all that is wrong with the ethics of medicine is that health care professionals have adopted the Homeric list while patients have been expected to adopt the Pauline character.[1] The very term *patient* conveys the notion of meekness and long-suffering passivity. Developing a theory of virtue for an ethic of medicine will be much more difficult and controversial than is often realized.

All too often virtue and the virtues in ethics – whether in medical ethics or more general reflections on good character – have been taken to be uncontroversial. Touting the virtuous life might be seen as gracious or even platitudinous, but hardly polemical. I shall suggest that this is a mistaken view of the innocence of virtue theory. The virtues and virtue theory in general, in at least some medical ethical systems, can miss the key point of ethics at best and, at worst, can be downright dangerous. I shall argue first that it is a mistake to assume that there is an obvious single set of virtues that constitute good character for all possible social roles; second, that even when describing the character traits of a particular role (such as that of the physician or patient) there is no obvious set of virtues that are ideal or correct; third, that an emphasis on virtue can lead to wrong conduct; and, fourth, that in certain medical contexts it is really not critical whether the actors are virtuous. Having argued for these unusual and, I take it, controversial, positions, I shall then, in the more constructive portion of the

Earl E. Shelp (ed.), Virtue and Medicine, 329–345.

essay, argue that there is a kind of health care – one that is almost non-existent today – in which a virtue theory and apreement on the proper virtues for specific professional and lay roles would be essential.

Before developing this argument, I must be precise in what I mean by virtue theory and by virtues. I mean by a virtue, a dispositional trait of character that is considered praiseworthy in general or in a particular role. I mean by a virtue theory a systematic formulation of the traits of character that make human behavior praiseworthy or blameworthy. My use of the term I take to be compatible with the usage of Gert and Pincoffs, but in some of the other essays in this volume these terms have been used in much less precise ways. Virtues are treated as synonomous with values or principles of right conduct, supererogatory actions, altruistic acts, or some other more general category referring to human conduct rather than human character. By contrast I am limiting my use of the term to character, inclinations of the will, persistent motivation, those qualities that make human actions praiseworthy or blameworthy, not right or wrong.[2]

The contrast is seen most clearly in the terms benevolence and beneficence. Benevolence is a characteristic of the will, hence the ending 'volence' from the Latin *volens*, wishing or willing. It refers to a will to do good. By contrast, beneficence is a characteristic of action, hence the ending 'ficence' from the Latin *facere*, to do or make. Using the terms this way one can make meaningful statements such as the following: *X* was a benevolent character who always failed to be beneficent, that is, do the good. One could by contrast say, '*Y* was always malevolent, but because of a terrible fear of the law and/or a terrible failure to anticipate consequences always managed to be beneficent in spite of himself.'

FOUR PROBLEMS IN VIRTUE THEORY IN MEDICAL ETHICS

It has upon occasion been suggested that ethical theory in medicine has undergone a major conceptual shift within the past fifteen years. It has been argued that traditionally medical ethics articulated by professional groups and by individual health care professionals emphasized the virtues, but that the new generation of medical ethics – that offered primarily by medical lay people, including philosophers, theologians, and lawyers – has abandoned the virtues in favor of an ethic of right conduct [8]. It is possible that there is some truth to this historical claim. A careful historical study would probably reveal that this is at best a generalization. The Hippocratic corpus, for example, emphasized norms of right conduct as well as character traits.

It spoke of applying dietietic measures for the benefit of the sick and not giving a deadly drug to anybody and not using "the knife, not even on sufferers from stone", as well as the virtues of purity and holiness ([14], p. 6). Moreover, recent work in medical ethics has not been totally oblivious to problems of virtue, as is illustrated by the present volume as well as other recent writings in medical ethics [7, 10]. I do not wish to deal with the historical problem, however. I wish to deal with the normative problem of whether virtue theory is either necessary or sufficient as a medical ethic. I shall argue that in the normal world of medicine it is not and shall state my case by making four points.

That the Proper Virtue Set is Not Obvious

The first point should by now for readers of this volume be a simple one. Some writers seem to assume that there is some obvious set of virtues that together make up virtuous character. The theologians writing within a particular tradition can be excused if they automatically assume that faith, hope, and love are the virtues that constitute good character. Even here, as Amundsen and Ferngren have shown us, there are variations within the tradition. Paul, in places other than I Corinthians, adds compassion, kindness, forgiveness, and patience. Other New Testament authors will provide slightly different lists. The beatitudes find blessedness in being poor in spirit, meek, merciful, and pure in heart as well as in mourning and in hungering and thirsting for righteousness. The problem really gets serious, however, if one wishes to work in a more secular vein. We have already seen how radically antithetical the classical Homeric virtues are to those of the Christian tradition. To this list we might add the Confucian virtues of humaneness and compassion (which resemble somewhat those of the Christian list) and also filial piety (which seems to have little resemblance) ([16], p. 200). The gentlemanly virtues of a Victorian ethic are hardly compatible with the egalitarian virtues of a liberal feminist ethic.

A more severe problem arises when one realizes that from the time of Plato it has been recognized that different virtues received different emphasis when attached to different roles. Courage is the virtue of the warrior; wisdom of the philosopher. Any theory of medical ethics will have to take an agreed-upon list of virtues and apply it to the various medical roles. The dominant role in medical ethics has been that of the chief professional, the physician, but a full medical ethical theory would have to determine the character traits of other health professionals as well – nurses, pharmacists, social workers, and so forth. Moreover, it would also have to identify the virtues of the lay actors – patient, family, friend, policy maker, and so forth.

That the Proper Set of Virtues for a Particular Role Is Not Obvious

That particular virtues are proper to a particular role is a point that is easily conceded by many defenders of virtue theory. What they often do not realize, however, is that even within a particular role it may be very controversial what constitutes virtuous performance. Alasdair MacIntyre ([9], p. 56) discusses at great length how in the classical Aristotelian tradition excellence in a particular role is thought to be derivable from our conception of that role. Thus excellence in being a man is derived by understanding of the good internal to being a man just as the concept of a 'good watch' or a 'good farmer' or a 'good harpist' is inherent in our understanding or watch, farmer, and harpist. Once one understands a practice such as farming of harp playing, one, according to this view, immediately has an understanding of the traits necessary to demonstrate excellence in the practice.

Among virtue theorists it is common to consider medicine 'a practice.' MacIntyre is explicit about this ([9], p. 181). Pellegrino, in his essay in this volume, for instance, says that "the virtuous physician on this [Aristotelian–Thomist] view is defined in terms of the ends of medicine." Erde, in his discussion in this volume, not only labels medicine as a practice, but equates that to the work of physicians and "the orientation they need in order to provide excellent health care."

We must at least challenge the idea that medicine can be reduced to physicianing. Medicine, if it is a practice at all, is a practice that is engaged in by all manner of people, professionals and lay people, nurses, and parents, patients, and healthy rational adults striving to promote their continued health. If medicine is a practice, it is one that virtually all humans engage in just as they engage in education or labor or recreation.

More critically, however, it is not clear whether we can assume that practices are singular, easily defined activities. Is it clear, for example, that one and only one set of characteristics could possibly constitute excellence in playing a harp? What, for example, if different skills were needed to play symphonic music and jazz? It rapidly becomes clear that there is no Platonic ideal of a harpist. In fact, here may be several different images of harpist, each with its own traits of excellence.

The same problem arises in a virtue theory in ethics. Some who have attempted to apply virtue theory to medical ethics seem to be confident that they can identify a set of virtues that apply to the role of physician. My colleague William May, for example, assigns to a professional such as a physician virtues such as benevolence, discretion, and justice without worrying very much about whether these are controversial or not [10].

Even for a particular role, such as that of the physician, it may be the case that there are several radically different competing images of the role. A gentlemanly physician of Percival's England may have little in common with a public health officer, an Orthodox Jewish practitioner, a feminist health collective staff physician, or a physician who conformed to the ancient Confusian ideal of practicing medicine as a family member on one's relatives and not as a mere professional [16]. If medicine, even more than harp playing, is not one practice but many, then it will be impossible to define one set of virtues for a particular role such as that of physician.

Even a role as specific as that of physician may in fact really be countless roles, each with its own set of virtues. The selection of the particular description of the role (and therefore the virtues that attach to it) can never be a philosophically neutral task. One will have to make important normative judgments before the type of physician role can be formulated.

A survey of the virtues assigned to the physician role reveals the enormous variation among competing ethical theories. The Hippocratic Oath, as we have seen, identifies *purity* and *holiness*, two virtues appropriate for an ethic of a Greek mystery cult in which the physician is a kind of priest who is in possession of powerful, secret knowledge. By contrast, Islamic tradition assigns to the physician the character traits of being grateful, humble, modest, merciful, patient, and tolerant ([12], p. 387). The Confucian medical ethical literature takes up the classical virtues of humaneness, compassion, and filial piety for the physician [16]. The Japanese medical ethical literature of the Ri-shu school in the sixteenth century identifies kindness and devotion as the most important characteristics ([5], p. 8). By the end of the eithteenth century in England when the ethics of the gentleman dominated medical ethics, Thomas Percival [11], the writer of the medical ethical codification that dominated Anglo-American professional physician ethics for a century-and-a-half, could present not only tenderness and steadiness, but condescension and authority as the virtues of the physician, a formulation taken up verbatim by the American Medical Association in its original code of 1847 [1]. It was not until the 1980 revision of the AMA Principles of Medical Ethics that that organization adopted much more modern and secular virtues of compassion and respect for human dignity as its cornerstone virtues ([2], p. ix).

The difference is truly startling. It is not going too far to say that one could have virtually any set of virtues one wanted for the physician by simply picking the cultural tradition and time period properly. Any virtue-based medical ethical theory is going to have to resolve these disputes in order to have a comprehensive and defensible medical ethic.

That Virtue Theory Can Lead to Wrong Acts

What has been argued thus far, of course, does not demonstrate that a theory of virtue in medical ethics is unnecessary or insufficient or dangerous. It merely shows that the task before the defender of a virtue theory of medical ethics is much harder than might have been expected and that real matters of substance are at stake in picking which of the many competing sets of virtue might apply. It is, however, the case that virtue, at least by itself, can be extremely dangerous. It can lead systematically to wrong acts by health professionals and lay people. How can that be?

A virtue theory, when defined in the precise way I am using the term, spells out a set of character traits, habitual motivations, upon which people ought to be acting. When presented for a particular role, such as that of the physician, it spells out a set of character traits by which a physician ought to be acting. When he or she is so motivated, to that extent his or her actions will be praiseworthy or blameworthy (but not necessarily right or wrong).

I have been asked to provide a deontological critique of virtue theory. To a deontologist, that is, one who holds that rightness or wrongness is characterized by certain right-making characteristics independent of the consequences – by justice or truthfulness or respect for autonomy, for example – the critical question in this context is what the relation is, if any, between virtuous character and right conduct. The problem is essentially similar for utilitarian and other consequentialist theories or right action. What I have to say here generally can be applied to consequentialistic positions as well.

It is not obvious that adopting the deontological position on the question of what make right acts right has any direct bearing on virtue theory. At most, it will be relevant to one who is concerned about the impact that adopting or promoting a virtue ethic will have on right conduct. The relationship may well be a subtle one. There are a number of potential answers to the question of the relation of virtuous character to right acts. For example:

(1) Only virtuous character assures right acts.
(2) Virtuous character is one thing that assures right acts.

(3) Promoting virtuous character tends to promote right acts.
(4) Promoting the proper virtues tends to promote right acts.
(5) Promoting virtuous character risks facilitating wrong acts.
(6) Promoting virtuous character always produces wrong acts.
(7) Only promoting virtuous character produces wrong acts.

I take (1), (2), (6), and (7) to be false and (3) probably to be false as well, while (4) and (5) I take to be in the right ball park. Answer (1) is clearly false; the most despicable characters can be forced to act correctly if watched carefully and given appressive positive or negative incentives. Answer (2) is also false. People of sterling character can do wrong acts by miscalculations or misunderstanding the situation. Answers (6) and (7) are too implausible to debate. That leaves the propositions in the middle of the list. The correctness of (3), (4), and (5) will depend on two critical questions of fact, what I shall call the problems of 'naked virtue' and 'the wrong virtues'.

The naked virtue problem can be stated as follows: does having a virtuous character trait standing alone without ethical principle or moral rules or institutional checks to govern one's acts tend to promote or discourage right acts?

Consider first a simple ethical system in which there is only one virtue, benevolence. Virtuous physicians in such a world would be expected to act only out of the will to do good. Is it sufficient simply to produce physicians with the will to do good, i.e., be benevolent?

That would depend upon the extent to which we believed the actor was capable of using his or her own intuitions to produce good results, as well as on how important we think good results are to right acts. We live in a world where humans are fallible creatures. It is well recognized that the will to do good is not necessarily sufficient to actually produce good results, let alone right results. Even in a utilitarian world where one determines rightness by consequences, the problem remains. If the actor is sufficiently fallible, then he or she may rather routinely produce less good results or even harmful results by acting on naked good will than by using some rules or guidelines for right conduct. It may turn out that better consequences can be obtained by some other strategy than that of promoting benevolence.

Instilling in an actor (such as a physician) a strong and persistent determination to do only the good can have detrimental effects as well as good ones even if producing the good is the only objectives. It can lead to overconfidence, a sense of hubris, a disinclination to submit to peer or public monitoring, and, if others are convinced of the virtuous benevolence of such a person, lax regulation of conduct. Some of the more serious ethical problems in human subjects research, for example, have arisen from researchers who were (correctly) very confident that their only intention was to do good. Such persons are, in fact, often offended when an institutional review board dares even to look at their work. Naked virtue can produce a messianic complex in those who are not necessarily messiahs.

Realistically, it seems to me an open question whether instilling benevolence will produce beneficent acts. While there are serious problems of the sort just described, there are tendencies in the other direction as well. It is hoped that in at least some cases benevolence can increase rather than decrease the probability of beneficent act. It is by now well recognized that the *moral* virtues (character traits such as benevolence, veracity, fidelity, and the virtue of justice) correlate with right-making characteristics (the principles of beneficence, truth-telling, promise-keeping, and justice, respectively). If the naked virtue problem were the only one, we might be able to view a specific virtue as instrumentally useful in producing the right conduct associated with it: in the example we have been discussing, we might hope that benevolence would tend to produce beneficient acts, though given the potential for slippage between intention and actual production of right conduct, we should be concerned even here.

Matters get much complex in a world where there is more than one possible principle of right conduct and more than one associated moral virtue. Deontologists are particularly likely to opt for multiple principles (for example, W. D. Ross [13], James Childress [3], and my own work [17]), but rule utilitarians can as well (for example, Beauchamp [4]). The most one can realistically expect of a virtue is that it produce conduct with its correlative right-making characteristics.

A virtue theory will tend to produce wrong conduct when the moral virtues it includes are the wrong ones. Given the enormous variety of virtues and the lack of any systematic methods of resolving conflicts over which moral virtues should be included in a proper virtue set together with the earlier problem of slippage between virtuous intention and doing the right thing, the probability of right conduct resulting from a general promotion of virtuous character is not great. Looked at from the subjective perspective of a lay person in a health care setting, the probability of a health professional doing what the patient considers to be correct, simply because the professional has developed a virtue-centered approach to ethics, is infinitesimal. Both lay person and professional have an enormous range of virtues from which to choose. Getting them to pair up in a random pairing in a secular, modern medical facility is very unlikely.

Consider, for example, a physician whose theory of virtue led him to will that he would always benefit the patient. Even assuming that he could properly calculate the good and even produce good results, he might well still systematically produce the wrong act. He might, for example, benefit the patient by violating her autonomy. This can and has been done by benevolent physicians for centuries when they withhold information they consider disturbing from a terminally ill patient or a research subject. He might, for another example, benefit the patient by distributing resources in an unfair manner. In either case, the virtue of benevolence would have led to the wrong act even if the physician could correctly calculate the act that would produce the best consequences for the patients.

This might be adjusted by deriving a theory of virtue from a theory of action – the sort of move that Lawrence McCullough, in his essay in this volume, attributes to John Gregory in eighteenth-century Scotland. In that case, an action theory that included the principles of autonomy and justice would also have to possess the correlative virtues of respect for persons and the virtue of justice. As long as the virtues simply mirrored the principles of action, presumably right acts would result. They would result, that is, unless the emphasis on virtue so freed the actor from the constraints of the norms and rules of right conduct that the problem of the gap between willing the right and doing the right emerged once again just as it did with the virtue of benevolence.

Naked virtue together with the wrong virtues may well lead to wrong acts even though the intentions of the actor may be well-meaning. In such a case we might have to say that the actor did the wrong thing from a good motive or that the act was wrong, though of praiseworthy motive. To the extent that a robust virtue theory frees up actors to 'sin bravely', to act on the conviction that their motives are good, it may actually facilitate wrong acts. To the extent that emphasis on a free-standing virtue theory dissociates the actor from the proper principle of right conduct, a virtue theory is likely to lead to wrong conduct. This problem may be dealt with by forcing a theory of the virtues to be simply derivative from one's formulation of the principles of right conduct, but one doubts that that is what those urging a new emphasis on the primacy of the virtues have in mind. Barring such a move that makes the virtues derivative, one is led to reject proposition (3) in favor of the pair of conclusions that promoting virtuous character risks facilitating wrong acts, while at the same time if the naked virtue problem can be overcome and the correct virtues are identified, promoting those virtues tends to promote right acts.

That Virtue Theory Is Unnecessary in Stranger Medicine

Thus far, I have established that it will be extremely difficult to come up with a set of virtues and especially difficult to come up with a set for a particular profession. Moreover, if we are able to come up with such a set, it will not necessarily lead to right conduct, and, in fact, might, if it gives the actor a false sense of moral confidence, actually wrong conduct even if the set of virtues correlated with acceptable principles of right conduct are selected. At most, a set of virtues might be instrumental in promoting right acts. That leads us to the question of whether we need a virtue theory in medical ethics. Virtue theory may turn out to be extremely difficult to establish and dangerous, but nevertheless necessary for a full theory of moral action in medicine. In this section, we need to face this question head-on.

In order to tackle this question, I need to make a distinction between two very different kinds of medicine I shall refer to as 'stranger medicine' and 'communal or sectarian medicine'. I introduced the notion of stranger medicine in an earlier volume in this series [18]. I mean by it medicine that is practiced among people who are essentially strangers. It would include medicine that is practiced on an emergency basis in emergency rooms in large cities. It would also include care delivered in a clinic setting or in an HMO that does not have physician continuity, most medicine in student health services, VA hospital, care from consulting specialists, and the medicine in the military as well as care that is delivered by private practice general practitioners to patients who are mobile enough not to establish long-term relationships with their physicians. It is, for the most part, the health care of contemporary, urban America. This is to be contrasted with medicine that is delivered with a community – a *Gemeinschaft.*[3] In such a community, the health care provider and the lay person would have a relationship analogous to that of friendship or fellow members in a common cause. There would be a personal knowledge of and sharing of beliefs and values. It might be the medicine of what is now little more than the romantic image of the small town physician, but more importantly it would be the medicine that is practiced within sectarian communities that favor unorthodox medical systems, such as that of Christian Scientist practitioners or those within a feminist health collective or those getting their health care at Oral Roberts Hospital in Tulsa. All of these share the characteristic that lay persons and health providers share a common sense of beliefs and values upon which health care decisions might be made. In a real sense, they are part of a common community.

I think it can be argued that in the world of stranger medicine, what we are really concerned about in ethics is right conduct, while within a *Gemeinschaft*, we may be much more directly concerned about personal character traits, motivations, and virtues.

The question can be put quite simply: when we enter an urban hospital in a strange city, are we more concerned about having the physician do the right thing or act with the right motive? I think it can be reasonably said that in the world of strangers, we are much more concerned about conduct then virtuous character. If we could be assured that the physician would do the right thing, we would not really be concerned about motivation. At best, a concern about virtuous character is a concern that virtue will be instrumental in producing right conduct. Right conduct in such a context will, of course, include treating the patient with what appears to be courtesy. It will include communicating with the patient adequately. Virtuous character in the world of stranger medicine is at best a luxury and at worst a deterrent to right conduct. The problem is particularly acute in the world of strangers since there is no reasonable basis for assuming that the stranger with whom one is randomly paired in the emergency room will hold the same theory of virtue as oneself.

Critics of the position I am arguing here might be quick to point out that it may be just about as hard to come up with agreement about principles of right conduct as it is about the virtues of the various lay and professional roles in medicine. This may be true. Actually, I doubt it. The literature in action theory seems to have substantially more convergence than the literature in virtue theory. That is really not the point, however. Even if agreement in the two areas were equally hard, the need for agreement would not be the same. When one enters the world of stranger medicine – a big city hospital or the office of a specialist to whom one has been referred – it is crucial that there be some mutual understanding of what conduct is to be expected. A common theory of the principles or at least agreement on the rules of right conduct is essential before two human beings who are strangers can interact with each other. The same is not true regarding character traits. Patients can receive decent health care from health professionals whose idea of the virtuous life they do not share *provide the behavior of the professional conforms to the principles and rules of right action*.

Consider the plight of a deeply committed Methodist woman who is on the board of trustees of a secular city hospital. Suppose that two physicians who were for all practical purposes of equal technical competence were being considered for a position. One of the pair had a personal character that was aggressive, achievement-oriented, eager for rapid advancement and personal gain. Being a shrewd individual, however, she calculated that the best way to advance her personal objectives was rigorously to engage in conduct that was impeccable by some commonly agreed upon standard. She always got adequately informed consents, demonstrated what appeared to be courteous behavior, told the truth to her patients regardless of personal cost to herself, etc. She has a long track record that demonstrated a persistent ability to do the right thing time and time again.

The second of the pair had a radically different character. He was a Methodist who had absorbed the Christian virtues of love, faith, and hope as well as the beatitudes as much as any human being could, but was, perhaps in part because of this set of character traits, not as clever at calculating the rightness of his conduct. He was generally right in his conduct, but, upon occasion, admitted he erred, a fact documented by occasional disciplinary proceedings against him.

Which of the two physicians should the member of the board of trustees hire? I think it is clear that in the meeting of the board of the secular hospital she should opt for the first physician. Her job is to assure right conduct. That includes courteous, respectful behavior but not necessarily any particular set of virtues. In fact, were the trustee to opt for the second physician, she could justifiably be accused of imposing a particular religious conception of the virtues on secular patients, some of whom may in fact prefer more Homeric, Machiavellian, or Nietzschean views of the virtues.

The point here is not that the trustee should be satisfied with mere technical competence; far from it. Rather, in the world of stranger medicine, she has a primary concern with predictable morally correct behavior. Forcing agreement on virtues is unnecessary and virtually impossible to achieve. Forcing agreement on some minimal set of behavioral standards may be virtually impossible to achieve, but it is necessary if patients and professionals in the world of stranger medicine are to know what to expect of each other. Since coming up with a theory of virtue agreed upon by all is going to be extremely difficult, may further wrong conduct, and is not really necessary for the medical encounter in the first place, it is hard to see why virtue theory in the world of strangers should be given any prominence in medical ethics.

VIRTUE THEORY IN SECTARIAN MEDICINE

If virtue theory is to play such a minor role in secular medicine practiced among strangers, what is its role in medicine if it were to be practiced within a community among those who share a common set of beliefs and values? I have argued that it will be extremely difficult to come up with a universally agreed upon set of virtues, even for those in one particular role, such as that of physician or patient. In fact, there may be no definitive set of virtues independent of a more sectarian communities of beliefs and values in which all the actors share a common set of commitments.[4]

I have argued that within the world of stranger medicine, it is necessary to have a common set of action principles or rules of conduct simply to be able to transact medical business among strangers. Virtue theory, on the other hand, in the world of stranger medicine, is certainly not necessary and may actually be harmful.

If medicine were to be practiced within a sectarian community, however, there might be a much more prominant place for an agreed upon set of virtues. I have argued that in the world of strangers, one would prefer that one's physician engage in the right act for a suspect motive than engage in the wrong act for the right motive. Within a community, however, that is not obviously the case. It might be, for example, that within a family one would prefer that one's child engage in the wrong act for the right motive rather than the right act for the wrong motive. Within such a community, one is intrinsically interested in promoting good character according to some community-accepted standard of good character as well as or even rather than right conduct. There may be limits on such a priority for good character. For example, I would probably rather that my son refrain murder solely out of fear that he would be punished rather than have him commit murder for benevolent, but incorrectly reasoned, motivation. But, within certain limits, I would quite possibly prefer right motive to right conduct. Just the opposite is likely to be true in the world of strangers. From the man selling newspapers on the corner, I would rather have the right change for the wrong motive than the wrong change for benevolent motive.

An extension of our previous example is enlightening. Let us suppose that the same Methodist woman who selected the right-acting, poorly motivated physician for her secular hospital was also on the board of turstees of a Methodist hospital that had recently undergone a spiritual renewal so that the beliefs of Methodism were taken much more seriously than would normally be the case in nominally Methodist health care facilities. If she also had to choose between the two equally competent physicians for a position in her Methodist facility she would have every right, perhaps even an obligation, to opt for the physician whose virtues were much more in tune with those of the sponsorship of the hospital even it if meant occasional slips in right conduct.

If virtues seems to be highly dependent upon the religious or philosophical tradition in which one finds oneself and if virtues are much more important when one stands within a community that reflects such a tradition, then it would seem to follow that virtue theory would be much more important for a medical ethic within such a tradition. Thus, one might expect that a Catholic medical ethic would play out the Thomistic mix of Christian and Greek virtues that is so important to the theology of that tradition. Likewise, a Protestant medical ethic might elevate to highest priority the medical implications of the Christian virtues of faith, hope, and love.

In the same vein, other sectarian communities, such as the feminist health collective, might find it extremely important not only to determine what constitutes right conduct for patients and professionals within the feminist health care system, but to go on to ask what traits of character are crucial in their particular perspective on the world. Just as Percival's gentleman emphasized a proper condescension in the responsibility that is borne for those of lesser status, so the physician employed by a feminist health collective might be selected because she gave special attention to the virtues of respect and humaneness. In fact, it probably makes sense to say that planners of a feminist health care system would insist on competence and morally correct conduct in addition to the unique character traits that are considered important within that community, but, if forced, would sacrifice some of the former to gain the latter. Likewise, a hospital sponsored by a religious community might insist that certain virtuous character traits be present in its staff when the same insistence on idiosyncratic virtues would be totally out of place in a secular hospital setting. A secular hospital has every right to demand right conduct by its physicians. It is not at all clear that is has the right to expect particular traits of character.

Where does this leave a deontologist reflecting on virtue theory in medical ethics? The conclusion seems clear. Agreement on a set of virtues as universally acceptable traits of character will be extremely difficult both because different cultural groups seem to have radically different conceptions of the virtuous life and because different character traits seem to need different emphasis for different roles. Even for single roles within the health care system – those of physician or patient, for example – it will be extremely hard to get agreement on the ideal virtues.

Not only that, it is not clear whether promoting the virtuous life will lead to right conduct. A deontologist normally acknowledges many right-making characteristics of acts. Unless the list of moral virtues is simply read off the list of right-making principles of action, the chances of selecting the right virtues to produce right conduct is small. Moreover, even if the right virtue is selected, it is not clear that promoting it will promote right conduct.

Fortunately, in the world of stranger medicine, agreement on the virtues may not be necessary. What we are really interested in, so I have argued, is right conduct. At most, virtues are instrumental in achieving right behavior.

Rather, it is in the world of sectarian community medicine where groups are organized around commonly held beliefs and values – around traditions – that the virtues become critically important. Fortunately, it is in that world that agreement on the content of the virtues will be much more easily achieved. The task now is one of encouraging the reestablishment of sectarian medicine in which professionals and patients alike share a common tradition, a common set of beliefs and values, and a common set of virtues.

NOTES

[1] It is significant, for example, that in Karen Lebacqz's creative, activist image of the three virtues are from the Greek cardinal virtues while only one is from the list of Christian virtues.

[2] The distinction I am making is now standard in the literature of philosophical ethics. See Ross ([13], p. 7); Brandt ([6], pp. 465–471]); Beauchamp ([4], p. 150). Following Ross I use the term *act* to refer to the thing done and the term *action* to refer to the motivation.

In order to make a deontological critique of virtue theory clear, it is necessary to insist on this definition. The ambiguous, sometimes expansive, use of the Greek term often translated as 'excellence' permits virtue theory to slop over into the realm of action theory as well as disposition or character. One can refer to an excellence as doing a practice well in addition to having a disposition to do it well. Thus, some of the papers, often appealing to the Aristotelian tradition, refer to virtues as if they were inherently linked to right conduct. Pellegrino, for instance, properly quotes Aristotle describing virtue as that state of character which "makes a man good and which makes him do his own work well" (Pellegrino citing Aristotle's *Nichomachean Ethics*, II:6, 1106a, 22–24). Elsewhere, he speaks of a virtuous person as "someone we can trust to act habitually in a 'good' way" and "someone who will act well." At one point, he refers to virtue as implying "balance between noble intention and just action".

It seems clear that this broader usage is among those accepted in contemporary discussion of virtue, but using the term this broadly leaves no room for a deontological critique of virtue theory. If virtue includes right conduct as well as proper disposition, then deontology becomes a part of virtue theory. In order to avoid this confusion, I shall use virtue in a more restricted sense. There is support in Aristotle for this when he says that "the virtues cannot be capacities, either, for we are neither called good or bad nor praised or blamed simply because we are capable of being affected" (*Nichomachean Ethics*, II:5, 1106a).

I shall consistently use virtue in this narrow sense of the disposition or persistent inclination to act in a particular way rather than the actual capacity to act in that way. I have also refrained from adopting Pellegrino's and Lappé's notion that a virtue ethic is a 'higher' ethic or one that makes supererogation more essential. I would see the questions of rigor and supererogation as applying to both ethics of duty and ethics of virtue. One could apply a supererogatory standard in evaluating what behavior is called for as well as what virtues are appropriate. In fact, as we shall see, in some medical environments it might be appropriate to apply a rigorous standard of conduct (duty) and a lax standard of virtue (character).

[3] I take the term from Tonnies [15], but I use it in a slightly broader sense. While he limited community to those more intimate, face-to-face interactions of the kind found in a small group, I include all social groupings around commonly shared systems of belief, value, and moral commitment. I use Troeltsch's term *sect* and *sectarian* metaphorically to refer to medicine practiced within such a community. When I do so, however, I have in mind health care that might include some delivered by institutions that have associational characteristics, by what he would refer to as institutions that have the characteristics more of churches rather than sects.

[4] MacIntyre [9] stresses that images of the virtuous life are to be found within traditions, a point emphasized nicely by Neilsen in this volume. In fact, as Pellegrino noted, MacIntyre has linked virtue to local communities, something akin to what I am referring to as sectarian communities. Neilsen is asking too much of MacIntyre if he expects him to come up with a universally acceptable characterization of the good life and the virtues which constitute it.

BIBLIOGRAPHY

[1] American Medical Association: 1848, *Code of Medical Ethics: Adopted by the American Medical Association at Philadelphia, May, 1847, and by the New York Academy of Medicine in October, 1847*, H. Ludwig and Company, New York.

[2] American Medical Association: 1981, *Current Opinions of the Judicial Council of the American Medical Association*, American Medical Association, Chicago.

[3] Beauchamp, T.: 1982, *Philosophical Ethics: An Introduction to Moral Philosophy*, McGraw-Hill Book Company, New York.

[4] Beauchamp, T. and Childress, J.: 1983, *Principles of Biomedical Ethics*, Second edition, Oxford University Press, Oxford.

[5] Bowers, J.: 1970, *Western Medical Pioneers in Feudal Japan*, the Johns Hopkins Press, Baltimore.

[6] Brandt, R.: 1959, *Ethical Theory: The Problems of Normative and Critical Ethics*. Prentice-Hall, Inc., Englewood Cliffs.

[7] Hauerwas, S.: 1981, *Vision and Virtue*, University of Notre Dame Press, Notre Dame.

[8] Jonsen, A. and Hellegers, A.: 1974, 'Conceptual Foundations for an Ethics of Medical Care', in L. Tancredi (ed.), *Ethics of Health Care*, National Academy of Sciences, Washington, D.C., pp. 3–20.

[9] MacIntyre, A.: 1981, *After Virtue*, University of Notre Dame Press, Notre Dame.
[10] May, W.: forthcoming, 'The Virutes in a Professional Setting', *Soundings*.
[11] Percival, T.: 1927, *Percival's Medical Ethics, 1803*, C. Leake (ed.), Williams and Wilkins, Baltimore.
[12] Rahman, A., Amine, C., and Elkadi, A.: 1981, 'Islamic Code of Medical Professional Ethics', in *Papers Presented to the First International Conference on Islamic Medicine Celebrating the Advent of the Fifteenth Century Hijri*, I. El-Sayyad (ed.), Kuwait Ministry of Public Health, Kuwait, pp. 386–392.
[13] Ross, W.: 1939, *The Right and the Good*, Oxford University Press, Oxford.
[14] Temkin, O. and Temkin, C. (eds.): 1967, 'The Hippocratic Oath: Text, Translation and Interpretation', in *Ancient Medicine: Selected Papers of Ludwig Edelstein*, The Johns Hopkins Press, Baltimore, pp. 3–64.
[15] Tonnies, F.: 1957, *Community and Society*, trans. and ed. by Charles P. Loomis, Michigan State University Press, East Lansing.
[16] Unschuld, P.: 1978, 'Confucianism', *Encyclopedia of Bioethics*, W. Reich (ed.), The Free Press, New York, pp. 200–204.
[17] Veatch, R.: 1981, *A Theory of Medical Ethics*, Basic Books, New York.
[18] Veatch, R.: 1983, 'The Physician as Stranger: The Ethics of the Anonymous Patient-Physician Relationship', in *The Clinical Encounter: The Moral Fabric of the Patient-Physician Relationship*, E. Shelp (ed.), D. Reidel Publishing Company, Dordrecht, pp. 187–207.

Kennedy Institute of Ethics,
Georgetown University,
Washington, D.C.,
U.S.A.

STANLEY M. HAUERWAS

ON MEDICINE AND VIRTUE: A RESPONSE

Why virtue? Why virtue in relation to medicine? For many, medical ethics already seems complex enough, so why muddy the water by introducing the subject of virtue? It is no secret that virtue has not been a popular topic in recent philosophical and theological ethics. Therefore, it would seem to be the better part of wisdom not to introduce the subject of virtue into the discussion of medical ethics, as we stand the chance of becoming even more lost in the intricacies of ethical theory and as a result of throwing little light on the subject of medicine.

Yet I think this volume has rightly raised the issue of virtue and its relation to medicine. For as the historical essays make clear, the notion of virtue has been a crucial element in the ethics of the past. Indeed, since health was assumed to be a virtue, there were no hard distinctions between medicine and ethics. Moreover, it would seem that no account of the ethics of medicine as practiced today is adequate unless the virtues are given their due. In fact, it may be that one of the difficulties with the development of medical ethics has been the attempt to give an account of the ethics of medicine, with no or little regard for the significance of the virtues.

To put the matter more strongly the emphasis on the virtues may be the first indication that moral philosophy and theology are finally beginning to learn from their recent encounter with medicine. For on the whole the development of medical ethics over the last twenty years has done little to change the way moral philosophers work. Philosophers and theologians when they turned their attention to medicine did not change their assumptions about how either should be done, they simply changed their examples. We're taught that ethical decisions should conform to teleological or deontological normative theories, that autonomy is an overriding ethical issue, and that ethics primarily consists in responding to quandaries of this or that sort. Physicians were often satisfied with this result, as it seemed to bring some coherence to the bewildering challenges with which they felt themselves faced. Like a patient that is relieved to be told he has cancer, physicians were relieved by being able to name their ethical problem, even though that naming did little to help them understand, much less to do anything about, their moral unease.

Earl E. Shelp (ed.), Virtue and Medicine, 347–355.

Yet what physicians may well have overlooked is that their own best moral sense had been distorted, or at least not adequately articulated, by an ethic devoid of virtue. Medical ethics primarily became a discussion of cases but such discussions never seem to get anywhere in terms of helping or understanding morally what medicine is about. Since medical care so often deals with cases this kind of ethics seemed to make a good deal of sense. For a decision oriented ethic seems to underwrite the physician's impatience with theory when someone's life is on the line. Yet still something seemed to be missing in this account of medical ethics that is essential to the moral presumptions necessary to sustain the activity of medicine.

It would be a mistake, I think, to say that an account of the virtues is what has been missing. For I do not think that the emphasis on the virtues can or ought to be but a supplement to the kind of medical ethics that has largely ignored the virtues for the moral life. To reclaim the significance of the virtues is not first of all to offer a supplement for medical ethics, but a diagnosis of what is wrong with contemporary ethical theory and, in particular, what is wrong with the way medical ethics has been developed.

What must be recognized is that the kind of ethics that philosophers and theologians brought to the task of developing a medical ethics was an ethic that reflected a whole set of social institutions and assumptions. It was an ethic that assumed that we no longer suffer or benefit, depending on one's point of view, from living in societies that have a good in common. All our societies can assume we share in common is that each individual desires to delay as long as possible death. Ethics for such a society consists in an attempt to find rational means to settle disputes and conflicts that inevitably occur between autonomous units of desire. The important question for ethics is, 'What ought I to do?' not 'What ought I (much less we) to be?' The latter question inextricably requires an account of the good that modern ethical theory assumes is unanswerable in a public manner. Medical ethics, therefore, was made to conform to this general set of presuppositions.

Yet it did so with a good deal of resistance. Physicians continued to complain that their relation to the patient was more complex than the strict alternatives between paternalism and autonomy allowed. Moreover, the very nature of medicine as a profession did not seem adequately accounted for on the model of ethics that assumed the virtues at best were secondary to an account of obligation. For as Leon Kass suggests:

Being a professional is rooted in our moral nature and in that which warrants and implements making public professions or avowals of devotion to a way of life. It is

> a matter not only of the mind and hand but also of the heart, not only of intellect and skill but also of character. For it is only as a member of a community and as a being willing and able to devote himself to others and to serve some higher good that a man makes a public confession of his way of life. To profess is an ethical act, and it makes the professional qua professional a moral being, who prospectively affirms also the moral nature of his activity ([1], p. 1370).[1]

It is not surprising, therefore, that those within medicine and some ethicists who are trying to understand the ethics of medicine should be led back to a reconsideration of the virtues. For in a sense medicine continues to represent a community and set of activities that institutionalize the prerequisities for an ethic of the virtues. It does so because physician and patient alike are participants in a common endeavor not simply to preserve life but to maintain goods necessary for the community as a whole. For again, as Kass reminds us, as patients we are

> willing to expose ourselves – not only our bodies, but often certain intimate details of our lives – to this relative stranger, first, of course, because he has professed his ability and willingness to help (and we believe him), but also because we expect that he understands the meaning of our presenting ourselves for his examination and counsel. The physician, though he is healthy, shares the patient's awareness of vulnerability and mortality, not only because the *techne* of medicine rubs his nose in the multifarious fragilities of the flesh, but also because of his own humanity. The physician too is capable of self-consciousness, and recognizes himself in the other. He sees not only the one-sided relationship of benefactor and recipient appropriate to the patient's appeal for knowledgeable advice, but also the equal relationship grounded in their *mutual* and equal participation in the ever-precarious, necessarily finite, yet daringly aspiring and hope-filled venture called human life. Although he must keep his private feelings to himself, it is his own silent self-awareness that enables the physician to synthesize the abstract scientific knowledge of bodily workings with the concrete experience of living in health and in sickness. As a result, he acquires the kind of understanding that permits him not only to fight disease but also by helpful speech to mediate between the patient's understanding and its own mysterious and silent body ([2]. p. 1309).

Patient and physician, therefore, form a common community as they pursue goods of the whole community, some of which require certain virtues, and some being the virtues themselves.

Rather than being but another occasion to apply ethical theory derived from other contexts, medicine in respect to the virtues becomes a resource to challenge as well as reconstitute how we think about the moral life. For as many of the essays in this volume make clear, medicine, from the standpoint of both the physician and patient, presupposes the virtues. In short, there has been more moral substance in the practice of medicine than our moral theory has been able to articulate. Therefore, by turning our attention to the virtues we attempt to enrich not only our understanding of medicine but our moral theory.

But there is just the rub. For as I noted, philosophers and theologians have paid little attention to the virtues over the past few centuries. As a result it seems, at least judging by many of the essays in this book, almost everything remains to be done. For there simply is no agreement about how best to answer some of the most basic questions raised by introducing the language of virtue into ethical analysis. Yet I think this book has helped at least to begin the task by locating those questions, as well as proposing some extremely interesting suggestions as how they might be answered. In the interest of furthering the discussion begun here, therefore, it might be helpful to discuss what I take to be the agenda of questions we must continue to discuss if, in fact, medical ethics is to be further illuminated by recourse to the language of virtue.

The basic question, of course, remains, 'What is virtue?' Even that may be putting the matter wrongly, as we might better ask, 'What is a virtue?' The former question may be misleading as it assumes a far too unitary account of the virtues or that the virtues all partake of a similar nature. The latter assumption may well be justified if virtue is understood in the most general sense as an excellence for the perfection of a potential that increases the poser of its subject, but such general definitions do little to help us understand the nature of the virtues and their place in our lives. The resolution of this issue obviously involves the question to be discussed below of how the virtues are individuated as well as related.

A virtue, however, clearly seems to entail some decisive relation to the self, but the subject of the virtue remains anything but clear. It seems unobjectionable to say that a virtue is a disposition, but then not every disposition is a virtue, so a virtue cannot be simply a disposition or set of dispositions. The language of habit seems to be of some help in this respect as a virtue is a disposition we have by habit, that is, in a manner that it is not easily lost. Yet it makes sense for us to say that we discover some virtues that we did not know we had. In that sense some virtues seem to be closer to temperament than habit.

Moreover, some virtues may be inherent in a set of practices that are in a sense not in the self, at least not in the manner that the language of disposition suggests. Of course a person who performs such practices may discover that the self has been formed virtuously, but then it must be asked if they genuinely are virtuous. For example, they may be considerate only when they are in certain roles, etc. This issue involves the further question concerning the acquisition of the virtues, for example, is it possible for a person to be virtuous without having all the virtues? Can a person learn to be trustworthy in certain roles without also being courageous, truthful, and so on? Or put differently, does acquiring any one virtue entail opening the self to formation by the other virtues or at least some of the more important virtues? But then how are we to distinguish those that are important from those that are not?

Put concretely, an essentially selfish person may well become a physician or nurse. Yet inherent in the practice of medicine is a regard for others' welfare inconsistent with the vice of selfishness (assuming that most believe that selfishness is a vice). Does an account of the virtues require that the virtues entailed by the practice of medicine be carried over to other aspects of the physician's life? Of course such a carry-over may occur, but must it occur? Moreover, does the movement from selfishness to generosity require the acquisition of a whole range of virtues necessary to sustain the habit of generosity? In some ways it would seem that medicine is a particularly bad example for the life of virtue, as the practice of medicine may entail the overemphasis of some virtues at the expense of others. Thus the medical practitioner may exhibit a one-sidedness that we think incompatible with a well-lived life.

Connected with these issues is the troubling question of whether a virtue can only be the qualification of an individual. Though Bernard Gert maintains such is the case, Aristotle steadfastly argued that friendship is a virtue and yet friendship is a relation. Friendship is not simply a quality in each friend, though it entails certain qualities, but friendship itself is a quality that could not exist apart from the relation. This issue is particularly important for questions of medicine, since it may be that the virtues appropriate to the practice of medicine are more like friendship according to Aristotle's understanding of it. Exploration of this issue might well help us understand better why physician and patient are morally dependent on one another.[2]

Yet perhaps what these questions reveal is that our original question, 'What is virtue?,' may not be as wrong-headed as it first appears. If, as Aristotle maintained, we can only acquire the virtues as a person of virtue would, then perhaps an account of the virtues presupposes some more fundamental quality of the self. That is, perhaps, a virtue can only be a virtue if it is possessed as a person of character would possess it. But then what does such character involve? Can it be acquired all at once or is it a process of growth in the virtues? Can we have character and yet not have all the virtues? Or can we exhibit character in some of our activities and yet seem to be decisively deficient in others?

Obviously connected to but distinguishible from these questions is the issue of how the virtues are individuated. Pincoffs has reminded us that the list of virtues is long indeed and they do not all seem to have the same significance. Yet how are we to determine the individual virtues and their relative importance? Surely one tempting way is to distinguish the moral virtues from those that are amoral, but on reflection such a distinction seems much easier to make in the abstract than in the concrete. Certainly all attempts to suggest that only those virtues that involve our relations with others are moral is doomed to failure. For on reflection almost any virtue gains part of it significance in reference to our relations to others.

Surely one of the strengths, or what some regard as the weakness, of attending to the virtues as the distinction between etiquette and ethics, manners and morality, is rendered problematic. For virtues necessarily pervade all aspects of our activity, as otherwise they do not have the character of being genuine virtues. Therefore, traditional concern with patient etiquette may well have more moral significance than it is generally given credit.

Yet even if one refrains from pressing a hard and fast distinction between moral and non-moral virtues, the question still remains how the virtues are to be individuated as well as understood. For what unselfishness means in one community may be quite different in another. Are there some virtues that are invariant to the human condition? Justice, courage, truthfulness seem to be ready candidates for claims of universality, but all attempts to specify the nature of justice, etc., seem inevitably to require a particularly that cannot be justified by human nature in general. Moreover, are there virtues peculiar to the practice of medicine and if so, what are they?

Many of the authors in this book have responded to MacIntyre's suggestion that the virtues are correlatives of practices [(2), pp. 177–189]. The latter are similar to Aristotle's understanding of activity, as they are done as goods in themselves needing no justification beyond themselves. Such practices are virtue producing exactly because the self or character is formed through our involvement in them. They are in a fundamental sense self-involving. It is MacIntyre's contention that many of our virtues can be individuated by attending to such practices. Medicine, on such an account, seems a particularly fruitful area of analysis since, if as MacIntyre contents, the virtues still exist, they do so exactly in those areas of our lives that most nearly approximate a practice.

Yet MacIntyre also argues that a full account of the virtues requires more than the analysis of such practices. For the kind of character, or constancy, necessary for the acquisition of the virtues cannot be derived from the individual practices. Rather a narrative is required that places the self in relation to a good to which the virtues are intrinsic. That is, the good sought is constantly transformed, or our understanding of the good is transformed, by the virtues we acquire along the way. Therefore, the practices are finally dependent on their role in such a narrative.[3] We may, for example, learn to be courageous through military training, but such courage is not complete until it is given a role within a larger story. Therefore, the way we learn to individuate the virtues is finally dependent on such a narrative account that is shared by a people through an ongoing tradition.[4]

But what if such a narrative or tradition is missing? Indeed, that is the conclusion of many of the essays in this volume. Does that mean we must seek to develop our ethics without recourse to the language of the virtues? But what if medicine in particular, as I suggested above, requires an account of the virtues? Or as Erde suggests, what if the virtues that once sustained medicine in fact become vices when devoid of their original social setting? Do we ask too much of medicine if we ask it to embody an ethic of virtue in the absence of societal support for such a commitment? Or in fact is such support there, only overlooked by the insufficiency of our current conceptual machinery in ethics?

Inherent in the question of how the virtues are individuted is also the question of whether the virtues conflict. It was Plato's assumption, unhappily continued by Aristotle, that the virtues if rightly understood and acquired were one, or at least according to Aristotle could not conflict. However, such an assumption is by no means self-evident. Indeed, I suspect that is one of the reasons that MacIntyre sees so clearly the need for a narrative, as in the absence of any automatic harmony of the virtues a tradition is required to help us assess the virtues' relative significance and appropriateness.

This is an issue I think particularly important for medical ethics, since too often the problem is not that we are doing the morally wrong thing in medicine, but that we are not allowing for the possibility of other competing virtues. What is required is the development of an ethos sufficient to support the often conflicting virtues inherent in the practice of medicine. If we lack such an ethos, then it may well be that we are in deeper trouble than we had anticipated if we are to learn to practice medicine morally. In other words, it is not sufficient simply to recognize that medicine as an activity entails certain virtues, but that those virtues require still other virtues necessary to sustain the activity of medicine itself. Such virtues must obviously be carried by a community that protects medicine from being perverted by goals not intrinsic to its task. Whether such a community exists is, I suspect, one of the major issues confronting contemporary medical ethics.

These are the kind of questions we must discuss if this turn to the virtues for medical ethics is to be fruitful. Some may well feel that such a discussion, rather than confirming the usefulness of the language of virtue for medical ethics, has exactly the opposite effect. For the above agenda seems to present insoluble problems. Indeed, I have not yet even dealt with questions of how virtues are acquired or justified. Moreover, the issues all seem interrelated in a manner that makes one unsure where one can even begin to unravel some of the complexities involved in a turn toward the virtues. Would it not be better to go back to questions of what we ought to do in *X* or *Y* quandary?

Many will certainly think such is the case. However, I hope many will see the agenda developed by this volume not as a problem but as an exciting opportunity. For I hope by displaying some of the issues raised by the essays in this book that we will confirm that this way of approaching medical ethics offers a fresh and enlivening perspective on medicine as a moral endeavor. If that much has been done, then we are deeply in debt to the authors and editor of this book.

NOTES

[1] Ironically, because this aspect of medicine is so often ignored by those doing medical ethics, the moral nature of the training that medical students receive is overlooked or misdescribed. For medical training may in many ways be one of the last schools of virtue we have left in our society. Of course, the language and form that training takes is not explicitly moral, but that does not make it any less morally demanding. The young physician, without noticing it, is trained in virtues of honesty, unselfishness, faithfulness necessary for good patient care. I suspect part of the problem with the moral nature of medical training is not, as is so often alleged, that it is amoral, but that we have failed to acknowledge and articulate the kind of moral training involved. As a result, the ethos of medicine and medical training is distorted as we misdescribe what medicine morally entails. One of the difficulties of introducing ethics courses into medical school curricula is that it gives the impression that medical training is otherwise morally neutral. The first task of a course in medical ethics therefore is to try to help the medical student become aware of the morality, the virtues, that they are acquiring through their initiation into the profession of medicine.

Lest I be misunderstood, I am not suggesting the morality currently taught in medical schools is without difficulties or that reform is not needed. Rather I am suggesting that such reform has little chance of success as long as we fail to understand how medical schools do not so much lack a morality but rather are swimming in a morality. I think it is the case, however, that physicians as well as their critics often fail to notice the morally substantive connection and virtues inherent in the practice of medicine.

[2] For the best treatment of this issue see May [3].

[3] MacIntyre maintains "that unless there is a *telos* which transcends the limited goods of practices by constituting the good of a whole human life, the good of a human life conceived as a unity, it will *both* be the case that a certain subversive arbitrariness will invade the moral life *and* that we shall be unable to specify the context of certain virtues adequately" ([2], p. 189).

[4] The implications of this for medicine are immense. For if MacIntyre is right that we can be genuinely courageous or truthful only on occasion, then it is not enough that a physician be only a 'good doctor', he or she must also be a 'good person' ([3], p. 185). That is, the physician must have character that enriches the pratice of medicine but is not limited to that practice. The scientific and technological character of modern medicine has obscured the importance of the character of the physician, but it remains no less important. Of course a technically proficient but morally shallow person may give competent medical care to many patients. But if character is crucial it means such a person will lack decisive moral prerequisites for the practice of medicine over a life-time. We have a sense, however, of our contemporary moral difficulties just to the extent we want to avoid making judgments of character for determining those who should and should not be granted the privilege of being called a doctor.

BIBLIOGRAPHY

[1] Kass, L.: 1983, 'Professing Ethically: On the Place of Ethics in Defining Medicine', *Journal of American Medical Association* **249** (10), 1305–1310.

[2] MacIntyre, A.: 1981, *After Virtue*, University of Notre Dame Press, Notre Dame, Indiana.

[3] May, W.: 1983, *The Physicians Covenant: Images of the Healer in Medical Ethics*, Westminister Press, Philadelphia.

The Divinity School,
Duke University,
Durham, North Carolina,
U.S.A.

NOTES ON CONTRIBUTORS

Darrel W. Amundsen, Ph.D., is Professor of Classics, Western Washington University, Bellingham, Washington.

Tom L. Beauchamp, Ph.D., is Professor of Philosophy, and Senior Research Scholar, Kennedy Institute of Ethics, Georgetown University, Washington, D.C.

Martin Benjamin, Ph.D., is Professor of Philosophy, Michigan State University, East Lansing, Michigan.

Joy Curtis, M.S., R.N., is Associate Professor, College of Nursing, Michigan State University, East Lansing, Michigan.

Allen R. Dyer, M.D., Ph.D., is Assistant Professor of Psychiatry and of Community and Family Medicine, Duke University Medical Center, Durham, North Carolina.

Edmund L. Erde, Ph.D., is Associate Professor, Department of Family Practice, University of Medicine and Denistry of New Jersey, Camden, New Jersey.

Gary B. Ferngren, Ph.D., is Professor, Department of History, Oregon State University, Corvallis, Oregon.

Bernard Gert, Ph.D., is Stone Professor of Intellectual and Moral Philosophy, Department of Philosophy, Dartmouth College, Hanover, New Hampshire.

Stanley M. Hauerwas, Ph.D., is Professor of Theological Ethics, Divinity School, Duke University, Durham, North Carolina.

Marc Lappé, Ph.D., is Adjunct Associate Professor of Health Policy, Social and Administrative Health Sciences, University of California, School of Public Health, Berkeley.

Karen Lebacqz, Ph.D., is Professor of Christian Ethics, Pacific School of Religion, Berkeley, California.

Laurence B. McCullough, Ph.D., is Associate Director, Division of Health and Humanities, Department of Community and Family Medicine, School of Medicine, and Senior Research Scholar, Kennedy Institute of Ethics, Georgetown University, Washington, D.C.

Gilbert Meilaender, Ph.D., is Associate Professor, Department of Religion, Oberlin College, Oberlin, Ohio.

Kai Nielsen, Ph.D., is Professor of Philosophy, The University of Calgary, Alberta, Canada.

Edmund D. Pellegrino, M.D., is Director, Kennedy Institute of Ethics, and John Carroll Professor of Medicine and Medical Humanities, Georgetown University, Washington, D.C.

Edmund L. Pincoffs, Ph.D., is Professor of Philosophy, University of Texas, Austin, Texas.

Earl E. Shelp, Ph.D., is Associate Professor of Theology and Ethics, Institute of Religion, and Assistant Professor of Medical Ethics, Department of Community Medicine, and Member of the Center for Ethics, Medicine, and Public Issues, Baylor College of Medicine, Houston, Texas.

Robert M. Veatch, Ph.D., is Professor of Medical Ethics, Kennedy Institute of Ethics, Georgetown University, Washington, D.C.

Dietrich von Engelhardt, Dr. Phil., is Professor, Institut für Medizin – und Wissenschaftsgeschichte, Lübech, West Germany.

Marx W. Wartofsky, Ph.D., is Distinguished Professor of Philosophy, Baruch College, The City University of New York, New York, New York.

INDEX

The Philosophy and Medicine Book Series

Editors

H. Tristram Engelhardt, Jr. and Stuart F. Spicker

1. **Evaluation and Explanation in the Biomedical Sciences**
   1975, vi + 240 pp. ISBN 90-277-0553-4
2. **Philosophical Dimensions of the Neuro-Medical Sciences**
   1976, vi + 274 pp. ISBN 90-277-0672-7
3. **Philosophical Medical Ethics: Its Nature and Significance**
   1977, vi + 252 pp. ISBN 90-277-0772-3
4. **Mental Health: Philosophical Perspectives**
   1978, xxii + 302 pp. ISBN 90-277-0828-2
5. **Mental Illness: Law and Public Policy**
   1980, xvii + 254 pp. ISBN 90-277-1057-0
6. **Clinical Judgment. A Critical Appraisal**
   1979, xxvi + 278 pp. ISBN 90-277-0952-1
7. **Organism, Medicine, and Metaphysics**
   Essays in Honor of Hans Jonas on his 75th Birthday, May 10, 1978
   1978, xxvii + 330 pp. ISBN 90-277-0823-1
8. **Justice and Health Care**
   1981, xiv + 238 pp. ISBN 90-277-1207-7
9. **The Law-Medicine Relation: A Philosophical Exploration**
   1981, xxx + 292 pp. ISBN 90-277-1217-4
10. **New Knowledge in the Biomedical Sciences**
    1982, xviii + 244 pp. ISBN 90-277-1319-7
11. **Beneficence and Health Care**
    1982, xvi + 264 pp. ISBN 90-277-1377-4
12. **Responsibility in Health Care**
    1982, xxiii + 285 pp. ISBN 90-277-1417-7
13. **Abortion and the Status of the Fetus**
    1983, xxxii + 349 pp. ISBN 90-277-1493-2
14. **The Clinical Encounter**
    1983, xvi + 309 pp. ISBN 90-277-1593-9
15. **Ethics and Mental Retardation**
    1984, xvi + 254 pp. ISBN 90-277-1630-7
16. **Health, Disease, and Causal Explanations in Medicine**
    1984, xxx + 250 pp. ISBN 90-277-1660-9

YOUNGSTOWN STATE UNIVERSITY
3 1217 00374 8756